NAIL YOUR NUMBERS

A Path to Skilled Construction Estimating & Bidding

David Gerstel

Latitude 67

Publisher: Latitude 67

Developmental Editor: Jackie Callahan Parente
Copy Editing: P&M Editorial Services
Cover Design: David Gerstel and Six Penny Graphics
Interior Design: David Gerstel

Library of Congress Control Number: 2017907365
ISBN-13: 978-0-9826709-1-0
Includes Index
1. Building—Estimates 2. Construction industry—management
3. Construction contracts
TH435

Printed and Bound in the United States of America
Published in Kensington, California

1 2 3 4 5 6 7 8 9 10 21 20 19 18 17

Please contact us at DavidGerstel.com for permissions and bulk purchases

An Important Note to Readers

This book is intended as a guide to the development of sound estimating and bidding procedures and practices. Readers are, however, solely responsible for developing their own procedures and practices, for determining the costs of their own projects, and for seeking expert legal and financial counsel when necessary. The author, publisher, and all other persons or entities associated with production of Nail Your Numbers shall have no liability to any person or entity with respect to any liability, loss, or damage caused or alleged to be caused directly or indirectly by information, concepts, strategies, or descriptions of materials, methods, or equipment described herein.

for Sandra

Contents

You, This Book, and Me

An Introductory Note from the Author

Let's get the bad news out of the way first and fast. Construction is a tough business. It can be tough to get started and tougher yet to stay on your feet, to keep your company up and running. Especially during recessions, construction companies wink out of existence by the thousands.

There are lots of reasons builders perish. Foremost among them is the failure to develop reliable bidding and estimating systems. Deep into their careers, too many builders are still winging it. The results are painful to watch. Companies collapse. Field employees and office staff are thrown into unemployment. Builders lose their businesses, their life savings, their sense of self-worth, and sometimes even their homes. Why? Because, for lack of sound bidding and estimating systems, they blew their numbers on a couple of big projects.

You've got a chance to do much, much better. The fact that you are picking up this book is evidence of that. You have recognized that you must build a strong bidding procedure and learn to estimate your construction costs closely. You are not willing to wing it forever. You are determined to nail your numbers.

If you are lucky (as I was), you are making that crucial determination during the start-up phase of your career. You may be a builder who has already built a successful operation but is always on the prowl for ideas you can use to make your company stronger yet. Or perhaps you are well into your career but find yourself struggling because, as you have realized, too often you are going after the wrong jobs or badly misestimating your construction costs. If you've been even less fortunate, you might be just coming out of bankruptcy—as was the case for one of the readers of my earlier book on running a construction company. As he was, you are determined to try again. You are determined to get it right this time. (He did. Drawing on ideas in the book, including ideas about how to move beyond bidding for free, he built a respected, financially strong company. Now he is able to spend much of his time tending his vineyards and hiking in the Himalayas.)

I have written this book so that it can be of value to all of you—start-up guys, start-over guys, and owners of successful companies. In each chapter, I first discuss the most basic aspects of an important component of estimating and bidding, and then move up through the more advanced features that mature, durable construction companies typically have in place. For start-up builders, I have provided two chapters that the experienced guys may choose to merely skim (if that). They include reviews of the essential math and a walk through the language of estimating and bidding that doubles as an introductory overview of the whole process.

I am confident that experienced builders

will likewise find much of value here. I talked with or formally interviewed hundreds of builders in preparation for writing this book. Repeatedly, I have seen even the most seasoned and successful of them take enthusiastically to one another's ideas (and to my own).

Whether you are in the early years of your career or a seasoned pro, I think you will find it worthwhile to give serious consideration to the estimating and bidding practices that have sustained us, including:

- Systematically qualifying projects for bids before investing valuable estimating time in them.
- Creating comprehensive records of your own cost experience so that you don't have to rely on guesswork or pricey cost catalogues and databases full of "average" costs that are likely to vary from your own to a dangerous degree.
- Moving beyond bidding for free and moving up to getting paid for your estimates at a professional rate.
- Building a powerful electronic estimating and bidding program of your own rather than relying on off-the-shelf construction industry programs that, over time, can burden you with substantial overhead expense even while producing unacceptably imprecise approximations of the costs you actually incur at your jobsites.

Of course, the path to skilled estimating and bidding that I propose is not the only valid path. There are others. But, if I have observed correctly during my research for this book—during interviews, at builders' conventions and meetings, at web sites, and while reading a pile of articles and books—they don't vary all that much from the path I have traveled and will suggest to you. They include the same essential steps. They avoid the dubious shortcuts taken by the guys who are going to go over a cliff.

If you aim to nail your numbers—to do good work and get paid for it, or to put it another way, to get paid enough to do good work—the path I have suggested should get you well along your way toward realizing your goal. Certainly the wisdom and know-how of the savvy builders I have woven into my narrative should prove valuable.

It will take determination and persistence to put the practices we suggest into play. But the payoff can be huge. Your efforts will reduce the chances that you will be one of those builders who will wink out of existence. The estimating and bidding systems you produce will lower the financial stress that can drain the pleasure out of building your projects. You will be not only more likely to stay in business, but also to prosper and enjoy your work.

Good luck,
David Gerstel

PART ONE

RAMP UP

1 The Heart of Our Business

2 Get Ready

3 Get Set

CHAPTER 1

The Heart of Our Business

Why estimate? Why bid?

When Kent Thompson, a veteran builder in the San Francisco Bay Area, makes his first sales call to potential clients, he usually requires that all owners be present. If it's a residential remodel he's going out to discuss, he wants both husband and wife there. But for one client he felt he had to make an exception. The wife was in D.C. for her regular meeting; and Kent anticipated that it would be difficult, given her responsibilities, for both her and her husband to be available together any time soon.

As he completed his conversation with the husband, Kent promised to work up some preliminary numbers for the project. On the way out the door, he smiled and said, "Please say hello to your wife; tell the chairperson of the Fed that she and I are in the same business. We are both trying to keep people working."

Hustling work for your crew, whether it's just yourself or dozens of people, is a major reason we do the hard work of estimating and bidding. But it is not the only reason. One general contractor, who is also an architect, says that he views estimating as a way of getting a deeper understanding of his projects. It's his way, he explains, of putting together "the complete set of parts needed for a job" even as he costs them out and determines whether the project can be practicably constructed as he had been envisioning it. "Estimating," he says, gets the project "ready for delivery." Likewise, a builder who runs a design/build shop charges for her estimates as an extension of design. She calls the estimating process "specifying." And she charges for it at the same rate as she does other design work.

Paul Winans, a builder who has served as president of the National Association of the Remodeling Industry (NARI), maintains that all construction is really design/build—especially, Paul emphasizes, if the plans originate in an architect's office. He resists even using

COMING UP

- Why we estimate and bid
- The heart of your business
- Excuses for bad estimating and bidding
- Sloppy shortcuts
- Estimating vs bidding: The difference
- The first principle of estimating
- Hard work
- Outside help
- The reward

Lifeblood

Knowledgeable people across the construction industry understand that to be successful, a builder must be skilled at estimating and bidding.

The Associated General Contractors (AGC), an association for builders of large commercial and public projects from shopping malls to bridges, declares: "The lifeblood of a contractor is detailed estimating. [Without] very detailed estimates a contractor cannot compete, cannot complete a project, and cannot stay in business."

Here is the word from *Fine Homebuilding*, a favorite magazine of guys who build residential projects ranging from decks to kitchens, additions, and houses: "Of all the business tasks confronting a builder, accurate bidding is the most important."

The Journal of Light Construction declares: "Few skills are more important to the success of the small builder than cost estimating. Fewer still are more undeveloped and undervalued."

the words "estimate and bid." Given what he experiences as the chronically incomplete state of plans when they reach his office, he characterizes the process of estimating as "preparing a set of plans for production and attaching costs."

Other builders see estimating and bidding as an extension of marketing—though that is a practice, as I have learned from hard experience, that can be dangerous. If you let your desire to close the deal excessively influence your estimating, you can tempt yourself into understating the costs for a potential project. On the other hand, as the owner of a respected plumbing company emphasizes, creating "comprehensive" estimates that are "proactive and include everything called out in a set of plans and also what is not called out" is a great way of servicing clients. It is a way to win their respect and their trust. It is a way to win their business. Or at least it can be during boom times. During busts, as the plumber lamented during the Great Recession, it is generally only the lowest price that wins appreciation.

However you view estimating and bidding, whether as an extension of design, as marketing, or just as a pain in the butt, *you have to do it*. As Kent Thompson says, "If you want to get work and build things, you have to estimate. You have to give clients an idea of what their project is going to cost if you want them to work with you."

Not only that, you must also—and better sooner than later—travel the path to skilled estimating and bidding. When you are young and just starting your business, you can, as I did, make up for inadequate estimates and bids with brute force. You can put in such long hours at the job site and at your desk that you will manage to squeeze enough income to live on even out of projects that have been woefully mispriced. But if you want the long-term satisfaction, freedom, and prosperity that a well-run operation can bring, you must learn to efficiently nail your numbers.

A while back I got an e-mail from a young builder. On the advice of his attorney, he was asking me to serve as a consultant for his rapidly growing company. At our first meeting he told me he needed help with his "back-office work."

"What's that?" I asked.

"Well, to start with, estimating and bidding," he replied.

Startled, I answered, "No, no! That is a dangerous misconception! Filing, preparing the mounds of paperwork required for public works projects, and filling out insurance forms—that's back-office work. But estimating and bidding—no, that is not back-office work. That's the heart of your business."

Rationalizations and work-arounds

Many builders never do travel the path to skilled estimating and bidding. Their estimates and bids are like a sloppy framing job with nails sticking out all over the place. Of course, any framing job will sport an occasional shiner. But in a good job, they will be few and far between. When we say "nail your numbers," we are not seeking perfect estimating, any more than in framing we expect every nail to be a sixteenth of an inch from its ideal target. But we also don't buy the rationalizations for completely missing the target over and over. Many builders do miss by far too much, and they have invented all kinds of excuses for why the numbers in their estimates and bids differ substantially from their actual financial performance on projects

One excuse you'll often hear is "You just can't find good help these days." The problem is not with the estimate. Oh no. It's those bums goofing off at the job site and not getting the work done in the estimated time—as the boss figures he surely could if he were out there with his bags on.

Iris Harrell, founder of a residential remodeling company in Silicon Valley legendary for its financial success, rebuts that excuse. Speaking at a builders' conference, she admonished the audience, "When a job comes in over budget it usually has to do with the estimate, not with the carpenters who produced it. You either have to hear that [truth]," or you can choose to ignore it "until you go out of business because your estimates are so bad."

A commonplace excuse for weak estimating is "Estimating is not calculating, it is an art." Ah yes, an art! I do not know exactly what "art" is, but I am acquainted with accomplished artists in different media. All do their work with discipline, executing every note, phrase, or brush stroke with practiced focus. A good estimate and bid is likewise artful, for every dollar in it is honestly and knowledgeably figured. None are just splashed on the canvas.

Another much favored excuse: "A construction project is like a lawsuit. You can't know what it will cost till it is over." A construction project may get you into a lawsuit, especially if you've created a sloppy estimate and bid and then try to make up for it by cutting corners and piling on extra charges during construction. But producing reliable estimates and bids is nothing like pursuing a lawsuit. Costs in a lawsuit are driven by genuinely unpredictable factors—people who are so angry or so fearful that they will do irrational things. There are rarely, if ever, costs in a construction project that are so unpredictable. The great majority, as we will see, can be predicted with great accuracy. Except in a few exotic situations, the remaining costs can be projected within narrow ranges.

To make up for the alleged impossibility of accurate estimates, many builders resort to work-arounds. One favored strategy calls for randomly inserting additional charges here and there in an

Excuses

- You just can't find good help these days.
- Estimating isn't calculation; it's an art form.
- A construction project is like a lawsuit; you just can't know how much it will cost till it's over.
- You have to throw in lots of fudge because accurate estimating is impossible.
- Time and material (T&M) contracts are the most fair because the customer pays what the project really costs.

Yes, but...

At times, when I debunk the excuses for sloppy estimating, a builder will respond with something like, "Yes, I could be doing a better job, but you know it's really impossible to know what a job will cost." So why bother attempting to nail your numbers, he's saying.

My answer: He is letting the perfect become the enemy of the good. It is true that it is not possible to know exactly what a project will cost before it is completed. Owners and designers regularly do introduce changes during projects. And every job—not just remodeling projects—can generate extra costs because of hidden conditions discovered during work.

But for two big reasons, that is no excuse, any more so than any of the nonsense about "art" and "lawsuits," for giving up on estimating and bidding.

First, the goal in estimating and bidding is to closely capture your costs and determine your price to the owner for the work that is known or can reasonably be inferred before construction begins. That's all. The estimator is not obligated to see the invisible nor to read what might be in the owner's or designer's mind at some point in the future. Only when the invisible does become visible, when hidden conditions are revealed or the owner declares a new wish, does the builder generate numbers for those extras. Until then, his primary goal is to nail down numbers for work he already has been told about. That is what nailing your numbers means. It is not about gazing into a crystal ball.

Second, builders who will take the trouble to develop good estimating and bidding skills can even provide owners with reasonable projections of additional costs that may be encountered during the project. They can suggest reasonable amounts to hold in reserve for hidden conditions and/or for requests for additional work the owner may think of during the job.

estimate. Often dignified by the term "contingencies," they are more honestly called "fudge," as in fudging. Are you suddenly feeling uneasy about your labor costs? Okay, double up the labor charge for each sheet of plywood you will be installing. Not sure what your baseboard installation will run? Okay, crank up the waste factor on the material and cost it at retail rather than your contractor's price. Still not feeling confident? Add a line for "additional trim work." And so on.

Targeted and restrained use of contingencies is legitimate. It can even be necessary, as we will discuss in Part II and Part III. Overuse, especially of roughly guesstimated contingencies, can bring trouble; for when you have plugged in too many, you have no idea whether your estimate is reasonably on target, or much too high, or still way low. As one industry consultant emphasizes, "An estimate that is 20% high is just as bad as an estimate that is 20% low." Both are lousy estimates that carry consequences. With estimates that run too high, you don't get the jobs you bid and get a reputation for being too expensive. With those that are too low, you may get the jobs, but you feel pressured to cut corners during construction, thereby risking damage to your pocketbook along with your reputation.

Of all the work-arounds for weak estimating and bidding skills, the most favored is the use of

Four builders and their estimating shortcuts

"Kent," as I will call him here, is respected for the high quality of his work and his loyalty to his employees. But after 40 years in business, he still has not recorded costs from past jobs that he can use to accurately estimate costs for new projects. "We still miss our framing numbers by 50%," one of his project managers told me with chagrin. Kent relies on time-and-material contracts to make up for his weak estimating.

One of Kent's competitors, "Dave," avoids T&M but also disdains the practice of doing detailed estimates. He generates estimates for the second-story additions that are his specialty by calculating their square footage from plans and multiplying it by a dollar figure that he guesses to be about right for the work to be done by his crew. Then he adds on the bids from his subs. Then he adds on another rough figure for the cost of operating his company and for profit. Viola! In a few hours he produces an estimate and bid for a large job.

"Daniella" loves "helping clients realize their dreams." But she hates crunching numbers. Rather than mastering estimating and bidding, she decided to hand off the work to an amiable ex-contractor-turned-project-manager who walked into her office one day and applied for a job. A fourth builder, "Antonio," builds his bids using labor and material costs from a database he leases from a national vendor of estimating software. He knows the data is suspect and that his estimates are often way off, but he has never tracked and recorded task-by-task the actual costs of his crew's work in order to provide himself with a reliable data.

Here are the results of their shortcuts:

When the last economic downturn struck, Kent could not compete for jobs with his usual approach. The customers for construction were largely people looking for the bargains made available by the recession. They wanted hard numbers, not T&M proposals. But Kent could not generate reliable hard numbers. Determined to keep his people employed, he instead generated hopeful numbers. He lost money on job after job, burned through his company's working capital, and eventually had to borrow half a million dollars to keep it afloat.

Dave got by with his square-foot estimating until an addition came along that, at a glance, appeared similar to all the others he had built, but turned out not to be. During construction, Dave discovered that it had structural requirements far beyond the usual, and his costs greatly exceeded his estimate all through the foundation and framing work.

Daniella's amiable ex-contractor proved to be not much of an estimator. In fact, he had given up on running his own business and gone to work for other builders because he had not been able or willing to develop the skills necessary to succeed on his own. Daniella did not go over his work closely enough to see, until her projects were already under construction, that the first four estimates her estimator had generated were going to bring her to the brink of bankruptcy.

Antonio has managed to stay in business using costs generated by the software and data he leases. The estimates are, however, regularly off by 25%, usually to the low side. He shrugs off those results, for by larding the software-generated estimates with extra costs here and there, he is able to make enough out of his jobs to provide a good living for his family. Having somehow convinced himself years ago that the software and data package were the last piece of his business puzzle, he sticks with them.

time-and-material (T&M) contracts. They are tempting. But Michael Stone, a veteran construction industry educator who is in contact with a large number of builders across the U.S., has observed that T&M contracts, as compared with fixed-price contracts, generate two to three times the number of lawsuits. In my book *Running the Successful Construction Company*, I have described a series of T&M contracts that are in use in our industry. Ending with the most sophisticated, I have explained how each successive version derives from the previous inadequate one—and why even the most evolved version strikes me as snake oil.

You can nail your numbers!

As a last-ditch rationalization for poor estimating, some builders cite the dictionary. By definition, they insist, "estimating" means "approximating." Yes, but there are sloppy approximations and there are close approximations that are useful.

Plumb, level, and square are approximations also. But if you are a good builder, your frames are close enough to dead-on plumb, level, and square so that they facilitate the work that follows. You are not battling with the framing when you hang and case doors, set cabinets, or install trim; finish work fits cleanly and without undue fuss. Your framing is true. Your estimates and bids can be true, too. Then they won't stress and distort the construction of the project that follows.

Especially if you have recently transitioned from the trades, you may find the task of learning to create true estimates a bit intimidating. It should not be. Estimating is not as difficult to master as carpentry, plumbing, or electrical work. It is less challenging than planning and coordinating a remodel project or constructing a new home or restaurant, less challenging than rewiring a house. You learned how to build one step at time. You can learn to estimate and bid a step at a time, too.

In fact, if you are experienced in construction, you already have a good start. Estimates are put together in the same order as buildings. Do you understand how the construction of a building is organized and accomplished? Then you will quickly grasp the organization and principles of estimating. Fernando Pagés Ruiz, a builder and the author of *Building an Affordable House* and the *Fine Homebuilding* business blog, says, "When I am estimating, I am studying the plans carefully and building the structure in my mind." He thinks of estimating as a dress rehearsal for the real performance.

I don't mean to suggest to you that estimating is easy. It is demanding work requiring "meticulous attention to detail," emphasizes Ruiz. "It is labor intensive, and it is costly," adds Wayne Del Pico, whose books are used in the training of estimators of large commercial and institutional projects. "*Fast* and *accurate* estimating is an oxymoron," Dennis Dixon, an Arizona builder of custom homes, told the attendees at a construction conference in Las Vegas.

To begin with, reliable estimating requires adherence to several basic principles. They are:

- *Don't miss anything* on the plans, in the specs, or implied by the plans or specs. Capture all the detail in your estimate for a project. As Paul Cook laconically notes in his great book on estimating (see Resources), "Estimators tend to break a project down into more detail than nonestimators." They note all of it. When they estimate the cost of a slab they don't miss anything: Not the grading, the off-hauling, the dump fees, the compaction. Not the fine grading, the drain rock, the membrane, or the sand. Not the rebar, the concrete, or the finishing. And, not the waste, standby time, setup time, or cleanup. All that and more, if necessary, is captured in the estimate.

- *Don't miss anything* when it comes to attaching costs to the items in your estimate: Not wages nor wage increases that will come about during a project. Not any of the burdens on wages such as social security and unemployment taxes. Not the crew supplies, the consumables like blades, that are used up during a project. Not tax on material nor charges for delivery. Not the mobilization time needed to set up for different tasks. The list goes on, and in Part II we will discuss ways of creating and organizing a checklist that supports you in catching every detail of a project.

- *Don't miss anything* when, after calculating your costs for labor, material, and subs, you figure the additional charges necessary to cover the costs of running your company during the project, i.e., your "overhead." And don't miss anything when you survey the risks a project entails in order to set an appropriate additional charge for "profit," a fascinating and much misunderstood issue that we will discuss in Part IV.

In short, be thorough, even knowing that you will inevitably miss something minor here and there and that you will need to include in your estimate specific, well-placed contingencies to cover your oversights. Be thorough in estimating your direct costs for labor, material, and subs. Be thorough and thoughtful when you expand your estimate into a bid—understanding that estimating and bidding are two separate processes, as the sidebar on the next page, "Estimating vs. Bidding," lays out. Then you will produce reliable estimates and bids. Aim to nail your numbers and you will. It is possible; it takes hard work, but it can be done, as many successful builders—including me—have demonstrated. If you have gotten far enough in construction to need this book, pretty obviously you can handle hard work. And, good news here, you won't have to do the work entirely on you own.

A well-built frame is plumb, level, and square. It is "true." That does not mean it is perfect. The framing shown here was dead-on. And the whole frame for the two-story house of which it was a part stayed that way until we stood up the walls for the last exterior corner of the second story. There we ran 5/16" out of plumb. To this day, I have not been able to figure out why. A well-built estimate is "true" also, but that does not mean it is perfect, and sometimes you may not be able to figure out just why it is off by a bothersome but not ruinous amount.

Outside help

As you work to create a well-organized estimating and bidding system and to create comprehensive estimates and bids, you'll encounter help from several sources. The most significant "outside help," as I like to think of it, is the law of probability, or law of averages. The Associated General Contractors' (AGC) handbook on estimating and bidding describes it accurately, if somewhat obscurely: "The actual costs of the elements (i.e., items of construction in a project), when estimated

Estimating vs. bidding

In the 1970s, a professional estimator named Paul Cook wrote two magnificent books, *Estimating for the General Contractor* and *Bidding for the General Contractor*. It was from those books that I learned about estimating and then about bidding. As a result, I have always understood them as separate processes.

When I interviewed other builders for this book, I was surprised to hear that the concept was foreign to many of them. One said, "I don't even know what you mean by separating estimating from bidding."

I mean that it is useful to think of estimating as a management task sandwiched between two stages of bidding, both of which are leadership tasks. During the first bidding stage you are evaluating a project. Is it well conceived and worth building? Is it right for my company? Will it usefully expand my network of business relationships or my company's capabilities? Or will the project overwhelm me and burden my business with too much risk?

Whether you are operating a compact business that does small remodel projects or a sizeable one that builds large new structures, you are thinking strategically when you ask such questions. You are thinking as a leader. You are deciding where to next take your business.

If you decide to go after a project and begin estimating its costs, you have moved to a management task, for estimating involves consideration of tactics. What, for example, will be the most cost-effective way to get those 10"x18"x34' glulams into place? Should I estimate for a crane? Likewise, for this project do I need my top subs, or can I move down a step?

With the estimating of on-site costs done, you return to bidding. You move away from tactical and back to more strategic thinking. Having surveyed market conditions and the project's risks, you determine a markup for profit. Again, you are making leadership decisions, deciding with what financial arrangements you can march your company into a project.

At large construction companies, the distinction between estimating and bidding is made so sharply that the two tasks are done by different people working in different rooms. In smaller companies, where the owner does both the estimating and bidding, the distinction is often blurred, and eagerness for a project leads to wishful estimating. Even in the small company, however, there are ways to build a wall between estimating and bidding. We will discuss them in later chapters.

on a reasonable basis, will average close to the average of the elements (as estimated), even though each individual element will have been estimated too high or too low."

Ugh, yeah, very obscurely. So let's translate: To begin with, the costs you include in your estimate have to be reasonable. They have to come from reliable sources. They must come from trustworthy subs and suppliers; and for labor they must be based primarily on reliable records of your crew's productivity on past jobs. If you are using reliable sources, you may still run high on some items (or "elements," as the AGC calls them) and low on others. But the errors will tend to balance each other out. For example, on the other end of the seesaw from a modestly high estimate for concrete will be the moderately low one for wall sheathing. Each somewhat imperfect number will tend to make up for the error in another.

The averaging out will happen within estimates for individual jobs. It will also happen across jobs. For example, on an addition you built for a residential customer, your estimate of costs ran low; but on a tenant improvement for The SuperChocolate Corporation it turned out to be high. The one provided little profit; the other put more money in your company's bank account than you had projected. In other words, the low and high estimates more or less canceled each other out. Across many estimates and many jobs, assuming you are building estimates *from reliable sources and not missing much of anything*, the averaging should shrink the difference between bid prices and actual performance to minor amounts.

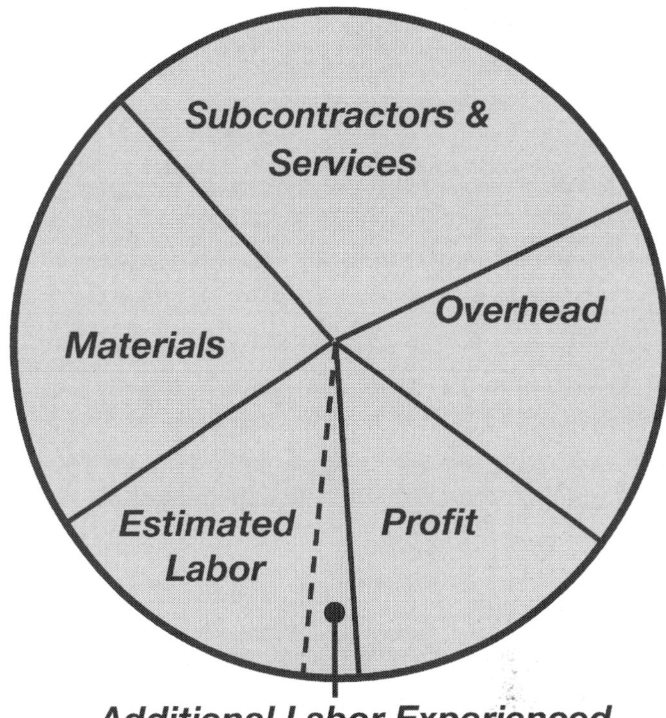

The charges included in a typical bid for a construction project fall into half a dozen categories. Of these the most difficult to nail down is labor. However, even if the charge for labor turns out to be too high or too low by a fairly sizeable percentage, that translates into a relatively small portion of the entire bid.

Additional help will arrive in the form of accurate quotes from vendors and subs. As the accompanying pie chart suggests, a typical estimate includes half a dozen categories of cost. For most of the categories, you will get help nailing down the costs. For materials, if you figure your quantities of material correctly, your suppliers will provide you with reliable figures for the costs of those materials. Likewise, if you accurately spell out the scope of their work to good subs, you will get back reliable quotes for their trades.

Similarly, the costs of running your company—liability insurance, office supplies, Internet service provider (ISP) and phone fees, equipment, etc.—are provided by the vendors you deal with. Your insurance broker tells you what your liability policy will run you. Your office supply company tells you the price of your printer and ink. Of course, you have to keep track of such overhead costs—using straightforward accounting methods such as I have described in *Running a Successful Construction Company*. And you have to recover, with each project you build, an appropriate proportion of the overhead costs. But you don't have to figure out costs for items. Your vendors take care of that.

Even when it comes to figuring profit on a project, you are not entirely on your own. In good part, the profit you include in your price for a job will be dictated by three forces: Market conditions in your locale at the time you are bidding the job. Your position and reputation in that market. Your need for work. You are responsible for sizing up such conditions, but you are not operating in a vacuum. The conditions are real and observable, and we discuss them in detail in Part IV.

The toughest challenge in estimating, the one you must face with virtually no reliable outside help, is calculating the cost of labor to be done by your own crew—including yourself if you are still wearing your bags—rather than by subs. Here you are on your own. William Mitchell, who wrote a thought-provoking book on how to make it as a builder after he experienced bankruptcy, maintains that it is really impossible

Life at the job site is a lot more fun if you are building with a reliable estimate and bid at your back.

to estimate labor accurately. I do not agree, and unlike Mitchell, I have been spared going broke. Instead, I have prospered as a builder for decades. In Chapter Seven, I will describe the system that I use and recommend for nailing down your labor costs for a project.

That said, I do recognize that even with the best systems and most diligent work, labor costs can prove elusive. Events beyond your control, such as bad weather, can affect the hours your crew will need for tasks from deconstruction through finish work. There are, however, practices you can adopt to protect yourself even against such events. For example, you can develop and rely on records for work done in different kinds of weather. Then you will be able to construct costs for work in an upcoming project from records of past projects done under the conditions most likely to be similar to those for the coming job.

Developing a file of your own labor productivity records is vital. But it's also true that even significant errors in labor cost projections may not wreck an estimate if all the other costs are figured reliably. Even if you are wide of the mark for labor, that does not mean that your entire bid will be hopelessly off. For in most bids (unless you are in a niche where you use few subs and your crew does virtually all the construction), your charge for labor will be only a modest fraction of your overall charges for a job. As a result, a sizeable error in your labor estimate translates, as suggested in the pie chart on the previous page, into a tolerable error in your total bid.

For example, say 20% of your charge to a client for a project is for labor, and your estimate of costs for that labor turns out to be off by 15%. That results in a 3% error in your overall bid: (20 × .15 = .03 = 3%). Three percent errors are not trivial. But over the course of multiple jobs, the law of averages may tend to rescue you from such errors, with the low and high estimates canceling each other out.

With an estimate complete and a project successfully bid, you will soon be out on the job site. Personally, I love it out there, whether I am working with the tools or just stopping by to support my crew and subcontractors. I love the smell of freshly excavated earth and newly cut lumber, the drama of a concrete pour, the sculptural power of a frame rising upward, the camaraderie between workers. But I also know, from having produced a few, that low estimates and bids can drain the fun out of a job. You worry whether you are going to lose your ass, whether you will have to dig into your savings and investments to pay your bills, even whether you will be able to support your family.

Bad estimating and bidding can cause real anxiety. But good estimates and bids let you sleep at night. They have your back.

CHAPTER 2

Get Ready

Your mission and your self

A few years back, so-called "mission statements" were all the rage in the business world. The fad has faded. When I ask my fellow builders about their mission statements, they generally can't remember them. I don't recall mine either, and of all those I have heard I remember only one: "Listen, Design, Build."

Even so, the creation of mission statements generated benefits. The effort moved us into strategic thinking and out of the tactical preoccupations—the organization of workflow, the ordering and buying and hiring and firing, the sales calls—that tend to consume a builder's days. Creating mission statements helped us step back, take stock of our companies, and dial into our hopes for the future.

In reality, most of us don't create our mission. We don't invent it. We find it or it finds us. We discover a direction or are nudged onto a path that takes us to a good place. I set out in my youth to become a carpenter/novelist. Fortunate events modified my ambitions and pushed me over to becoming a builder/business writer. Similarly, a builder I interviewed for this book set out to become a high-end residential remodeler. A severe recession pushed him into the construction of restaurants and cafes—the only work available. Once there, he saw the opportunity to work his way up to the larger commercial projects where the bonding companies screen out the flakier competition. He has made it his mission to grow a company that can compete at that level.

Whether you invent a mission or find a direction, knowing where you want to go can help you determine your estimating and bidding priorities. One builder in my area has found his way to residential remodeling with his crew doing virtually all the trades. For his estimating and bidding, he concentrates on gathering accurate records of the hours needed to construct the many items that his crew installs—from concrete foundations to finish plumbing.

COMING UP

- Your mission
- Making the Big Shift
- Nice-guy-itis
- Your essential skill set
- An estimator's space
- Making the move to estimating software
- Pitfalls in the plans
- Hazards in the specs
- Estimating & bidding math
- Estimating terminology

Another builder, operating in a hyperwealthy coastal community, found his opportunity in the construction of "estate homes"—sprawling houses on large and intensely landscaped lots. The estate builder has no field employees other than project managers and a two-man utility crew; he subcontracts all trade work. He compiles no labor-hour records but instead concentrates on refining systems that ensure every item of work in his projects has been anticipated and bid by one subcontractor or another.

Getting your own estimating and bidding ducks in a row will require that you take stock—not only of your company, but also of your personal tendencies and technical capabilities. I will be brief about the personal stuff, exploring only one key question: If you are about to start your own company—or for that matter have been in business for a decade but are still struggling—are you clear in your mind that you must make, or should have already made, the "Big Shift"?

The move from working as a tradesman, or even a project manager, in a company owned by someone else to running your own business operation is huge. It's a giant step even if, at the outset, you are the only person employed by your company. The move requires a big shift in thinking. You encounter a whole new set of responsibilities and challenges. You have the opportunity to enjoy a whole new set of rewards, including increased earnings and control of your time. You acquire work, get work done, and get paid in entirely different ways; and at the heart of those changes is your new responsibility for estimating and bidding.

Many tradesmen who start their own company fail to appreciate the magnitude of the move. They open a business without ever making up their minds to learn business thinking. They continue to think like wage workers, valuing and selling their time and labor at the same hourly wage they were paid by their former boss.

Worse, the wage they charge for on-site work is actually less than what it cost their former boss to put them on a job. They forego all the benefits that the boss paid—such as contributions to social security, workers' compensation insurance, and unemployment insurance. Worse still, they never ask for pay for the work of running a business or for covering its expenses: office and shop space, office supplies, phone and Internet, and much more. Worst of all, they charge nothing for taking on the long-term liabilities that come with each project, small and large.

In short, when they go into business for themselves, they do not grasp that they need to be paid for all their work at a rate at least equal to what it would cost someone else to keep them on payroll. And they do not recognize that their company needs income above and beyond wages or salary they pay themselves to cover its operating costs and risks.

The failure to make the Big Shift and learn business thinking is often rooted more in personal neediness than in lack of ability. For some, "business," especially "profit," is equated with greed. We do not want to think of ourselves or be perceived by others as greedy. We prefer to be admired. As one builder said of herself, we want too much to be liked. I call the excessive desire to be liked "nice-guy-itis."

The trouble with nice-guy-itis is that it does more harm than good. Low bids may please your customers. As the builder Paul Eldrenkamp has wryly observed, when owners seek competitive bids, the low bidder is often seen as a good fellow, and the other bidders—unless their bids are very close to the low one—as thieves. The problem is, low estimates and bids weaken your company so much it will soon not be able to support you or your family or serve your community.

Your skill set

Along with evaluating your personal tendencies and correcting any that may undermine you as an estimator and bidder, you would do well to take an inventory of your education. Do you have sufficient knowledge of the trades for which you will be estimating production costs? Can you read plans and

Bob's Problem

Bob is a builder afflicted with especially severe nice-guy-itis. Bob operates a one-person kitchen and bath business in a prosperous metropolitan area where he works for well-off people living in million-dollar homes.

He is a capable designer and an outstanding craftsman. But when it comes to generating estimates and bids, Bob includes material and subcontractors at his cost. He figures in his own labor at a rate roughly half of what the guy he used to work for as a project lead now charges for leads with skills inferior to Bob's. For overhead and profit combined, he apologetically asks for 10%, or about one-third the going rate in his area. All told, his bathrooms and kitchens are sold at one-half or less of what his competitors charge.

With his low pricing and good reputation, Bob gets eight out of 10 jobs for which he generates bids. But Bob recoils from the idea of raising his charges to the point he would get only six out of 10 jobs, which is the win rate enjoyed by a nearby competitor who charges more than twice as much as Bob. Being turned down by four out of 10 potential customers is a level of rejection greater than Bob can bear to consider. In fact, Bob wants to "improve" his winning percentage. He wants to "win" nine out of 10 of his bids. Savvy old-time builders not afflicted with nice-guy-itis have a phrase to describe Bob's kind of winning. They called it "losing."

follow specifications? Do you know the language of estimating? Do you understand the necessary math?

I will not attempt here to catch you up on trade knowledge. The chapters that follow are written with the assumption that you are far enough along in your construction career so that you recognize virtually every product available at a lumberyard or building supply house. You understand lumber grading, know the difference between nominal and actual dimensions, can tell type M from type L copper pipe, and recognize the load capacity of electrical wire by its color. You thoroughly understand how the materials to be installed by your crew are fitted together during construction.

You also know the sequence in which work is done. Knowing the sequence is critical, because effective estimating is done in the same order as building. If the order is embedded in your mind, you will instinctively sense when you leave something out of an estimate and correct the omission.

What I will do in this chapter is:

- Emphasize the need for a space where you can work on estimates and bids without distraction.
- Take a preliminary look at computer estimating (a subject covered in detail in Part V).
- Explore challenges that plans and specs pose during estimating.
- Survey the math essential to estimating and bidding.
- Review estimating and bidding terms in a way that will help you better understand the process even as you absorb its language.

I urge you, please, to not attempt to travel the path to skilled estimating and bidding without the necessary knowledge of plans, specs, math, and language. The necessary learning is well within your reach; if you were able to learn to build, you will certainly be able to learn the skills that underlie good estimating, taking them on in digestible chunks, one at a time. You may find the learning stressful at times. But it's a lot less stressful than never acquiring the needed knowledge and banging away at construction projects without reliable estimates and bids in place.

Your space, your computer

Estimating and bidding, says Fernando Pagés, require "a calm heart." You need to be focused and paying close attention. To achieve the focus and calm, you need a quiet space insulated from distractions. You need a place where you can close the door, turn off the phone, and bear down.

Your estimating space does not have to be large. Readers of my book on running a construction company may remember that my first office, which doubled as my estimating and bidding room, was a walk-in-closet. For today's computer-based estimating, even less space is required, as you can see in the photo of Black Creek Builders estimating setup.

While I suppose it is still possible to stay with paper and avoid using a computer for estimating, a computer with a large screen can greatly speed the work. Computers can virtually eliminate the tedious aspects of estimating—the endless multiplying and addition. They will take care of all that for you, almost instantaneously. Meanwhile, they leave to you the interesting aspects of the work—imagining how the project will be put together and how to build it most efficiently and effectively.

However, the speed at which you move to computerized estimating and bidding is an important judgment call.

One reader who looked over an early outline of this book was surprised that the chapters on computer-based estimating came at the end rather than at the beginning. I explained my reasons: Learning *what* you need to do to create sound estimates and bids and *why* you need to do it is not the same task as learning *how* to do it. They are separate tasks, and learning the what and the why should precede learning the how-to, whether by computer or other means.

As the Associated General Contractors (AGC) emphasizes, "estimators must understand the underlying theory, principles, and process of estimating before relying on any software estimating program." In other words, you do not want to be building an estimating system and learning to use computer software for estimating at the same time, any more than you would have an apprentice start cutting plates and laying out a wall frame while he is still learning to measure and safely use a Skilsaw.

If you are already skilled with computer software—meaning that at a minimum you word process with ease and know your way around Excel® or another electronic spreadsheet program—you may be able to develop your estimating and bidding systems and procedures directly on your computer, never creating a paper system. Otherwise, get your estimating and bidding chops down first. You do not want to find yourself tangled up in an unfamiliar computer program as you are working to strengthen your estimating procedures.

At Black Creek Builders in Oakland, CA, estimators can work efficiently while occupying only a corner of a small office space. A large computer screen displays plans while the smaller one is used to produce the estimate.

As a first step into the world of computer estimating, if you have not already done so, you should learn ten-finger typing. In response to that suggestion, you may be thinking, "I bought this book to strengthen my estimating and bidding skills. Now this Gerstel guy wants to send me off to secretarial school! I am going to ask for a refund."

Well, before you do, please consider this: Over the course of a career as a builder you likely will, even if you are a skilled typist, average—at a minimum—several hours a week banging away on your keyboard. You will be answering e-mails, creating contracts, writing up the assumptions that will accompany your bids, and recording notes for use during the actual construction of a project, even while you create your estimate for it.

Each year you are likely to spend a couple of hundred hours on the keyboard. Over the course of a 40-year career, you will easily pile up 8,000 hours (40 × 200 = 8,000). That's if you are a skilled typist. If you are a hunt-and-peck typist, your time on the keyboard could double from 8,000 to 16,000 hours. The extra 8,000 hours equal an extra 200 workweeks (8,000 ÷ 40 = 200) spent pecking away. Those 200 weeks amount to *nearly four years*! On the other hand, learning to ten-finger type, via a printed or online program, should take roughly 40 hours, or one workweek. It will save you 199 weeks. So take your choice. Hunt and peck away. Or take 100 extra two-week vacations over the course of your career (and don't forget to send me a postcard with a thank-you note from your favorite beach or mountain trail).

Once you can type, your next step, if you have not already picked up the skill, should be to learn word processing—either by playing around on your keyboard while consulting a manual or via an online or community college course. Whether you are still crunching numbers with a calculator or have moved to computer-based estimating, producing the text documents that must accompany bids on your computer will greatly speed your work.

Comprehensive plan reading

Settled into your quiet workspace with your mind in business mode, you are ready to open up a set of plans for a project you have decided to bid. What you will be looking at, as Sal Alfano, long-time editor of the *Journal of Light Construction*, once wrote, is "a story about an imagined place," told in a number of drawings and notes. It is also a story told in a specialized language. Even the printing is different, with all letters capitalized. And much of the story is told in a swarm of symbols, terse references, abbreviations, and lines whose different weights and formations indicate a range of meanings.

From this weirdly told story you must, as an estimator, be able to visualize every system, assembly, and item that goes into the construction of the "imagined place." If you visualize badly, you will leave something out of your estimate. You may overlook an entire system, as did estimators who have overlooked heating, venting, and air conditioning (HVAC)—perhaps because HVAC is often only hinted at in the plans by tiny boxes representing registers. On the other hand, visualize well and you can produce an estimate that is very close to the actual costs of construction. You can nail your numbers.

Early in your career the plans you estimate and bid from may be simple. The plans for my first kitchen remodel consisted only of a single page showing views of the floor and the walls as they would appear at the end of construction. The challenge here was not understanding the plans, but reading into them all the trade work necessary to remove the existing kitchen and replace it with another.

From that remodel, however, the length and complexity of the plans I dealt with ramped up to dozens of pages for large projects designed by architects and engineers. Plans created by design professionals can be overwhelming to a startup builder. One way or the other, however, if you wish to maximize your opportunities, you must learn to read all the various components of plans listed in the sidebar to the right. (If you need help with plan reading, see Resources for a good book by Wayne DelPico.)

No matter what stage of your career you are at, you need to be alert to the challenges and hazards plans present to estimators. To begin with, note that plans may be sent to you in various stages of development:

- *Schematics*, conceptual drawings that provide no more than a sketch of a building's shape and intended functions.
- *Preliminary drawings*, produced after further design development, and typically including roughly half the detail of the final drawings. Sometimes, particularly if the clients change their minds about what they want, several iterations of the preliminary drawings may be sent your way.
- *Working drawings*, which should be fully enough developed so that they will pass muster at the building department and enable you to build the project without undue requests for clarification.

You may be called upon to produce estimates for all the stages of the drawings, and often for no pay or for a pittance, a trap we will deal with in Chapter Sixteen. The repeated recalculation of costs can consume huge amounts of time. Moreover, it may foster errors in your estimate. You may not notice the extent to which an item, assembly, or system has changed from one iteration of the plans to another. As a result, you may use quantities and costs figured for a first estimate in a revised version and end up significantly off in the revision. For example, if you do not notice that the structural engineer has, between the preliminary and working drawings, increased the depth of a foundation footing from two feet to three feet, your costs for excavation, off-hauling, backfill, and forming can be off by a third or more.

When you have received the working drawings and the plans are ostensibly complete, they are not likely to be actually complete. In fact, it is fair to say plans are never complete. For as one architect declared, it is virtually impossible, given the

Components of plans for a project

Site plans

- Boundaries
- Topography
- Plants and streams
- Other natural features
- Drainage
- Building setbacks
- Paving
- Utilities
- Landscaping

Structural plans

- Foundation
- Frame
- Sections
- Details

Architectural plans

- Roof plans
- Floor plans
- Reflected ceiling plans
- Sections
- Details
- Exterior elevations
- Interior elevations
- Cabinet elevations
- Door & window schedules
- Finish schedules

Mechanical plans

- HVAC
- Plumbing

Electrical plans

Estimating challenges in the plans

- Estimating required at three levels: Conceptual schematics. Preliminary design. Working drawings.
- Even at working drawing level plans far from complete
- Different scales used on different pages of the plans
- Inaccurate or inconsistent dimensions
- Critical details missing or wrong
- Contradictions between one area of the plans and another
- Work not shown in plans but for which builder is liable
- Obsessive, blinding detail
- Changes in plans sent during estimating, so that the estimate must be repeatedly adjusted
- Details that are unnecessarily costly to construct

complexity of the way we build these days, to call out in the plans everything needed to construct a project.

That is not to suggest, however, that you can leave an item that is missing from the plans out of your estimate and bid. The fact that the designer failed to provide for an item does not necessarily mean that you will not be held responsible for building it. For example, at a large residential project, the builder's crew did not install flashing at the skylights, because the plans did not show them. The skylights leaked. A lawsuit ensued. Under a legal doctrine known as "standard of care," which requires "reasonable caution such as a prudent man would have exercised," it was the builder, not the architect, who was held responsible for the costs of retrofitting the missing skylights with flashings and for repairing the damage caused by the leaks.

Even for less obvious blunders than failing to install skylight flashing, you *may*, without compensation, have to build work that was omitted from the plans. Note, please, the emphasis on "may" here. There's a lot of gray area around standard of care. While a builder may be clearly responsible for some items not explicitly called out in a set of plans, about others there can be much dispute.

A good estimator, however, picks up on virtually everything that will be needed to build a project, even if the designer did not. Whether all items will be included in the bid submitted to the owner is, however, another issue. If you are operating in the hardball world of competitive bidding and being exploited for free estimates and bids, you may elect to leave the owner in the dark about certain of his designer's oversights and look forward to charging for extras during construction. More about that controversial subject in Chapter Sixteen.

Here we will move on to another challenge. Even as you are well into estimating a project, collecting bids from subs, and working up costs for work to be done by your crew, changes can flow in from the designers' offices. They may be the result of new ideas that have occurred to the owners or designers. They may result from other bidders raising concerns.

Even a seemingly minor change can impact costs across multiple trades. For example, inside the cloud a designer has used to alert you to a change, you might spot a wall-hung water closet (W.C.) replacing a previously indicated floor-mounted one. Now you will have to produce an adjusted cost for plumbing, for framing, for finish flooring—and then, if the stud bays behind the W.C. were to contain wire or ducts, for electrical and HVAC as well.

Subcontractors' responses to the inevitability of changes can also pile on work. To avoid repeatedly redoing their bids to account for changes, they may sensibly wait till they have received what they guess to be the final iteration of the plans before submitting their bids. As a bid deadline nears, builders may find themselves hurriedly comparing and choosing sub bids to determine which to include in their own submittals. Or, if the builders do not get all needed sub bids in time, they may be forced use so called "plug figures," for the costs for a trade they do not know well enough to price accurately.

Beyond the perpetual incompletion and changing of plans, estimators face challenges arising out of the use of a variety of scales in plans, precedence of one area of the plans over another, designers' desire to avoid redundancy, and constructability. A set of plans is, of course, a scaled representation of the imagined place. A ¼" or ½" segment of a drawn line can represent a foot of construction. But scale often varies throughout a set of plans. The civil engineer who drew the site plan may have used one scale while the structural engineer uses another. Meanwhile the architect may use several scales: ¼" for the floor plan, ½" for the cabinet drawings, and a full inch for certain details which are of special concern.

The estimator must make sure she is using the appropriate scale at each drawing when she is figuring quantities. Say, for example, she has been using 3/16" to calculate quantities of work drawn by an engineer. Then she moves to figuring the amount of exterior wall frame represented on an architect's floor plan at ¼" per foot. Distracted by a phone call, the estimator stays with 3/16" as she estimates framing material from the floor plans. The result: She scales 10" on the plans as 53'4" rather than the 40' of wall actually shown (I will leave it to you to pencil out the math here).

Along with appropriately shifting between scales, an estimator must be aware that in case of conflict between two drawings, the drawing using the larger scale takes precedence. An architect may, for example, first provide a floor plan at ¼" scale, leaving out costly detail, then subsequently show that detail at a larger scale. The incompleteness of the smaller-scale drawings can result from a designer's preference for drawing details once, not multiple times. Redundancy takes time. In addition, if they draw a detail more than once, a contradiction may creep in, and that can lead to conflicts and confusion during construction. Regardless of the designer's motivation, an estimator is responsible for capturing every detail no matter which drawing shows it.

> ***Regardless of the designer's motivation, an estimator is responsible for capturing all detail no matter which drawing shows it.***

Likewise, to avoid redundancy and conflict, if drawings by one design professional display a detail or assembly, another designer may elect to omit it from his drawings. Say a civil engineer shows drainage lines on his site plan. Noting that the civil has provided for the drainage, the structural engineer omits it from her drawings. That decision shifts responsibility to the estimator. He must pick up the work from the one drawing where it is shown, even if it is not where he expects to find it.

Finally, an estimator might reasonably assume that a project can be built as shown on the plans. Constructability is not, however, a sure thing. I learned that lesson most vividly when my company built a master bedroom and bath in the unused space beneath a steep gable roof. The plans were beautifully drawn, displaying—among other exquisitely rendered details—an archtop window shown at the exterior south elevation of the project. Unfortunately, the beautiful details were not accompanied by a section accurately locating the window relative to the level of the interior floor. During construction, my crew discovered that if the window were located as suggested by the exterior view, it would extend below the floor line on the inside so that its sill would have been visible a foot below the ceiling of the living room. Oops.

For an estimator, verification of constructability matters because if constructability is not possible, then it will also not be possible to put up the building for the estimated costs. Likely they will run more. The builder will pay the difference or risk an adversarial encounter with the owners as well as the designer if he decides, however legitimately, to insist that the cost of adjustments is their burden to bear. In the case of the archtop window, the owners graciously offered to pay me for any extra work when I showed them the problem. Not so graciously they threw the architect off the job; the window snafu was just the last straw for them. Of course, he never asked me to build one of his projects again, so my estimating oversight was not without cost.

As challenging as plans can be, there is another construction document that can contain even more obscure and obnoxious obstacles to accurate and reliable estimating. That is the "project manual," the "specs" as it is informally called, though only one of its several sections actually focuses on the technical specifications.

Vigilant spec review

Plans and specs have different but complementary functions. "If the drawings are the quantitative representation of the project shown in graphic format," writes Wayne DelPico, "then the specifications provide the qualitative requirements of a project in a written document." During estimating, you take your quantities off the plans, say 90 yards of concrete for the footings. From the specs, you take the strength and mix requirements for the concrete—PSI 3,500, 50% fly ash, for example.

For small projects, such as those designed by kitchen and bath designers or interior decorators rather than architects, specifications will likely amount to nothing more than minimal notes in the margins of the plans. The designer will be relying on the builder to provide specs for nearly everything except selected finishes. You sometimes see minimal specs even for much larger projects. I have seen plans for huge estate homes cranked out at architects' offices that are virtually specification-free. However, a fat project manual crammed with detailed requirements is likely to show up on most large projects, especially commercial jobs, and most certainly public works projects.

When you encounter your first full-blown project manual, you may feel, as I did, as if you have been handed the *Iliad* written in ancient Greek and asked to translate it. I had never seen specifications even remotely as detailed or as thick with legalisms and specialized language. To this day, I find reading specs irritatingly tedious. I think I'd rather form up a retaining wall in the mud during a rainstorm. But read them we must; every requirement in a project manual must be accounted for during estimating, because it can burden you with substantial cost. And reading them does deepen your understanding of the project you hope to build.

The basic divisions of project manuals include:

- *Bidding Requirements*, which may call for specific bid forms and set out strict bond requirements.
- *Contract Forms*, typically an American Institute of Architects (AIA) contract, which—*of course!*—is designed to move responsibility for a project from the designers to the builders.
- *General Conditions*, which extend the basic contract and can include required procedures for meetings, submittals (such as cabinet shop drawings), and testing—for example, of concrete strength.
- *Supplementary General Conditions*, which layer on additional requirements specific to a project.
- *Technical Specifications*, not to be confused with "specs," which is just an informal designation for the entire project manual.

Technical specifications sometimes resemble a junk drawer, appearing to contain whatever the designers feel like throwing in—including cover-their-ass requirements from manufacturers of materials with which they have little or no experience. They may be poorly written, unclear, and even contradictory. Incredibly, they may even contain requirements that have nothing to do with the project for which they are intended. They've just drifted over from another project as a set of technical specifications created for that earlier project was sloppily modified to govern a new one.

Recently, for example, a buddy of mine was reviewing plans and specs for a large renovation project that included a small amount of landscape construction. He opened the specs and came across a requirement for ultra-high-strength concrete reinforced with epoxy-coated steel. My friend rang the architect. Politely, with his teeth gritted, he asked, Was the concrete spec just maybe a little over the top? After all, the only concrete in the job was for an eight-foot-square patio.

Not to worry, the architect told the builder. Don't pay any attention to that spec. It was just left over from the last time the firm had used the project manual, namely for the construction of a new public school with intense requirements for earthquake resistance.

Along with requirements that are irrelevant to the project at hand (but for which you may nevertheless be held financially responsible), project manuals can trip up estimators in other ways. First of all, the specs may, or may not, take precedence over the plans. In bygone years, construction industry convention had it that the specs always took precedence. Now, either the specs or the plans can. Precedence goes to the document that calls for the higher quality or provides for the greater benefit to the project. The upshot: The specs may ask for construction that seems reasonable to you. You go with it and do not notice a note on the plans that calls for something more exotic, far costlier, and that you are going to be required to provide during construction.

Of related concern is the use of the term "or equal" in the specs. Your sense of equality within a particular class of products may differ from the designer's. His "equal" may be much costlier than the one you provided for in your estimate. And the

Estimating challenges in the project manual

- Unclear writing
- Contradictory requirements
- Boilerplate requirements that do not apply to the project
- Requirements left over from a manual for a previous project that do not apply to the project being estimated
- Conflicts with requirements in the plans
- Vague "or equal" requirements
- ASTM (American Society for Testing and Materials) numbers that can refer to several different grades
- Ambiguous requirements for "highest quality work"
- Unrealistic requirements to "match existing finishes"
- Prevailing wage requirements
- Requirements for costly meetings, submittals, & testing
- Punishing holdbacks (retentions)
- Bid submission requirements
- Liquidated damage burdens
- Requirements for insurance and bonds

contract may well empower him to decide which "equal" is to be used. You take the hit.

Similarly, ASTM numbers, i.e., standards produced by the American Society for Testing and Materials, may be used in the specs to indicate the quality of products to be installed during construction. For estimators, the problem is that an ASTM number is a standard that may cover a range of specific products with widely differing prices. Again, it may well be the designer who is empowered by the contract to determine which grade is to be installed. If you have included a lower-priced choice in your estimate, your company absorbs the upcharge for the higher-grade product if the designer or owner insists you install it.

Along with "or equal" and ASTM numbers, specifications lay a trap for estimators with their use of the term "highest quality." How is it possible to determine what is the highest quality? Who determines it? The term sets up the designer (and the owner) to put the squeeze on you by demanding during construction that you redo work until it matches up to some farfetched standard.

The use of "highest quality" is particularly problematic when included in the specs for projects being put up for bid by multiple builders. Some bidders may ignore the highest quality requirement while the designer may have no intention of enforcing it, may not even realize the phrase is included in his project manual, or may not have been hired to do any work beyond providing plans and specs and therefore will not be on the job to enforce the highest quality requirement. But you take the requirement at face value. You put much work into producing a bid that assumes all work will be done by your best crew and subs. As a result, your bid is out of the running. I once lost a bid for a new home for just that reason. I did not feel less regretful about my wasted some 80 hours of estimating time when the owners told a mutual acquaintance that they realized they had made a mistake by going to a cheaper builder to save 10% on the job.

Yet another troubling requirement: "Match existing." Often it is used on remodeling projects to refer to finish work that has aged in place. Such work often cannot be matched. It is made of material that is no longer available or has acquired a patina that only time can produce.

Fortunately, there are ways you can defend your estimate and bid against the use of such terms as "or equal," "highest quality," and "match existing." We will take a look at the defensive measures in Chapter Fifteen.

In addition to the hazards arising out of requirements for the actual construction, a project manual can pose other kinds of problems you must be alert for during estimating and during bidding. So that owners and their architects can more easily compare bids, the manual may require that bids be submitted in a particular format and/or that your charges be broken down in a particular way. The manual may call for pricing of numerous alternates, which is especially time-consuming work. Fail to use the format or submit the alternates and your bid may be declared "nonresponsive" and tossed out. All the costly labor that went into producing it goes down the drain.

Particularly if you elect to try your hand at public works projects, you must be alert for a "prevailing wage" requirement. John Martin, a builder in a major metropolitan area, decided to test the waters beyond his usual residential and private commercial projects by bidding for the construction of an annex to a park welcome center. At the bid opening he saw that he was low. He saw, in fact, that he was low by a scary margin. With a little digging he learned why: The project required that "prevailing wages" be paid all workers. For public works in his area, "prevailing wage" referred to a union compensation package, which was 50% higher than the compensation that "prevailed" in his usual niche. John could have been required to build the project, even if it would have bankrupted him. But he got lucky. The authorities let him withdraw his bid.

For a public job almost certainly, very likely for a commercial job, and for many residential projects as well, the project manuals may impose still another hazard—namely a holdback or "retention" from each of the progress payments you are to receive during construction. The retention can amount to a sum greater than your combined markups for overhead and profit. In that case, during the project you will be using your capital or borrowing money to keep your company afloat.

When the project reaches completion, the owners can decide that you have not performed according to the quality standards called for in the project manual. Or they can claim that you have reached completion later than the required date and owe "liquidated damages." You may disagree. You may feel your work is top-drawer. You may point out that the extras the owners called for extended the completion target date and that you were well within the revised target. But the owners have your money. They decide to keep your retention for themselves. "Don't contact me again," they tell you. "If you have a problem, have your lawyer call our lawyer. Goodbye. Click." That, more or less, was what one builder heard when he phoned a client to ask for his final $100,000 payment.

The builder could have called his lawyer, of course. He could have gone to court. But he understood the risk, that the lawsuit could have gone on for years, with bills from his $400-an-hour attorney landing on his desk monthly; and that he might well end up recovering less from his client than he would have to pay for legal services. So, he just walked away from his hundred grand. "Not worth the stress and anxiety to go after it," he said. His experience is not unique. I have heard similar tales from builders many times.

Like the plans, the project manual is subject to change—via addenda from the designers—in the middle of your estimating process. The changes require additional work. And they require that care be taken so that all the extras and deletions they imply are accounted for in the final estimate.

Yet, for all the challenges posed by plans and specs, builders do cope with them. Using the kinds of procedures we will cover in coming chapters, some even prosper, finding meaning and satisfaction in their work and enjoying good earnings. But, of course, to get to that good place you must be able to attach costs to the myriad requirements contained in project documents. That, in turn, requires facility with the kinds of basic math you should have been taught before you got out of high school...but, alas, may not have been. If not, you may find the following review helpful.

Trying to estimate and bid without understanding the necessary math is like trying to work off a ladder that is missing several rungs.

Basic math

Sometimes when I speak at builders' groups I am asked, "what is the secret to your success?" My main answer is "luck." My luck started early with my assignment to the sixth-grade class of the terrifying and aptly named "Miss Cross." By the time my nine months with her came to an end, I had been given all the math I would need for estimating and bidding construction projects: Counting. Addition. Subtraction. Fractions. Decimals. Multiplication. Division. Plane geometry. Rudimentary algebra. And I had been drilled so persistently at much of it that even decades later I could do it largely in my head.

Here I can't provide a math course and do for you what Ms. Cross did for me. I can only survey the needed math, dividing it into three groups: Basic Math. Key Estimating Math. Key Bidding Math.

You must have all of it. Trying to estimate and bid without at least understanding the necessary math is like trying to work off a ladder that is missing three rungs. You might manage to get up

Basic estimating and bidding math

- Addition
- Subtraction
- Multiplication
- Division
- Fractions
- Geometry calculations
  - Area of rectangles
  - Area of triangles
  - Pythagorean theorem
  - Area of circles
  - Volume of cubes
  - Volume of cylinders
- Decimals
- Percentages
- Proofing (Auditing)

and down a few times, but eventually you are going to break your head. If my survey suggests to you that your math needs improvement, please find your own Ms. Cross—in a community college, on the web, or in Resources at the back of this book.

Estimating requires counting, collecting, and calculating of quantities. If you are building a new structure and are framing it with your own crew, you must accurately count the number of hold downs, joist hangers, and other hardware items needed. For the sub trades, you must collect numbers, namely the subs' bids, for their portion of the work in a project—and, vitally important, you must make sure the numbers cover every item you assume to be within their scope of work. Otherwise, you may follow in the footsteps of Kent Thompson, whose company employs dozens of people, but who is nevertheless still a lousy estimator after 40 years in business, and says that he regularly finds himself splitting the difference with subs over items left out of their bids.

Other than the collection of sub bids, every cost that you will need to place in an estimate will require calculation using the math shown in the sidebar on basic estimating and bidding math. After looking it over you may be thinking, "Hey, this is the twenty-first century. I can rely on a specialized construction calculator or computer to do the math." You would be right. But please realize that overreliance on a calculator leaves you with vulnerabilities. For example, if you need to know the quantity of finish flooring for a living room that measures 24' × 22' (528 s.f.), and your finger slips so that you accidentally punch in 24' × 12' (288 s.f.), you will not look at the total and instantly sense the mistake as will someone drilled in mental multiplication. When you step up to computer estimating, your vulnerability gets worse. Then a typing mistake can be embedded in a formula that will be used, erroneously, on job after job.

At the very least, if you rely on a calculator or computer to do your math, you must understand the functions the machine is performing for you. Otherwise, you will not only fail to sense errors it produces. You will not even be able to feed your machine the right information in the correct sequence it needs to perform the calculations.

For the function known as addition, the key thing to understand with respect to sequence is that it possesses the "commutative property." That means that no matter the order in which you add up a series of numbers, you will get the same result: $4 + 3 + 2 = 9$. Likewise, $2 + 4 + 3 = 9$. Not all math functions possess the commutative property. Subtraction does not. Four minus two equals two; but two minus four does not equal two ($4 - 2 = 2$; $2 - 4 \neq 2$).

That may seem obvious beyond mentioning. But the commutative property comes crucially into play at the final step in estimating construction costs for a project, namely auditing (i.e., "proofing") your results. To perform audits, you add your numbers, such as

A fraction is a part of a whole.
The lower number of a fraction tells you how many parts the whole is divided into.
The upper number tells you how many of the parts the fraction is representing.
Here the fraction tells you the whole has been divided into four parts and that one of those parts is being represented by the fraction.

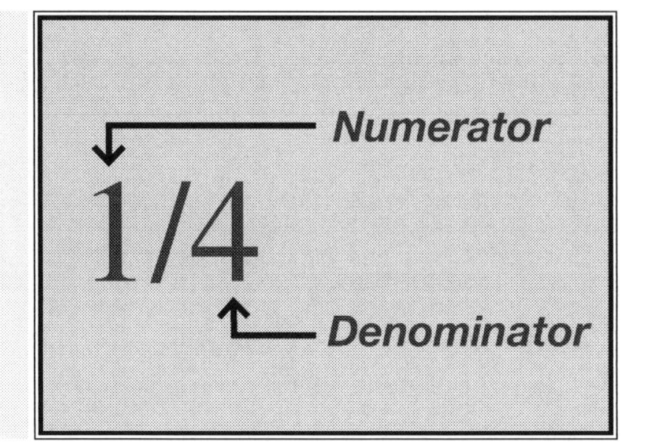

all the labor costs for all items in the estimate, in two different sequences. If you arrive at the same total, you've proofed your addition; your math is correct. If the two sequences do not produce the same result, then you know that you have made a mistake or multiple mistakes.

Multiplying is really a rapid form of addition of identical numbers. Instead of adding a column of six 32s, you multiply 6 × 32 and get the same result. Multiplying may be, for estimating, the most important of the basic math functions. You rely on it for figuring the largest quantities of work in projects—such as excavation, concrete forming and installation, framing, and wall and floor finish surfaces. Note that multiplication, like addition, is commutative: 8 × 32 = 32 × 8. It does not matter in which order you multiply a set of numbers; you get the same result.

Dividing, which can be thought of as reverse multiplication (8 × 32 = 256; 256 ÷ 8 = 32), is less often used in estimating. Unfortunately, quantities and costs in construction generally multiply rather than get smaller. But you do occasionally need division to complete calculations you begin with multiplication, as we will see in the next section of this chapter.

Often in construction we deal with numbers that are not whole numbers such as the "8" and "32" used above. Instead we find ourselves dealing with fractions, which, as you know from working in the trades, express or represent a part of a whole number. For example, as your tape measure tells you, a quarter inch is one of four equal parts of a whole inch.

Fractions are expressed with two numbers divided by a line. The bottom number is called the "denominator." It tells you how many parts the whole is divided into. The top number is called the "numerator," and it tells you the number of parts the fraction is expressing, i.e., representing. For example, with the fraction $\frac{5}{8}$, the lower number, the denominator, tells you the parts are one-eighth of a whole and the top number, the numerator, tells you that the fraction is expressing five of those parts.

Fractions are useful on the job site where you do a lot of measuring and cutting to a fraction of an inch, for example, cutting 2x6s into $14\frac{1}{2}$" lengths for use as blocking in 16" on center (o.c.) wall frames. However, you generally do not need to multiply fractions for *building* purposes.

In *estimating*, however, you often do have to multiply fractions. To simplify the task, you convert fractions into decimals. For any particular fraction, that is accomplished simply by dividing the denominator into the numerator. With fractions changed to decimal form, you will be able to easily figure quantities for your estimate.

For example, let's say that you must figure how much concrete you need for 20 foundation piers, each requiring $3\frac{1}{2}$ cubic yards (c.y.). To calculate the amount of concrete needed for all 20 piers: Convert the fraction to a decimal: $3\frac{1}{2}$ = 3.5 Multiply your 20 piers times 3.5 c.y. per pier to get 70 c.y.: (20 × 3.5 = 70).

Please note that when you multiply numbers with decimals you must properly place the decimal point in your result. How do you place it? You

Fraction Equivalents

1/16	.063	6%
1/8	.125	13%
3/16	.188	19%
1/4	.25	25%
5/16	.313	31%
3/8	.375	38%
7/16	.438	44%
1/2	.5	50%

Fractions are converted to decimals by dividing the denominator into the numerator. For example, 3 ÷ 8 = .375

Here decimals are rounded off to three places. For example, 7/16 is rounded off to .438 from .4378.

Decimals are converted to percentages by replacing the decimal with a percentage sign: .25 = 25%

Percentages in the table are rounded off to two places, which is as precise as you will typically need for estimating and bidding.

count the number of places from the right that it is located in each of the numbers you multiplied. Then, you move the decimal place by the total count. Thus, in our example above, you moved it only one place. Why? Because only one of the numbers that were multiplied, namely 3.5, contained a decimal point. And that decimal was only one place from the right.

Here's a more complicated example: 6.32 × 4.1 = 25.912. In the result, the decimal has been moved three places from right to left. Why? Because in 6.32 the decimal is two places from the right; and in 4.1 it is one place from the right; and two plus one equals three.

Key estimating math: rectangles and triangles

If the preceding section threw you, you will need a refresher course in elementary math to become a capable estimator. If you breezed through the basic math section, good! You can move on to reviewing further applications of the math for estimating purposes.

A calculation of area—namely the size of a two-dimensional figure such as the surface of a floor—is likely the one most frequently made in estimating construction costs for a project. Easiest to calculate are the areas of a square or rectangle and the quantity of material needed to cover them. To do the calculation, you use this simple formula: width by length equals area (W × L = A), wtih area usually expressed in the U.S. in square feet (s.f.) or square yards (s.y.). (Please note: If the area is horizontal, as in the case of a floor, you use W × L. If the area is vertical, as in the case of a wall, you might wish to change the formula to W × H, with the H indicating height. In the discussion that follows, to simplify the presentation, I have largely used L rather than switching back and forth from L to H.)

Of course, area alone does not tell you how many units of material, such as pieces of sheet goods or tile, you will need for your job. To get that number—to figure material quantities—you need to take another step, one that involves division. For example, to figure the area of a floor that is 20' wide by 40' long as well as the number of sheets of underlayment required to cover the floor area, take the following steps:

- Multiply 20' × 40' to arrive at a floor area of 800 s.f. (20 × 40 = 800).
- Multiply 4' × 8' to arrive at an area of 32 s.f. per sheet of underlayment (4 × 8 = 32).

- To calculate the total number of sheets needed, divide the total area of the floor (800 s.f.) by the total area of each sheet (32 s.f.) to arrive at 25 sheets (800 ÷ 32 = 25).
- In actual estimates, add for waste. (Note: Here and in other examples below I can't tell you exactly how much to allow for waste. That will depend on your experience, the conditions where you are working, and your skill at controlling waste. However, in Chapter Four I will provide a range of waste factors that I and other builders have found useful.)

If the material you are trying to quantify has an area that is not a whole number, but instead a fraction of a whole, likely expressed as a decimal equivalent, you must take an additional step. That step often will involve rounding off your fraction or decimal. Rounding off is just simplifying a number by shortening it while taking care to not significantly increase or decrease the quantity expressed.

For example, if you are figuring the quantity of 10"-square ceramic tile to cover your 800 s.f. floor, you go through the following steps:

- Convert one foot to inches: (1' = 12")
- Express the 10" dimension of the tile as a fraction of a foot, i.e., of 12": 10"/12".
- Convert the fraction of a foot to a decimal: 10" ÷ 12" = .8333', which you can round down to .83'.
- Multiply width by length (W × L) to get the area each tile covers and express the area in square feet: .83' × 83' = .6889 s.f.
- Round .6889 up to .69 s.f.
- Divide to get the total number of tiles you need for your 800 s.f. floor: 800 ÷ .69 = 1,159 tiles.
- Add a quantity of tiles for waste.

For more complex forms, to calculate an area and the materials needed to cover it, you may need to break down the area into components. Thus, breakdown is necessary for areas that are a combination of several squares or rectangles, such as a house with an L-shaped floor plan, as well as for certain triangles. It is also necessary—as you can see in the illustrations of geometric shapes and area formulas on the next page —for certain triangles and for areas that combine rectangles and triangles, such as a gable roof that intersects another roofline.

In construction estimating, after rectangles the shapes for which we most commonly estimate area are triangles. They fall into two basic groups. Right triangles—namely those with one right angle, i.e., a 90° and two 45° angles—are by far the most common. Occasionally we encounter triangles featuring acute angles (less than 90°) and an obtuse angle (greater than 90° degrees but less than 180°)—as in the Victorian apartment building in the photo below, which features both acute and obtuse angles.

Interestingly, the area of all triangles—whether right or acute/obtuse—is figured using the same formula: One half the width times the length (or height) equals the area (½ W × L = A) or (½ W × H = A). Why? Because, to begin with, if you take two same-sized right triangles you can put them together to create a rectangle, It follows 1) that the area of each of the triangles is one-half the area of the rectangle, and 2) that if the area of the rectangle is figured with the formula: (W × L =A), then for the triangle: (½ × W × L = A).

For triangular areas that feature obtuse and acute angles instead of a right angle, the visualization is more complex. But if you look at illustration #5, you will see

Geometric Shapes and Formulas

A - Area
L - Length
W - Width
H- Height
o - Degrees

4 Obtuse (Acute) Triangle

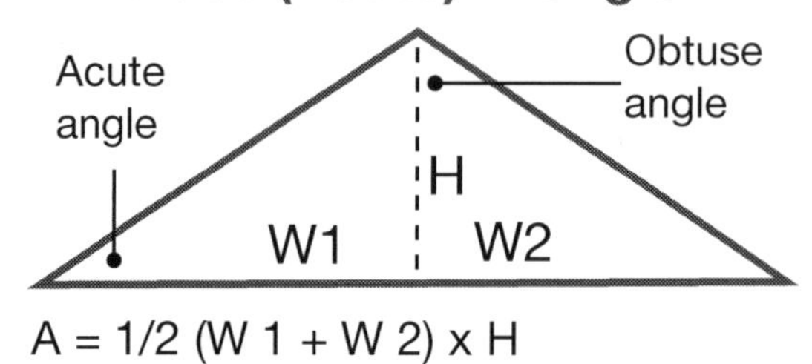

A = 1/2 (W 1 + W 2) x H

1 Square

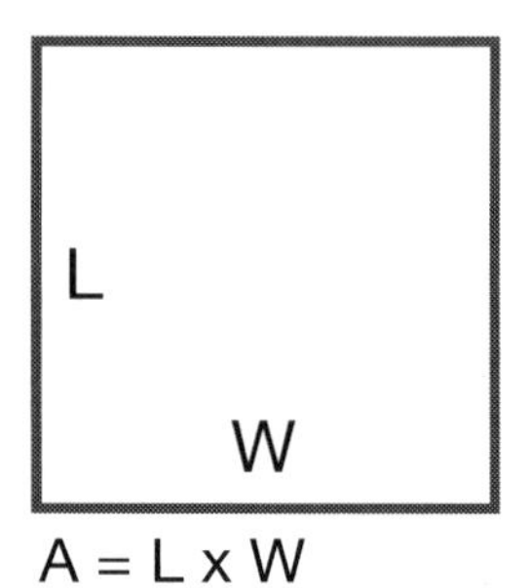

A = L x W

5 **Parallelogram** (formed by two triangles)

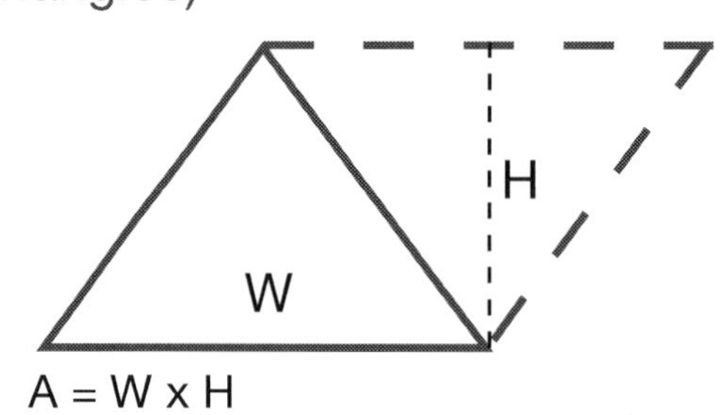

A = W x H

2 Rectangle

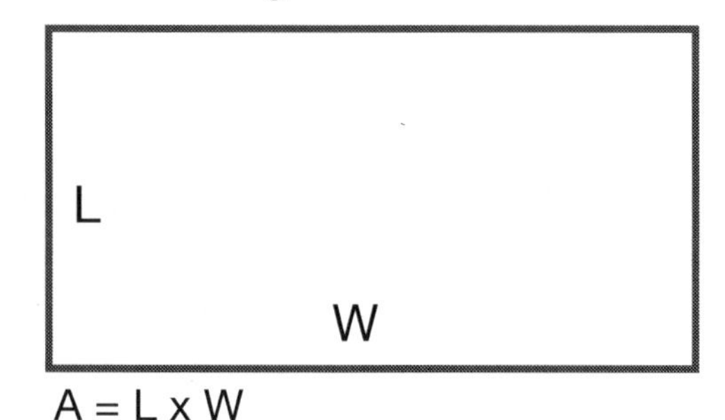

A = L x W

6 **Rectangle + triangle** (one side of a gable roof intersecting another)

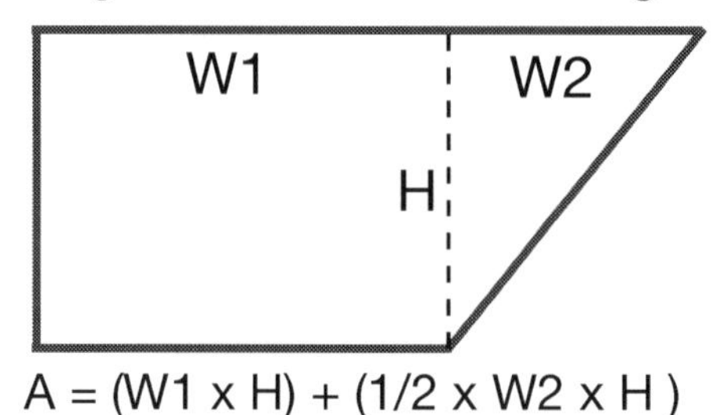

A = (W1 x H) + (1/2 x W2 x H)

3 Right Triangle

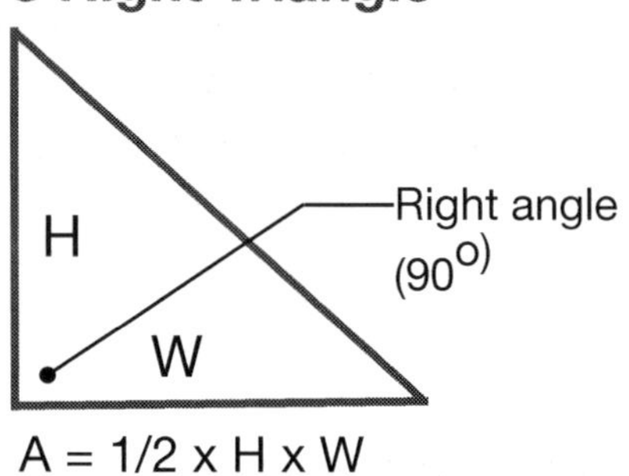

A = 1/2 x H x W

7 **Trapezoid** (floor of a bay)

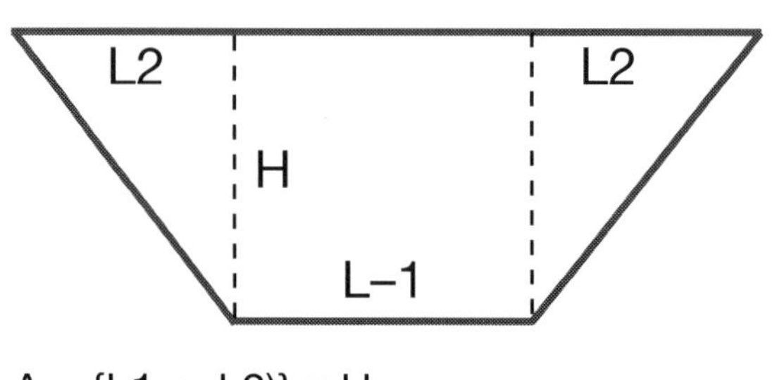

A = {L1 + L2)} x H

that if you take two of the

same-sized obtuse triangles and fit them together you get a parallelogram, which is basically a rectangle leaning to one side. If you imagine pushing the parallelogram upright till its angles are 90°, you will see in your mind's eye that you have reshaped the parallelogram into a rectangle. The area of the two triangles within the parallelogram did not change as it was pushed upright to become a rectangle. Just the shapes of the triangles changed. They became right triangles, and their area can be figured using the formula for the area of right triangles: (½ × W × H = A).

To figure the area of a single acute or obtuse triangle, such as the one shown at illustration #4, take the following steps:

- Divide your obtuse triangle into two right triangles.
- Figure the area of each one by using the formula for the area of right triangles: (½ × W × H = A1; ½ × W × H = A2).
- Add the results to get the total area of the obtuse triangle: (A1+ A2 = A).
- Alternately, take a shorter path to the same result. Just add the two widths together, multiply the total by the height, and divide by two: (W1 + W2 × L/2 = A).

To calculate the volume of concrete needed to fill this Sonotube, visible through the scaffolding, I used the formula: ($V = pi \times r^2 \times H$)

Occasionally you may need to quantify material for an area that combines a rectangle with a triangle. Again, breaking the shapes down into components simplifies the math. Thus, for one side of a gable roof (illustration #6) that intersects another roof plane, you can visualize the surface as breaking down into a rectangle and a right triangle. (If you are not seeing that, just go outside and find a house with a gable end roof butted into another one. You will see that each surface of the roof is made up of a rectangle and triangle). To get the area of the surface, simply figure the area of the rectangle and of the triangle and add the two results together ($A_r + A_t = A$).

As you gain fluency in estimating, you will find other ways to visualize areas to speed your work. For example, for a gable end roof that intersects another roof plane, the two sides of the roof can be joined in your mind's eye to form one long rectangle, and the area can then be calculated with a single application of the W × L formula. Similarly, with the bay floor (i.e., trapezoid, illustration #7) you can rearrange the three component forms into a rectangle and figure the area of the bay using W × L. I have not illustrated these visualizations here, but if you are working to strengthen your estimating math skills, you might want to grab a pencil and do so for yourself.

Key estimating math: circles

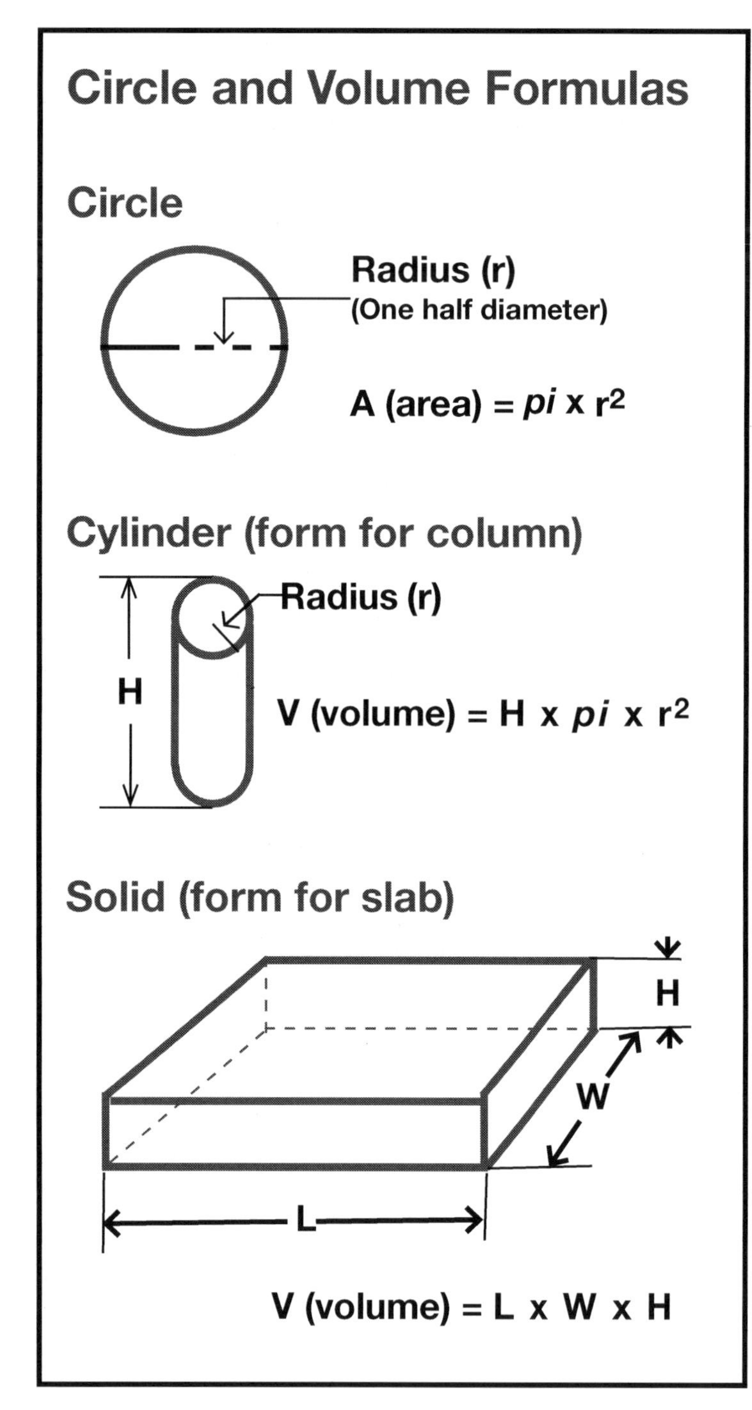

Circle and Volume Formulas
Circle
Radius (r)
(One half diameter)
A (area) = pi x r2
Cylinder (form for column)
Radius (r)
H
V (volume) = H x pi x r2
Solid (form for slab)
H
W
L
V (volume) = L x W x H

and cylinders

Figuring the area of circles and cylinders involves use of a special number called "*pi*." *Pi* equals circumference divided by diameter (*pi* = C ÷ D) for any circle. Yes, if you divide the circumference of *any* circle by its diameter, you get exactly the same result. For that reason, *pi* is known as a mathematical "constant."

Pi was first discovered 5,000 years ago and recently, with computers, has been calculated out to 13 trillion places after the decimal. Fortunately, for estimating we only need two of the places. We can assign *pi* a value of 3.14.

Using the value 3.14 and the radius (i.e., 1/2 the diameter of a circle) we can figure the area of any circle by multiplying *pi* times the radius squared (*pi* × r^2 = A). For example:

- Determine the diameter of a circle (Say, 10').
- Divide the diameter in half to get the radius (5').
- Square the radius, i.e., multiply the radius times itself: r × r = r^2 (5' × 5' = 25 s.f.).
- Multiply *pi* times r^2, in this case 25 s.f., to calculate area: (3.14 × 25 = 78.5 s.f.)

In your real construction life, you may never actually encounter a circular area for which you must calculate material and labor. Few buildings or rooms are circular. So why, you may be wondering, am I taking up your time going over the formula: (*pi* × r^2= A). The answer is that while you may never need *pi* to calculate the area of a circular room, you may well need it to figure the volume of material required to fill a cylinder, such as the sonotube form shown in the photo on the preceding page. Cylinders are encountered regularly in construction estimating.

To figure the volume of a cylinder, you need the formula for the area of a circle. Why? Because to figure the volume of *any* three-dimensional form, you first figure the area of a horizontal section of the form and then multiply that area times height (V = A × H). Therefore, to figure the volume of a cylinder, you figure the area of the cylinder's circular section and then multiply it by the height of the cylinder. (*pi* × r^2 = A; A × H = V)

For example, to estimate the volume of concrete needed for a column that is 3' in diameter and 10' high:

- Divide the diameter of 3' in half to get the radius of the circle formed by the column, i.e., of a circular section of the column: (3'/2 = 1.5').

Calculating rafter lengths

Calculating of rafter lengths involves the use of specialized math: square roots and the Pythagorean theorem.

Finding a square root is simply the reverse of squaring a number. For example: Squaring two gives you four ($2 \times 2 = 4$). Going in reverse, you find that the square root of four is two ($4 = 2 \times 2$). In other words, the square root of a number is another number that, when multiplied by itself, gives you the first number.

The Pythagorean theorem states that the long side, i.e., the diagonal side or "hypotenuse" of a right triangle, is equal to the square root of the sum of the squares of the two shorter sides: ($c = \sqrt{a2} + b2$). The length of a rafter is the hypotenuse of a right triangle. Therefore, if you know the run and rise of a roof, using the Pythagorean theorem, you can figure the length of a rafter from ridge to plate. Of course, to get the total rafter length you must add for the rafter tail.

In real-life estimating work, you are probably going to want to use a construction calculator to handle square roots and find rafter lengths using the Pythagorean theorem. Unless you happen to have become a square root master in high school, the math is just too involved and too tedious to do by hand. But if you understand square roots and the theorem, you will understand what the calculator is doing for you.

Pythagorean Theorem

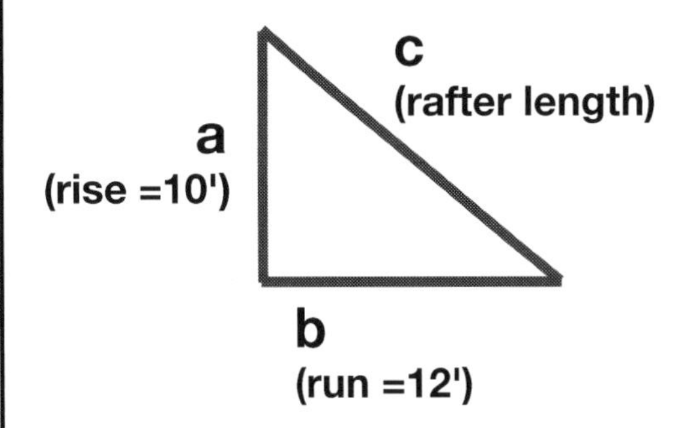

Note: The symbol $\sqrt{}$ instructs you to take the square root of a given number.

Theorem:

$$c^2 = a^2 + b^2$$

$$\sqrt{c^2} = \sqrt{a^2 + b^2}$$

$$c = \sqrt{a^2 + b^2}$$

Example:

$$\sqrt{c^2} = \sqrt{10^2 + 12^2}$$

$$c = \sqrt{100 + 144} = \sqrt{244}$$

$$c = 15'6\text{-}1/8''$$

- Use the formula for areas of circles ($A = pi \times r^2$) to figure the area of the column's circular section ($A = 3.14 \times 1.5' \times 1.5' = 7.1$ s.f.).
- Figure the volume of concrete by multiplying the area of the circular section by the height of the cylinder ($V = 7.1$ s.f. $\times$ $10' = 71$ cubic feet [c.f.]).
- Note that concrete volume is measured in cubic yards (c.y.), and that one c. y. contains 27 c.f., ($3' \times 3' \times 3' = 27$ c.f.).

- Convert cubic feet to cubic yards by dividing 71 by 27 (71 ÷ 27 = 2.63 c.y.)
- Round 2.63 up to 3 c.f. to allow for waste.

As mentioned above, volume is equal to area times height (V = A × H) not only for cylinders but also for any other three-dimensional form encountered in construction. Therefore, to estimate the volume of concrete needed for a slab 20' wide, 26' long, and 6" thick:

- Figure the area it covers: (20' × 26' = 520 s.f.)
- Convert the height—the thickness—of the slab to a fraction of a foot and then to a decimal: (6" = ½' = .5').
- Multiply the area times the height to get the volume in cubic feet: (520 s.f. × .5' = 260 c.f).
- Convert the cubic feet to cubic yards by dividing 260 by 27: (260 ÷ 27 = 9.63 c.y.).
- For an actual estimate, round 9.63 up to 10 or even increase it to 10.5 yards to allow for waste because the base of a slab is often uneven and extra concrete is needed to fill in the low spots.

Key bidding math

As we have discussed earlier, an estimate of your costs of construction is not the same thing as a bid. After completing an estimate of the direct, on-site costs of a project, you must roll it into a bid to arrive at the "selling price" you offer the client. The bid will include two additional numbers: 1) A charge for company overhead incurred during the project. 2) A charge called "profit" that you need as insurance against the risks of being in the construction business.

Some builders prefer to add overhead and profit in a single operation. In my view that is a mistake. I recommend adding overhead and profit in two distinct operations because overhead and profit are very, very different financial creatures. Overhead is a trackable, foreseeable experience of relatively fixed costs. Profit represents hope and serves as a buffer against unknown and unpredictable costs that can occur in the future. We will explore the two concepts in more detail and deal with the controversy about the best means of adding overhead and profit to a bid in Chapter Fourteen.

Here we will look only at the math necessary for adding on, or "marking up," for overhead and profit. One commonly used method for overhead is to calculate it as a percentage of the direct, on-site costs of construction. As you recall from the section on basic estimating math, a percentage is simply another way of expressing a decimal, which in turn is derived from a fraction. Thus, one quarter is expressed as .25 in decimal form, and .25 is the equivalent of 25% (¼ = .25 = 25%).

To determine the amount of overhead you want to include in a bid using the percentage method:

- Determine the direct, on-site construction costs of the job.
- Determine the percentage markup necessary to recapture your overhead incurred during the job.
- Convert the percentage into a decimal.
- Multiply the direct costs times the decimal.
- Alternately, if you are using a calculator with a percentage function, simply multiply the direct costs by the percentage. The calculator will make the conversion to a decimal for you.

For example, you are bidding on the renovation of a small playground. You have estimated that the direct costs for all labor, material, subcontractors, and services will run $42,000. You determine that to recapture the overhead costs that will be incurred during the project, you will need a markup of 22%. To determine the dollar amount to add into your bid for overhead, you multiply $42,000 × .22 and arrive at $9,240: (42000 × .22 = 9240). Or, alternately on your calculator: (42000 × 22 [% key] = 9240). To determine the amount you will add for profit, you perform a second such calculation—or you add direct costs to overhead and multiply the total times a percentage, such as 10%, for profit. For example, $42,000 + $9,240 = $51,240; and $51,240 × 10% = $5,124 for profit.

If you are new to construction, a 22% markup may strike you as extravagantly, unrealistically, unobtainably high. In fact, it is very much in the range of overhead markups used by well-established remodeling contractors in major metropolitan areas. And for profit, they typically add another 10%—or even more when markets permit.

But I am getting ahead of myself. I am digressing from the math and diving into the judgment calls necessary to create actual estimates and bids. Before we get into that material, I would like to take one last step here in Get Ready. Especially for those readers just starting out as builders, and also for those who have been at it a while but who are still struggling a bit, I would like to go over the terminology used for estimating and bidding.

The language of estimating and bidding

Every trade and profession creates its own specialized language. Words are given meanings that are not close to their everyday meanings. In law a "brief" is a lengthy argumentative statement. In basketball you "drain" jump shots from "the elbow." In the high-tech world, the "cloud" is no heavenly misty thing but a series of massive industrial facilities.

Estimating and bidding have their own language, too. As you go about learning the craft, unfamiliar terms can derail your attempts to understand crucial topics. I remember my own confusion when I first heard a builder use the term "overhead." I thought he had somehow switched topics and was referring to the storage bins above his seat on an airplane.

Here I will go through the basic vocabulary of estimating and bidding, not in alphabetical order, dictionary-style—as I do in the Glossary at the back of the book—but in the order that the terms come up during the work. My hope here is to provide you with an overview of estimating and bidding even as you acquaint yourself with the terminology. This section is intended first and foremost for readers who are new to estimating and bidding. If you are already fluent in its language, you might reasonably choose to only scan the section. Either way, you will notice that some of the terms covered here have come up earlier. They are included again for the sake of completeness and coherence.

Here I will go through the basic vocabulary of estimating and bidding not dictionary style — as I do in the glossary at the back of the book — but in the order that the terms will come up during your work.

Estimating and bidding for a project typically begin with a **lead**—here referring not to a project foreman but to a first contact from someone who hopes to build something. Builders in the world of heavy construction use the term **sponsors** to describe the varied parties, from public agencies to corporations, who contact them. In the light-frame world, we build largely for private **clients**, the people we used to call "customers" before we became so intent on creating a professional image for ourselves.

Clients often assume that you will be delighted to be contacted and offered a shot at providing them with an estimate and bid. They may not realize that even as you let yourself be interviewed, you are **qualifying** not only their project but also the clients themselves for your services. During really **slow markets**, the regular periods when construction work dries up, almost any client and project may qualify for a bid. During **strong markets**, you can be choosy and turn down projects that are boring or look to be financially marginal and clients that are offensive. You will be especially alert for the **grinders** who will maneuver to get as much as they can for as little as possible.

During the first contact with the clients you may record your initial impressions of them and their project on a **lead sheet**. You may even score different aspects of the project and the project as a whole in an attempt to bring a degree of numerical precision to your qualifying process. If the **lead sheet score** looks satisfactory, you may meet with the clients for a **site visit**. If the clients and their project continue to feel like a good fit for your company, and the clients likewise continue to be interested in you, you may accept their **invitation to bid**.

Your next step, if you are operating a **design/build** company that provides one-stop shopping for everything from initial conceptual drawings through working drawings and construction, is to draw up a **preconstruction services contract** for the clients. Once it is signed, you can begin drawing **plans** and writing **specifications (specs)**. Alternately, if the project is designed by others and is to be **awarded** to a builder via a **competitive bid** process, you will likely begin dealing with a designer chosen by an owner. For certain types of projects, you may never even meet the owners, but instead with an **owner's representative** or **project manager**, or with their designer acting as their representative. The designer may be:

- An **interior designer**, who would likely provide little more than drawings and notes that suggest the look of the project.
- A **kitchen and bath (K & B) designer**, whose plans are also likely to be thin on structural and mechanical details.
- An **architect** who will have passed a stiff licensing exam that allows her to legally call herself an architect, but who may or may not be a member of the **American Institute of Architects (AIA)**.

On occasion, even an architect may produce quite sketchy drawings barely complete enough for a building permit. But if the project is in a major metropolitan area, where building departments require highly developed **construction documents**, the plans may include drawings of the site developed by a **civil engineer**; foundation and frame details from a **structural engineer**; and from the architect at least the following:

- **Floor plans** showing the layout of walls, doors, windows, fixtures, cabinets, and other major installations.
- **Elevations**, views of vertical surfaces, especially walls.
- **Sections**, views created by cutting vertical slices or planes through the building.
- A **reflected ceiling plan**, a view of the ceiling as it would look reflected in the floor if the floor were a mirror.
- **Details** providing views, often at larger scale, of critical items.
- **Schedules** listing and describing dimensions, brands, model numbers, grade, and other detail about doors, windows, fixtures and finishes.
- **Electrical plans** showing the location for devices (switches, outlets, lights) and the panel.
- **Mechanical plans**, indicating the location of plumbing equipment and fixtures and HVAC (heating, venting, and air conditioning) equipment and registers.

If the mechanical plans provide only minimal detail, as is often the case, that indicates you or your subcontractors are expected to install the plumbing and HVAC on a **design/build** basis.

Accompanying the plans, you may receive a thick **project manual**, more informally known as the **specs**, dictating everything from construction quality to contract clauses.

Drawings and specs, aka **construction documents**, will likely be sent to your office not in paper form but in the computer file format abbreviated as **PDF**. If you are one of several builders vying for a project in a competitive bid, the documents you and the other bidders receive ideally would be close to complete (though in reality, that's often not the case) so that bidders are not making divergent assumptions about what to include and not include **(NIC)** in their bids. As you and your competitors proceed with your estimates and bids, you will receive notification of changes to the plans. They will be indicated by

clouds, i.e., cloud-like forms drawn around the areas of change, and/or by **color highlights**. Often a change is generated by new requests from the owner or by a **request for information (RFI)** sent in by a bidder who has noticed a contradiction or omission in the plans.

For some projects you may need to generate a series of estimates. First, you may create a rough projection of costs from a set of **conceptual drawings** (or **schematics**) that barely define the form and function of the proposed building. After further **design development**, you will generate from **preliminary drawings** an estimate that aims to be within 10% to 15% of the final costs.

Finally, you will take on the fully developed **working drawings** and really nail down your numbers—including **contingency** funds that you will advise the owners to set aside for additional work necessary to correct **hidden conditions** discovered during construction, or to provide for additional enhancements to the project. Along with your final estimate, you will create a detailed set of **assumptions** (aka **clarifications**) to compensate for the omissions that typically occur in plans produced by even highly competent architects.

Project: Sherman Addition **Date:** 11/21/17

Item	Qty	Lab Hrs	Labor Cost	Material Cost	Subs	Services
Gen Requir.		10	500	100		380
Foundation	30'	78	3900	900		
Floor Frame	5	4	200	180		
Sub Floor	4	2	100	120		
Wall Frame	26'	14	700	440		
" Sheath	7	4	200	150		
Roof Trusses	5	4	200	500		
" Sheath	6	3	150	80		
Plumbing		1	50	—	400	
HVAC		1	50	—	400	
Roofing		1	60	—	600	
Windows	6	12	600	2200		
Siding	260[illegible]	20	1000	520		
Ext Trim	10	5	250	160		
Subtotals		159	7950	5350	1400	380
TOTAL DIRECT COSTS: 15080						
		OVERHEAD	2600			
		PROFIT	1600			
PRICE (BID)			19280			

Notes: Shell only; all other work by owners

At all three stages of estimating, you will make use of that indispensable tool of estimating, the **spreadsheet**. Spreadsheets are often equated entirely with the electronic versions, especially Excel® from Microsoft. Handwritten spreadsheets were, however, used by builders long before computers were invented. A spreadsheet, as suggested in the accompanying illustration, is nothing more than a page or series of pages, whether paper or electronic, that are divided by vertical and horizontal lines into **columns** and **rows**, which in turn define small boxes called **cells**.

Across the top of your spreadsheet, just below the project description, you place column headings for: **Items and Assemblies. Quantities. Labor hours. Labor cost. Material cost. Subcontractors. Services.**

In the labor hours column, you place estimated hours for work to be done by **your own forces** (i.e., your **crew**). You then **extend** those hours into the labor costs column by multiplying the hours by an appropriate **labor rate**—your cost per hour for the employee who will do the work. The result, or extended number, is your labor cost. For example, as shown in the sample spreadsheet, you figure three hours for roof sheathing, assume a $50-an-hour labor rate, and extend into your labor cost column to get a figure of $150.

To estimate labor hours for a grade beam such as this one I formed with my crew, you could proceed item by item—excavation, forming, lining and bracing, rebar, mud sill bolts, hold downs, concrete installation, stripping and cleanup. For more efficient estimating, combine all the items into a single assembly and estimate labor cost for it by the linear foot.

In the **subcontractors** column, you place charges proposed by outside companies to do specialized portions of a project such as plumbing, electrical, insulation, and drywall. **Services**, a column I find useful, though it is not used by all builders, contains costs for such items as sani-cans and scaffolding that support work at a project but do not become a permanent part of it.

Often builders seek bids from several subs in the same trade, choosing the best bid, which may or may not be the lowest. Certain builders will incorporate one sub's proposal into a bid, proceed to win it, then shop around for a lower bid, and use the cheaper sub on the project in order to harvest the difference between the bids as extra profit. Such **bid shopping** is considered unethical. Savvy owners prevent it by requiring that bids include the list of subcontractors who will perform construction.

Labeling the rows in a spreadsheet is a much more demanding task than labeling the headings. A fully developed light-frame builder's spreadsheet can include hundreds of rows of items (sometimes called **parts** or **elements**) of work. Items are self-contained units—such as a mailbox, a sink, or a unit of studs—delivered to a project complete and ready for installation. During their start-up stage, builders sometimes instinctively estimate an entire project item by item, or **stick by stick**. For wall framing, for example, their spreadsheets display separately the costs for sills, plates, studs, and headers.

More experienced estimators, seeking greater efficiency, combine items into **assemblies**. For example, when estimating an exterior wall frame they would, at the least, combine sills, plates, and studs into an assembly. They might go even further and include headers, shear panel, wall wrap, and finish siding in the assembly.

All the varied terms—"items," "parts," "elements," "sticks," and "assemblies"—used to describe the content of rows in a spreadsheet can get a little confusing; at least they were for me when I began to learn estimating. The confusion worsens when the term "**unit costing**" is added in. After all, when you cost out the purchase and installation of a mailbox or wall assembly, are you not creating a cost for a unit? Are you not always costing out one kind of unit or another?

Well, yes you are. However, advocates of unit costing mean something more specific. They are suggesting that we cost items and assemblies with labor and material combined into a single number. For example, for subfloor they would have you record 600 s.f. for the subflooring in one cell of your spreadsheet and extend a unit cost that covers both the material and labor for that work, say $3,200, into the adjoining cell.

Beyond unit costing comes still another term for yet another process, **unit pricing**. Here you lump together not only your on-site construction costs for labor and material, but also your overhead and profit.

In my work and in this book I stick with the old reliables, "items" and "assemblies." I do not advocate unit costing and unit pricing. Unit costing denies you a separate look at your total labor costs for a project, for they are lumped together with material costs. That separate look, as we shall see in Chapter Thirteen, provides an invaluable check on the accuracy of your estimate.

Unit pricing, in my view, is a move still further in the wrong direction. With unit pricing, you are mingling together material and labor and then making matters muddier still by lumping in the costs of running your company and your hoped-for profit. That is no way to nail your numbers.

Along with developing an extensive list of items and assemblies for your spreadsheet, you will want to organize them into **divisions** of work such as concrete, framing, and plumbing. You may choose your divisions from those included in the **MasterFormat**™, published by the Construction Specifications Institute (CSI). Alternately, you may take MasterFormat as a starting point, but modify it to fit your project, or you may even develop your own sequence of divisions from scratch.

With your spreadsheet sufficiently developed, you are ready to drill down into the plans and specs for a project and produce a **takeoff**. The Associated General Contractors (AGC), an organization of companies that do larger-scale commercial and institutional projects, defines a takeoff more formally as a **quantity survey**. And in contrast to what the term "takeoff" would suggest in ordinary usage, namely a reduction, that is exactly what a construction takeoff involves, a systematic survey and quantification of all the items and assemblies in a project.

You can **take off** (the term can be used as both a noun, "takeoff," and a verb, "take off") quantities of material in a variety of units: **Linear feet, l.f.** for foundation forms and wall frames among other assemblies. **Square feet, s.f.** for sheet goods such as roof sheathing. **Cubic yards, c.y.** (or, for small quantities, **cubic feet, c.f.**) for concrete and gravel.

You can also use **board feet, b.f.** for taking off quantities of lumber. A piece of lumber that is one inch thick and measures one foot on each side (1" × 1' × 1') contains a **board foot** of material. A piece of lumber 1"× 1' × 10' contains 10 board feet (1 × 1 × 10 = 10). A plank 2"× 1' × 20 = 40 b.f. (2 × 1 × 20 = 40).

With quantification of labor, you encounter another choice of units. Estimators for industrial-scale projects may take off labor in **crew-days** or even **crew-weeks**. For example, three weeks for a crew of seven could be entered into the spreadsheet as "7-crew/3 weeks." In the light-frame world, **crew-hours** or simply **hours** are generally the unit of choice—for example, "3-crew/16 hours" for a framing crew, or "carpenter/12 hours" for a finish carpenter.

Your spreadsheet, with its pages-long list of items and assemblies, does double duty as a checklist that helps you to avoid missing work called for

If you have specialized in one area of construction and then move to another, your spreadsheet/checklist will likely not cover all items and assemblies in the new work. When a builder who specializes in kitchen remodels bid on the reconstruction of the structure partially shown here, he wisely subcontracted the concrete work, realizing it was beyond his expertise. But because he did not yet have a line on his spreadsheet for "under-slab insulation" it slipped by him. He did not include it in his bid either as work to be done by himself or by his subs and had to eat the cost.

in plans and specs. Over time, you will steadily build up your spreadsheet/checklist. It will increasingly help you to avoid overlooking what I like to call the "**slipperies**." (Items and assemblies that I have found to be elusive, i.e., slippery, are called out in Chapter Five.)

Especially for builders in their start-up phase, one entire division of costs can prove especially slippery. That is **General Requirements**, typically the first division of work covered in an estimating spreadsheet. General Requirements includes work and costs that occur steadily over an extended portion of a project rather than being associated only with a particular item or assembly. Examples include scaffolding, material handling and storage, and the big one—**project management** (also known as **lead person** or simply **lead**) **time**.

General requirements sometimes go by other names—**General Conditions**, **Direct Overhead**, or **Project Overhead**. I prefer "General Requirements" because the term "general conditions" is used as a title for the section of the designer's project manual that deals with contractual issues. Using the same name for a section of a spreadsheet is confusing. Similarly, "overhead" is the word used to describe the costs of running a company; using the same word to describe on-site costs of constructing a project can (and does) cause builders to erroneously categorize costs.

Of the various job site costs—that is, the **direct costs** for material, labor, subs, and services—labor is far and away the most slippery. To begin with, **labor rates**, i.e., the cost per hour for workers, vary greatly between persons with differing skill levels. If you assume the wrong skill level for installing an item or assembly, your estimate will be off. For example, if you estimate a day's labor for an apprentice for framing pick-up, and instead a full carpenter spends a day on the task, then the labor cost you put into your estimate will be low.

Labor rates, moreover, include not only the wages you pay your employees, but labor **burden**—all the many costs for items from health benefits through disability taxes to **consumables** (items such as saw blades, safety goggles, and tape measures) that must be incurred to keep an employee in the field and that are piled on top of wages. Burden can add up to half or more of a wage. If you are paying a crack lead carpenter a wage of $40 an hour, with burden added the carpenter will likely be costing you $60 or more per hour. Let burden, or just a single component of burden, slip by when you establish a labor rate and you can seriously throw off an estimate. The slippage will show up every time you use the wrongly calculated rate to figure the labor cost of installing an item, and that could be dozens or hundreds of times across an estimate.

To accurately determine labor hours and costs (as well as other direct costs) for their estimates, builders draw on various sources. Some turn to **cost books**—either paper books or the digital versions that are included with some estimating software packages. Others rely on the **job cost records** they keep to track the costs for the various divisions of a project while it is under construction: so much spent for the foundation, so much for framing, and so on.

Beware! Cost book numbers are likely to vary wildly from the costs you actually experience in the field. Job cost records also enable only rough approximations of actual costs. They are too broad. For example, overall foundation costs for one project do not accurately predict foundation costs on another. Builders intent on really nailing their numbers go beyond cost books and job cost records and develop **historical labor productivity records**, a term you can shorten to **labor productivity records** with no harm done.

A good labor productivity record includes descriptions of the assembly, the project, the crew, the conditions under which they worked, and their **productivity rate for the assembly**—for example, one-half hour per foot of wall frame. I have provided an extensive sampler of labor productivity records in Chapter Twelve.

For some items or assemblies, you may find costs so slippery that you just cannot nail them down. For those risky situations, you may want to include a **contingency** in your estimate, an added charge to buffer you against worst-case costs. Alternately, some builders attempt to move the risk, or some of it, to their clients by proposing in their bid to do part of or even an entire project on a **time-and-material (T&M)** basis. Rather than building for a **fixed price** or **lump sum**, they will charge the clients their actual costs plus a predetermined fee for overhead and profit. But as I have already mentioned, T&M contracts, appealing though they may appear, can bring on problems worse than those they eliminate. I view them as snake oil and think they should be avoided.

Along with calculating costs for work to be done by your own crew, you will also be collecting proposals from various subcontractors. To evaluate proposals from subcontractors, you can use a **matrix**, a sheet divided into columns and rows, such as the one shown in Chapter Eight. A matrix looks like a spreadsheet. But for estimating and bidding purposes it is used only to display numbers and not to multiply them and add them up as in a spreadsheet.

When your estimate for a project is complete, you must run the necessary **proofs** and **audits** to make sure your math is correct. Once it is, you can package your estimate together with a **recap sheet**. The recap sheet summarizes all the **direct costs** of construction—labor, material, subs, services—as well as the other charges to the owner, particularly **overhead** and **profit**, that are included in a **bid**.

Overhead, in more formal accounting language, is referred to as **Sales, General, and Administrative** expense **(SGA)**. In our world of construction, that's the costs for marketing such as job site signs and a web site (sales); the costs for maintaining an office, shop, yard, and equipment (general); and the costs of employing office staff (administrative). Tracking total company overhead is straightforward enough. But figuring out just how much of that total overhead to **recapture** on each individual project requires judgment calls.

Profit is even less straightforward than overhead. Most simply, and adequate for now, it can be described as a charge to cover the financial risks inherent in any construction job. Those risks

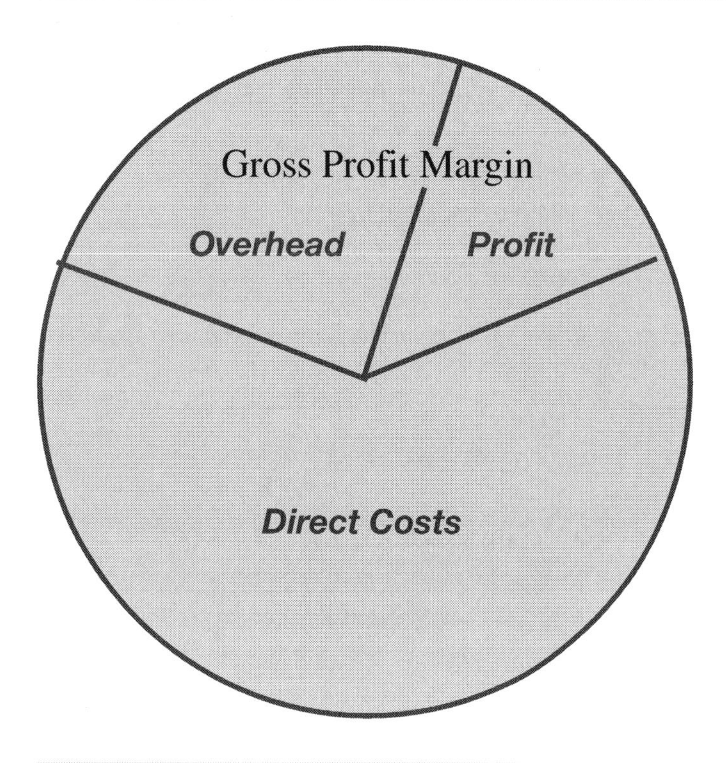

Gross profit margin includes overhead and profit. Selling price includes direct costs and gross profit margin (and also the costs of a bond, should the project require one).

include mishaps during the job, such as injury to a key worker. They include requests from the owner, i.e., **callbacks**, sometimes coming years after the project has been completed, to return and replace or repair defective work.

A handy definition covering both overhead and profit comes from the roofer David Haight. He thinks of overhead and profit as **company rent**—namely his charge to clients for mobilizing workers, materials, and services to roof or reroof a building and to take responsibility for defects in that work for years afterward.

Rather than "company rent," builders often use the more formal term **gross profit margin (GPM)** to describe overhead and profit combined. Some builders use a single calculation to figure GPM for a bid. Other builders prefer to calculate overhead and profit separately, using the method known as **marking up**. They mark up the direct costs of construction for overhead. Then they mark up the sum of the direct costs and the overhead for profit.

Whether you use the GPM or the markup method, for certain projects, you may need to add one more number to your recap sheet—the cost of a **bond**. A bond is a commitment from a financial services firm, known as a **surety**, to finance completion of a project on which you are bidding should you win the bid and then fail to **perform**. For typical residential projects, it is unlikely that you will be required to bring in a surety. Your word and your reputation will be the only bond your clients will require. For larger commercial jobs and for public works projects you will likely need a surety to back you up.

With direct costs, overhead, profit, and bond costs, if any, totaled up on your recap sheet, you have arrived at your **selling price**, the amount you will present to your client as your charge to build their project. But a bid can include more than a selling price. It can include written statements, particularly assumptions, aka "clarifications," that fill in the blanks you spot in the plans and specs and spell out what is **not included (NIC)** in the bid.

Builders are sometimes accused of **lowballing** their bids, even bidding at a price they know will be below their costs, to secure a job, then aggressively submitting **change orders** with extra charges for additional work in order to secure a profit. The practice is demeaned as **change order artistry**. It is, however, sometimes also the case that **bottom-feeding** designers or owners make it impossible for a builder to win a project at a fair price and thereby provoke change order artistry.

As you gain experience at estimating and bidding, you will develop a comprehensive master spreadsheet/checklist. From it, you can produce shorter spreadsheets for distinct categories of projects such as restaurants, kitchen remodels, tenant improvements, or new homes. You can save spreadsheets from specific projects and use them as **templates** for other projects of the same type. Creating templates is much easier once you have graduated from pen-and-pencil spreadsheets and automated your estimating with the use of a computer program. Then you will encounter additional specialized terms, but those terms are better left to Part IV, where we go into the use of computers, software, and electronic spreadsheets for estimating and bidding.

At this point, however, you can appreciate that if you master estimating and bidding, you have developed a valuable skill. Going forward you may elect to do the work of bidding without charge, as a "**loss leader**," i.e., as part of your marketing. But you may not. Many builders want to move beyond bidding for free. They don't like doing highly skilled work without pay. They insist on being paid either with a commitment from the client that they are moving toward a contract for construction or with a fee for their estimating work, or with both.

To achieve those goals, they have developed various processes. I call mine **cost planning**. Other builders use other terms such as **negotiated bidding**, **preconstruction services**, and **construction consulting**. From the architect's side of the industry, we are seeing a push in the same general direction that goes under the name **Integrated Project Management (IPM)**. Whatever its name, the service almost always includes **value engineering**, an analysis intended to turn up ways of bringing a project to fruition with the client's needs met and dreams realized, but also within budget. Whatever the name, builders bring value to the process only if they practice **transparency**—resist the temptation to conceal markups for profit in their estimates of other costs.

In Chapter Sixteen we'll go in detail into the subject of moving beyond bidding for free (one of my favorites). Meanwhile we'll turn our attention to the what, why, and how of the many processes our survey of construction terminology suggests. If you've got determination and have the necessary math and language under your belt you will, I am confident, master them steadily.

CHAPTER 3

Get Set

Choosing the right job to estimate

As important as knowing how to produce accurate estimates is knowing which jobs to estimate in the first place. Klay Gould, who started out as a carpenter and over several decades built a construction company that employs 100 people, believes that giving a client an estimate should be "the last thing you do, not the first. You want to move estimating much farther out." Ideally you want to move it out to the point where you have a commitment from the clients to work with you. At the very least, before beginning an estimate, you want to be sure you are a strong candidate for a project that is actually going to get built and that you are capable of building it at a profit. Estimating is just too expensive to give it away casually, for it is done by the most capable and most highly paid people in your company.

COMING UP

- Choosing the right job to bid and estimate
- Asking the budget question
- Qualifying the project and the designer
- Evaluating plans and specifications
- Evaluating other bidders
- Hitting winners
- Danger zones
- Saying no
- From bidding to estimating
- Site inspections

The same is true, of course, if you operate a more compact company than Klay Gould's and are yourself the estimator. Your time is valuable, and you can't be frittering it away on estimates that have little chance of bringing in business—the right kind of business. As Judith Miller, a consultant who has worked with builders across the country for decades, says, "before you head out the door" to discuss a project with potential clients you should know they "are good possibilities for you."

Usually your first sniff of a promising lead comes with a phone call or e-mail to your office. Someone who hopes to build something is contacting you to ask you for what they usually call an "estimate," though they really mean a bid with a selling price. Likely, they assume that you will be eager to provide them with one for free. Likely they do not realize how much work is involved for you. Even if they do have an inkling, they may believe what a prospective client once told me during our first phone conversation: "You are a contractor! You're supposed to give free estimates!"

From your perspective, however, the caller is, at this point, simply a prospect, someone to be listened to and responded to thoughtfully as you fill out a lead sheet. Your lead sheet likely will begin with a header for prospect's name (or names), project location, and other basic information suggested in the accompanying list.

If you are operating a design/build company and you learn that the prospective clients have already invested in design by others, you likely will decide to go no further with them. You politely ask them to give you a ring if they decide to move on from their designers. The lead sheet goes into the file you maintain for future marketing efforts. At some point down the road you want to remind the clients that you exist. Perhaps they will decide to call you first for their next project.

A lead sheet does not have to be filled out in any particular order. You don't want to put the prospective client through a question and answer drill, marching them down the list of concerns as if you were a medical assistant readying a patient for a colonoscopy. You are taking a first step toward establishing what may be a long-term, mutually beneficial working relationship.

You are tentatively bidding for the job, even if not yet estimating its costs. So listen attentively; don't interrogate. Let the conversation flow. Let the client know you are hearing them. Most of the concerns called out in the lead sheet will be covered spontaneously. Those that are not you can raise toward the end of the conversation.

You may have to be patient as the prospect describes the project to you. Clients are often naïve about the challenges and possibilities their project will present. My remodeling clients often fretted during our initial conversations about a necessary removal of an interior wall. They worried that their house might fall down or that the removal would be prohibitively expensive.

Lead sheet items

- Date filled out
- Prospect name(s)
- Prospects' contact information
- Prospects' relationship to one another
- Project location
- Property owned or not
- Project description
- Project suitability
- Stage of design
- Designer (if any)
- Engineers (if any)
- Clients' prior experience with building
- Reasons the clients are not calling prior builders
- Reasons clients are calling you
- Relative weight clients give to quality, schedule and price
- Other bidders
- Competitive bid? Or is client open to alternative project delivery methods?
- Schedule
- Budget
- Funding
- Client respectfulness and tone

They did not understand how straightforward replacing even a load-bearing partition wall with a carrying beam can be, or that the cost would be modest compared with other items in their proposed project such as foundation work, fire sprinklers, or cabinetry.

If the first contact comes from a person speaking not just for himself or herself but for others as well, you want to learn what their relationship is. Boyfriend/girlfriend? Girlfriend/girlfriend? Married couple? Business partners? Fellow investors or church building committee members?

You also want to learn whether the prospective clients actually own the property on which they are thinking of building. People seeking free estimates do call builders to determine whether a property purchase makes sense. A woman seeking feasibility information called two builder friends of mine during the Great Recession. Neither builder thought to inquire whether the client owned the property she wanted to build on. One, nevertheless, decided against investing the many hours he knew it would take to generate a reliable estimate from the woman's sketchy plans. The other, hungry for work to keep his crew going, plowed 200 hours into an estimate and bid. When the woman saw the costs of building her dream house, she decided against buying it.

If the clients do own the property and have engaged designers with whom you are not acquainted, you should get their names and check out their reputations with your fellow builders. Some designers are so disdainful of builders or do such sloppy work you want to avoid them. More likely you will learn that the designer is okay, though perhaps with an irksome tendency or two you will be glad to know about in advance. Now and then you will hear that the designer is such a joy to work with that you will want to really go after his or her project in hopes of initiating a long-term relationship.

When evaluating designers for potential bids, you will want to learn something about the quality and comprehensiveness of the plans and specifications they produce. Paul Winans, the former president of the National Association of the Remodeling Industry (NARI), insists that plans should be highly detailed, lest they cause misunderstandings during estimating and construction. I value completeness myself. However, I am also happy to work with sketchier drawings produced by others so long as the designer is responsive to my questions and respectful of my need to fill in the blank spaces he has left behind—and I get paid for the work.

David Bass, the founder of a highly regarded construction company and frequent speaker at construction business events, goes even further. He prefers working with interior designers, whose plans are notoriously free of the detailed information needed for construction. Why? Because the designers need him! They are focused on the look and feel of a project and "don't have a clue about how to build," says Bass. They need a capable and reliable builder to handle "the specifics of what's behind the look and the how-to of achieving those specifics." They need Bass's firm so badly they are willing to bring him into their projects as a collaborator rather than forcing him to bid against other builders.

As you develop your lead sheet and criteria for selecting projects for estimates, you will want to remain flexible. One builder recalls the propitious moment when he bent his usual rule that prospects must already own the property at which they hope to build. A woman had contacted him by e-mail. She was looking at a house that, coincidentally, he had hoped to buy himself but could not afford. During their first conversation, the builder recalled, the client gave him "a good feeling. She was real forthright. She answered all my questions straight up. She said she had seen my work. She liked it. Now that she knew I already had an appreciation for the house she wanted to buy and remodel, she was even surer I was the guy she wanted to work with. Sometimes you go with your gut, break your own rules. I gave her a lot of my time." The woman bought the house and hired the builder to design and build a million-dollar renovation.

Key concerns

As you talk to a prospect and fill out your lead sheet, you are trying to get a sense of whether the project is a good fit for your company. Is it within the capacity of your crew and subs? When construction time rolls around, will you have a lead available who can properly run the job? Is the project a type that you have found to produce good financial results? If not, does it offer other payoffs, such as a chance to build something exceptionally useful, beautiful, technically challenging, or all three?

A less obvious lead sheet topic is "client's prior experience with building." Even if the prospective clients have never worked with a builder, you may decide to spend an hour with them, especially during your start-up years. As my own company matured, however, I learned there was little chance that inexperienced clients would go with us or anyone else at our level. The costs of building with an established outfit were just too daunting; the rookie clients likely would go in search of a cheaper guy.. I often spent time with them anyhow. I enjoyed the conversations. And I found they could lead to the clients' calling me for their next project, not wanting to go through the frustration of working with a "bargain" builder again.

> ***"I am going to ask you something a little weird. We don't know each other very well yet. But may I ask, what is your hoped for budget?"***

If a prospective client does have experience, it is more than useful to inquire why they are not working with the builders that they had contracted with for other projects. Their answers may quickly disqualify them from further consideration and save you time. Chuck Tringali, a builder who works in the Hartford, Connecticut area, recalls a prospective client who phoned about a new and, at first glance, promising project. But during interviews, the builder discovered why the client had called him rather than his previous builders. One had been too expensive. Another had done shoddy work. "I thought there just might be a connection there," Tringali said. "The guy wanted good work but did not want to pay for it." He put the lead sheet in his dead file.

With the more matter-of-fact lead sheet topics covered, you may have developed a sense of comfort with the person on the other end of the phone line and are ready for crunch time. You are ready to ask, Why is it that the prospective client has called *me*? It is always nice, maybe even critical, to hear in response that you were strongly recommended by a previous client. One painting contractor in my area responds only to people who have been sent to him by his past clients. He has learned that it is not worth his while to try to sell a job to people who have not already been persuaded that his exceptional work is worth his high price.

Even if a past client has referred you, there are questions worth asking the prospective client. Just what was it in the recommendation that the client found appealing? Any answer may have implications, but if it's something along the lines of "a really good price," as opposed to a reference having raved about the quality of your performance, you should go on alert. There may have been special reasons why you gave a low price to the past client—a hole in your schedule that you had to fill, or just plain inexperience. You may not want or need to sell your work so cheaply again.

Especially if price does surface as the major consideration, you want to know how many other bidders the clients are talking to as they shop around. From time to time, the answers may amuse you. Iris Harrell, founder of a Silicon Valley remodeling firm, laughed as she recalled a prospective client who reassured her that Harrell Remodeling was only the fifth company she was inviting to bid on her project.

Experienced builders go beyond asking about the number of other bidders. They want to know the clients' criteria for choosing the winning bidder. And they want to know who the other bidders are. In response to the answers, they may accept the invitation to bid—or they may decline. One who specializes in foundation, drainage, and structural repair work says he draws the line at four competitors, and if he learns that one of the competitors is "Joe's Patio and Concrete Company," he declines to participate

The bottom-feeder

When prospective clients first call me, I usually take as much time as they wish to talk about their project. An exception was the guy whose first sentences to me were, "You a contractor? What is the lowest price per square foot you've built for?"

"Yes," I answered, "I'm a builder, and I am licensed as a general contractor."

"Good, good. What's the lowest square-foot price you build for?"

"Square-foot prices vary depending on the kind of project you are building."

"I want to know what is the lowest square-foot price you build for. I have a job. I need a price."

"I don't estimate by the square foot. It's not accurate enough. I keep cost records from past jobs, but not by the square foot. I probably could not…"

He interrupted me and asked again, with greater urgency, "What's the cheapest you build for?"

I wished him good luck, and put the phone back in its cradle.

You never know for sure, but I am fairly confident I was right to decide against spending further time with that gentleman. It was pretty clear; he was a bottom feeder, someone who would eventually dredge up some poor guy who would bid his project at below cost.

in the competition (though taking care to not denigrate Joe's outfit). He also walks away from people—typically, developers—who repeatedly seek multiple bids and always accept the lowest bid. He does not want to be competing on price alone.

Increasingly, as construction becomes more complex and the time needed to estimate its cost becomes greater, builders decline to bid for free against other bidders, especially during strong markets. One declares flatly, "I don't do competitive bidding. If they ask me and another guy to price it, I won't. I am not going to pour a lot of my time into a project if someone else might get it. I don't come out well on the numbers because I am really good at what I do. I see everything in the project, and I am going to include everything that is needed for construction in my estimate. I am not, like some other guys, going to keep a list of all the things that are left off the plans so that I can hit the client up for them later."

Another builder feels that for a public works project, competitive bidding can be okay. Then, he says, "the plans and specs are thorough, all the bidders have to be financially responsible, wage and insurance requirements are in place, and the authorities are going to hold everyone's feet to the fire and make sure the requirements are enforced. But for residential projects the plans and specs are typically so loose everyone has a different interpretation of them. And unfortunately, unqualified people get invited into the bidding process." A request for a free bid for a residential project, he adds, "sets his teeth on edge."

He explains all that to prospective clients. After hearing his spiel, the builder says, a prospect phoning him about a residential project usually ends the call. But about 10% stay on the line. And those folks, the builder figures, are good candidates for an estimate.

The lead sheet topic that many builders focus on more intently than on any other (except perhaps budget, which we will deal with in an upcoming section all its own) is client respectfulness and tone. One says, "My first question is, does the client want to be in a relationship with me, and vice versa." Another declares that what he likes to hear from a prospective client is, "We have seen your work. We are very excited about it and want to get together with you." He steers away from the people who ask, "Can you come over and give me an estimate," who see him as a source of financial information and construction expertise they can exploit without charge.

You can score a lead with numbers, colors (red for no, green for go, etc), or marks such as the pluses, checks, and question marks used here.

LEAD EVALUATION				
Client: *Don & Sue Aird* **Contact:** *634.7895 Aird@xmail*				
Proj. Address: *212 Hill Rd. Kensington* **Designer:** *To be selected*				
	Minus	OK	Plus	
Prior Bldg Experience			+	*Several projects for home and business*
Prior Builders		✓		*McNeil, retired. Other guy erratic*
Designer			+	*Client wants our recommendation*
Why Called?			+	*Strong reference from Doris Smith. Have seen our work.*
Project Description	???			*Professional grade photog studio. Upslope (steep?)*
Size			+	*800 s.f. one story—Guesstimate*
Commute			+	*Local*
Budget		✓		*$300K, could go higher*
Funding		✓		*Have cash "set aside"*
Prop Owned?			+	*Yes.*
Schedule			+	*Do right, no rush.*
Priorities		✓		*Willing to spend for quality. Want to use funds "effectively"*
Bid Process			+	*Want to work with builder can trust*
Our other Jobs			+	*Schedule open for estim. & build*
Crew Lead			+	*Fred available. Will be done with current proj.*
Crew			+	*Fred's crew available*
Subs			+	*"A" team available*
Suppliers			+	*"A" team available*
Client Tone			+	*Candid, open, listened, respectful*
Conclusion	*YES – PROMISING. Need to involve right designer, maintain budget control. Check access to upslope.*			

A third says, "I don't choose the project; I choose the clients. Was their first question about square-foot costs? Or was it about beauty, functionality, durability, and energy efficiency? Do they want to challenge the truth as I have experienced it? Or do they say, 'Yes,' when I ask, 'This is my program, is it right for you?' And do they value me enough to wait for my services? Or do they expect me to start on Monday?"

Follow up

To bring increased objectivity when evaluating prospective clients, some builders incorporate scoring into their lead sheets. Anita Wang, for example, records a numerical grade at each of her lead sheet topics. If the client registers a total score of less than 75, she dumps the lead. If they score higher, she schedules a visit with them. If they come in at 110 or higher, she schedules the visit optimistically, thinking to herself, "This job is ours."

I have used scored lead sheets and even included one as an illustration in *Running a Successful Construction Company*. Since its publication, I have realized they are not entirely reliable. They can mislead. Here's why: The lead sheet gives equal weight to all factors and uses a limited range of scores, say one to ten or simply a minus sign, check, or plus sign.

There's a limit to how low a score or how high a score any item on the lead sheet list can receive. As a result, a project might be exceptionally attractive in a few aspects. But even the high numbers they receive may not be enough to push the overall score up enough to let you know you want to go after the project. Vice versa, a project might be so potentially deadly in just one dimension—such as the clients' attitude toward builders—that it should be avoided. But the "2" it shows for client tone gets buried in its generally good scores, and so you go for it.

Usually a follow-up visit will help you spot problems, and also opportunities, that the lead sheet score does not adequately reflect. Wang vividly recalls a visit she made some years back. With a good-looking lead sheet score sitting on her desk after an initial phone contact from a client, she eagerly went out to the client's home. The meeting was awful. For three hours, Anita remembers, the client was "babbling away, unable to carry on a logical conversation."

Anita tried to back out of the meeting, but the client repeatedly cut her off and rambled on. Finally she escaped. Unpleasant though it was, at least the follow-up visit had given her the information she needed to downgrade the lead sheet score for the prospective job and reject it.

On the other hand, it is sometimes the case that a project might look barely okay on a lead sheet, but you go out for a visit anyhow and are glad you did. You begin to evaluate the project less by the numbers, and start to contemplate it more subjectively. You realize that it is going to be so architecturally striking you want to be the builder who brings it to your community. Or like Chuck Tringali, the Hartford, Connecticut builder, you see that a well-done renovation of a young family's dilapidated house is going to seriously enrich their lives. And you think that, all things considered, you are the builder most likely to deliver the project satisfactorily. So you decide to work with the clients despite budget constraints and other factors that drive down the lead sheet score.

Gordon Reed, a talented builder who started out as an apprentice in my company and has run his own for the last decade, does not put much stock in lead sheets. He believes that you get a much better sense of a person by meeting them at their property than by just hearing their voice over the phone. He likes to go out for a follow-up visit. "I am," he says, "always willing to give them an hour. I don't want to put myself out of the running too early. And, I want to see their body language. I want to see how their home is organized; is it in disarray or does it suggest stability. I want to see how they treat their kids. I am looking for graciousness," he says. "I want to know, do they see us as setting out to engage in a shared enterprise? Or am I merely there to do their bidding?"

The focus on tone and consideration has more import now than it did just a few years ago, and for two primary reasons. We used to say, back in the preweb twentieth century, that 10 times as many people would hear from an unhappy client as from a satisfied one. Now the unhappy clients can, via the web, etch their dissatisfaction permanently in a public record available to billions. In a word, they can "Yelp" you forever.

Secondly, because of the upward redistribution of wealth over the last several decades, builders now often find themselves solicited by people who have vastly more economic resources than the builders themselves. Some builders, understanding the power imbalance, prefer to not serve the super-rich, though their lavish projects may look tempting on the surface. Anita Wang stays away from them. "You will be like a Pluto to their sun for the rest of your life," she warns. "They pull you into their orbit, and they expect you to remain there, at their beck and call."

Of course, you cannot always tell from a phone conversation and a single visit just which client, super-rich or otherwise, might have malicious tendencies. Interviews only reveal so much. They are sort of like first dates; as a friend of mine says, "They are just brief encounters where a couple of people get together and lie to each other for an hour so." Everybody is on his or her best behavior. You can come away with the wrong impressions. Still, sometimes you can discern chronic anger or self-centeredness by letting the conversation drift away from the building project to other topics—neighborhood issues, politics, or sports. The way people express their opinions about such subjects—are they judgmental, harsh, and dogmatic, or tolerant, reflective, and appreciative?—can give you clues to what they will be like to work with.

If you sense too much turbulence you might want to back away. I have built a couple times for clients who turned out to be serial tantrum-throwers. The projects worked out okay—in part because I am comfortable with confrontation, and the clients backed off when I drew a line in the sand. But the truth is, had they decided otherwise, I could have been embroiled in a long battle, costly in dollars and peace of mind.

> ***If your business is in its start up phase you may be wondering, "how will I ever get any jobs if I make clients and projects pass all these tests?" It's a legitimate question. It has a straightforward answer.***

Paul Winans asks, "Ever work with a client from hell? It isn't worth it." Those guys are out there. Spot their tendencies if you can. Avoid them. As the builder Mark Kerson, author of *Elements of Building*, puts it, "Hire good clients."

Asking the budget question

"The super-rich do have this saving grace," one builder notes sardonically. "They can pay you if they feel like it." Don't take it for granted that all owners can, that they have either a reasonable idea of what their project will cost or the financial resources to see it to completion. They may not. Whether it is during the phone interview or a follow-up visit with the clients, you must find out if they are in touch with financial reality as *you* experience it. Are they envisioning a price that is in line with what you have learned you must charge (as opposed to what they have heard another outfit across town charges) to do reliable work, recapture your costs, and earn a decent living?

One way or the other, you are going to have to bring up the budget issue. Every experienced builder I have ever asked about it considers the "B" question crucial. They raise it with forethought, care, sometimes with a degree of delicacy, sometimes quite directly. Antonio Rodriguez declares, "I always talk turkey during the first phone call. Asking the budget question is like asking for contact information. I have to have it. I ask them the question, and then I shut up. That silence is usually rewarded with a number."

When I ask the question, I try not to spring it without warning. I tell the clients that I am going to raise a question some people find a bit uncomfortable but that we need to explore. In response to that gentle approach, clients usually do give me a figure. I reciprocate, giving them honest feedback about their budget: It could be ample (rare). Adequate for at least a restrained version of their project (typical). Too low for my company to do their job properly (occasional). Or—now and then—dangerously low, an invitation to unqualified builders or outright scam artists.

If it is the last case, and they are rookie clients still naïve about the actual costs of construction and disinclined to believe me, I might tell them about the young couple that budgeted $100,000 for a second story addition to their home. They signed up a guy they had seen working around the neighborhood who said he could build their addition for that amount. After he gutted the existing house and partially framed the second story, he told the couple he needed his next hundred thousand. They did not have it and had to abandon their house.

Other builders—typically people who have spent a lot more time in sales training than I have—use a more oblique and subtle approach. They probe, they pull back, they persist and probe again. One says that if the clients do not respond to his first cautious probe, he asks if he could summarize what he understands to be their hopes for their project. If they reassure him that his summary is about right, he tests them with a figure: "In our world," he says, "what you have in mind, delivered at a high-quality level, is currently costing around $120,000." Then he pauses. He waits. If the clients respond with, "You know, that's about what I'm expecting, that's what our friends paid to have similar work done," he considers the budget question answered.

On the other hand, if the clients give an unrealistically low figure, say $35,000, the builder answers gently, "I hate to say this, but good quality work is awfully expensive. What you are describing can run as high as $135,000." The conversation might end right then and there. But at times the clients respond with, "Well, we might be able to come up with $90,000." The builder thinks, "I've got a chance." The clients have just moved $55,000. They might move a bit higher even while the project scope is dialed back. Now that there's a chance for a contract for construction, the builder is thinking he'll invest more time with the client.

Klay Gould, the founder of the 100-employee company mentioned earlier, exercises the probing approach with even greater subtlety. During the first phone conversation, when he senses the moment is right, Klay interjects, "I am going to ask you something a little weird. We don't know each other very well yet. But may I ask what is your hoped-for budget?"

Klay does not drop the question once he has an initial figure. He comes back to it, testing for commitment and flexibility with other questions: "If the project goes over the hundred thousand you hope to get it done for, how will you feel about that?" Or "Have you considered the advisability of your budget relative to the value of your property?" Or "How is that kind of expenditure going to affect your life? Are you concerned that if you do all the work you have in mind that you might end up with a dream place but house poor?" The important thing, Klay emphasizes, is to come back to the "B" question at least three times.

Underneath the "B" question there's a second related question, obvious enough, but even more awkward to raise. It is a question, Klay observes, that "can be offensive." For that reason, he says, "many builders don't ask it and architects almost never do. But it should be the basis of everything." The question is, Okay, you have established a budget, but where is the money coming from? In other words, Do you have the dough now? If not, where and when are you going to get it?

It is even more important to ask that question for commercial than for residential jobs. More frequently in the commercial than in the residential world, owners are sometimes still trying to corral financing and even development rights as they begin seeking bids for construction. They may even sign a construction contract and ask you to begin work before all financing is in place. The Associated General Contractors (AGC) warns of the potential results: "Many general contractors have gone broke because they neglected to find out where funding was coming from."

At this point, you may have a question of your own. You may be wondering this: How do you know the budget information clients give you is valid? How do you know they are telling the truth? One old hand answers bluntly, "You don't, you just have to trust them." But he is also confident that if clients are not straightforward with him he will pick that up. "I've got a pretty good bullshit detector," he says. I believe him. And I'll venture to guess that almost all successful builders develop a detector over the years. If you don't you are likely to spend a lot of hours chasing jobs that won't happen.

Qualifying the project and the designers

Just the lead sheet issues we have discussed so far add up to a pretty difficult obstacle course for clients interested in obtaining a bid. If your business is in its start-up phase (or if you are battling through a recession), you may be wondering, How will I ever get any jobs if I make clients pass all these tests?

Qualifying projects: external factors

Designer
- Reputation with builders
- Responsiveness
- Fair mindedness
- Design talent
- Technical competence
- Environmental intelligence

Plans
- Pages (all present, in logical order)
- Clearly organized notes
- Site grade accurately shown
- New and existing on same page
- Adequate floor plans, sections, elevations
- Essential mechanical & electrical plans
- Accuracy of building dimensions
- Correspondence between architectural and engineering plans
- Conflicts between structural and mechanical systems
- Waterproofing details
- Adequate detailing
- Window schedules
- Finish schedules
- 3-D rendering

Specifications (Project Manual)
- Clarity and organization
- Number of alternate prices required
- Break down of bid required
- Time allowed for construction
- Sequence of construction
- Contract required
- Potential penalties if any
- Change order procedure
- Payment schedule

Other burdens
- Affirmative Action
- Prevailing Wages
- Bonds

Other Bidders
- Capabilities
- Reputation
- Need for work

That is a good question. It has a straightforward answer. You won't. You will not qualify the clients or their project so strictly when the market is poor or when you are starting out. You will build for a client you realize may be difficult. You will relax the budget test. You will work for less than you will during better times or further into your career. You'll do that to get established, to build a client base and references, or just to survive.

If you are just starting out, or just beginning to get serious about running your company in a businesslike way, you may create and begin using a lead sheet. That would be a good habit to get into. But you won't require scores as high as established builders with strong reputations do. When I first went

Many of the small projects—such as decks, tenant improvements, or this stairway I built—that come along during the start-up phase of your career are really design/build projects for which you need to create specs and plans. You can learn to charge for that work, helping clients understand that design needs to be thought through for their project to be a success, that it is time-consuming, and that you need to be paid for your time and/or the time of designers you hire. You will then have created a design/build option into your operation from the beginning.

into business for myself, I did not know what "lead," much less "lead sheet," meant. At the time I was a union carpenter. Along with 90% of the other guys at my hall, I had been thrown out of work by a severe recession. At my wife's suggestion, I got my general contractor's license and plunged into business for myself, taking every job that a neighbor, acquaintance, or friend of a friend offered me, never asking a single qualifying question. I was so eager for work and so grateful for it that I viewed everyone who asked me to do a job as a kindly person who could not possibly have any but the purest intentions. Fortunately, never, during that start-up period, did I encounter malicious clients or any so greedy they refused to pay me, and I never took on a project that was terribly over my head.

In time, though, I understood that I had been lucky and that prospective clients—and to an equal extent, their projects—should be scrutinized to determine whether they are a good fit for one's company. If a prospective project is one for which no independent designer has been or will be engaged, then there is little qualifying to be done. Only a few questions must be asked: Am I (or is my company) competent to do this job? Can we design it effectively, hire someone competent to produce the design, or connect the owner with an independent design firm? Can we build it? Is the job within our geographical range? Do we want it? If those questions can be answered affirmatively, and the client is okay too, you can sign a preconstruction agreement and move ahead.

If the project has come from an owner who has already engaged an independent designer, or directly from a designer representing an owner, then you move on to a more detailed project qualification procedure. You will be concerned with qualifications at two levels: *External,* the characteristics, conditions, and circumstances of the project. *Internal,* the capabilities of your company relative to the project.

The first external item to consider is the design firm. Start-up builders sometimes assume that they have hit the big time when professional designers begin to call them. They assume that the work and behavior of those professionals will, in fact, be professional. Oh, if that were only so! I don't think I have ever talked about designers with an experienced builder who has not expressed anger at some individual designers and/or dismay at the design professions, especially architects, as a whole.

There are good reasons for the frustration (as there are also good reasons for the dismay designers express toward builders), enough to fill another book. They include:

- Surprisingly low levels of training in how buildings are actually built. A popular book on architectural training lists some 30 areas of knowledge architects must engage with. Building is not one of them. It is not even mentioned.
- Disregard for builders' relationship with clients. Designers regularly disrupt builders' relationships with their clients, even when the builder has brought that client to the designer.
- Avoidance of responsibility. AIA (American Institute of Architects) contracts massively shift legal and financial responsibility for projects from architects to builders.

Fortunately, many individual designers surmount the dysfunctions of their professions. A few even know how to build. Even among those who do not, many pay attention as best they can to constructability and appreciate good construction. Those are the guys you are looking for.

To learn about the designer of a project you are considering for a bid, ask around among builders you have met at supply yards or at meetings of your builders' association. Haven't joined one? Please do. Now! Nowhere will you get better education and guidance than from your peers. You can ask them questions like these about designers you are evaluating:

- Is the designer generally well thought of among contractors? Would builders who have worked with him or her be willing to do so again?
- Is the designer responsive? When construction is underway and questions come up, can you get them answered, and fast, so that your crew is not spinning its wheels while awaiting a response?
- Does the designer recognize that no set of plans or specifications is ever complete, that extras do crop up during construction, and that the builder needs to be paid for doing them?
- Rather than trying to shift blame, will the designer own up to his missteps—like the architect who wrongly specified the dimensions of several windows ordered for a project; did not try to hide behind the standard phrase that appears in plans, "Builder to verify all dimensions in the field"; but instead paid for the windows out of his own pocket and incorporated them in a playhouse he built for his daughters?

Included in my own designer qualification checklist are a couple of other items: Does he or she produce designs I respect? I want to add good-looking and functional work to my community, not ugly buildings. And does she have "environmental intelligence"? I prefer designers who have put in the hard miles to develop a meaningfully green practice. They have gone beyond what I call "gadget green," the knee-jerk specifying of fancy equipment that is costly, perishable, and may well consume more resources than it saves. They have arrived at "green simple," the disciplined, step-by-small-step reductions of resource consumption for both the building's construction and operation. For example, do they call for wasteful traditional framing or for the "frugal framing," aka "advanced energy efficient framing" I call for in my book, *Crafting the Considerate House?*

Evaluating plans

Along with a methodical approach to sizing up designers, you need a systematic means for evaluating plans and specifications, perhaps even a checklist, covering such basic requirements as:

- *The pages of the plans should be sequenced in a logical order.*
- *Each page should be clear and easy to read*, with information arrayed so that it can be readily located. Plans from some designers seem to be modeled on minestrone soup, with information tossed into whatever open space was available. Other designers present information right where it is needed and in well-organized columns.
- *All pages must be numbered*, with no numbers skipped.
- *The architectural and engineering plans should be precisely coordinated*, with vertical and horizontal lines clearly labeled to facilitate easy navigation back and forth between the drawings.
- *The plans must be accurately and consistently dimensioned.* A wall broken into segments of 17'8", 6'4", and 10'3¾" will be shown correctly as having total length of 34'3¾". The dimensions are consistently shown from face of frame to face of frame or from a centerline. They are not shown from face of frame to face of finish.
- *Elevations must correspond to floor plans and sections are accurately drawn so that you can see that the levels of the building will connect as intended and that specified structural components will fit together as necessary.* Be suspicious of plans that do not include sections. They might not work. I once drew up preliminary floor plans for a two-story home and discovered only when I drew a section through the structure that there would not be adequate room for the stairway unless I extended the building with a bay.

In residential work especially, designers tend to provide little direction in their plans regarding HVAC, plumbing, or electrical systems. All you are likely to get from them is the location of equipment such as the furnace and of points of service delivery—namely location of registers, fixtures, and devices. But those should be indicated. Seriously important, *the suggested locations of equipment and points of service must not result in conflict with structural or finish elements.* For example, to get from a W.C. to the sewer connection, a waste line won't have to go through an 8" × 14" glulam intended to support the second story.

As with mechanical and electrical, you often don't get much in the way of *water management detail* (roof drainage, perimeter drainage, flashing, vapor barriers) in the plans. Ideally, what is shown will be correct. Unfortunately, it may not be, and you will have to step up and provide for it yourself. I was visiting with a builder friend, who is my go-to guy on water management, as he unrolled a set of plans from a prominent architect. "They already screwed up the waterproofing details," he noted matter-of-factly, but with not a hint of frustration. That was exactly as he had expected. It was the norm. He always has to provide drainage and flashing details.

Beyond such basics, you will be looking for a level of detail that works for you. As mentioned earlier, builders have different preferences here, ranging from extensive to sketchy detailing. Though I am happy with either, I tip my hat to the designer when I see fully detailed *window schedules and finish schedules*. While not a guarantee of all-around competence, they suggest that the designer is on top of her game and that the plans will turn out to be reliable during construction. Other experienced builders I know emphasize other items in their wish list for plans:

- *Existing construction and new construction shown on the same page* for remodel projects. That saves them from having to jump back and forth from one page to another to see the relationship between existing and new work.
- *3D drawings* that allow them to more easily envision how the proposed project will go together.

- *All details fully and clearly drawn* rather than vague references being made to "typical" details—particularly for modernist designs featuring difficult, and never merely "typical," details such as large expanses of glass butted right against dissimilar materials.

While it is useful to hit the high points mentioned here, and possible to do so in a few minutes once you are fluent in plan reading, it is not possible to fully vet a set of plans when you are merely qualifying a project for a bid. Paul Eldrenkamp, who operates Byggemeister, a highly regarded design/build firm in the Boston area, has created a thorough plan checklist for his own projects. It is 50 pages long. Time constraints will prevent you from putting plans through such a fine screen when simply considering whether or not to bid on a project. You can only hope that the designers have, and if they have not, that one way or the other you can be paid to spot their omissions once you qualify a project for an estimate.

Evaluating specifications

Specifications should be well organized, clearly written, internally consistent, and specific to the project. Badly organized and muddy writing is symptomatic of muddy thinking. You don't want to have to grope around in the mud and repeatedly go to the designer for clarifications. And you don't want to be stumbling across ludicrous specs, like the concrete spec described in Chapter Two that had been left in place when the specifications from one project were sloppily adapted to another.

The specifications must give you *adequate time to bid the project*. You do not want to be rushed and forced into a mistake. You want your subs to be given adequate time to get their bids to you so that you do not have to generate numbers for trades at which you are not expert. You will get those numbers wrong, like the remodeler who rushed to submit his bid, tossing in what he assumed to be an ample $4,000 for cabinets, only to find during construction that no shop would build them for less than $8,000.

Alternates, i.e., separate pricing for different areas of a project, such as a patio at the rear of a restaurant, should be limited in number. Each alternate requires a separate estimate and bid and will involve separate costing of labor, material, services, and sub trades along with separate markups for overhead and profit. Estimates and bids for projects with numerous alternates are costly to produce. And, as the AGC points out, "not only do alternates drastically increase the work of estimating; they increase the risks of mistakes exponentially." Why? Because parsing the work to be included in a base bid from the prices for the alternate work can be tricky. You can easily miss an item. You can mistakenly think it is in the base bid, therefore leave it out of the alternate, and thus capture it in neither.

Similarly, the specifications should allow you to *efficiently sequence construction*. They should provide that you be given access to the entire project from the beginning of demo and deconstruction to the completion of painting. If you are asked to renovate one section at a time while keeping others operating, you have really been asked to bid a series of projects, not one. You are in effect bidding alternates.

The specifications should ask for a *reasonable breakdown of the numbers*—framing, foundation, etc.—in a bid. For competitive bids, the preferred number of breakdowns is none, zero. You want to submit a total price to do the job, and that is it. I am not alone in my opinion. Other builders I have talked with resist participating in competitive bids if breakdowns are required. They are leery of owners comparing the breakdowns in the various bids, and using the lower numbers in one bid to lever down higher ones in another.

Savvy builders do not want to be sucked into that game. They understand that their bids are subject to the law of averages, as we discussed in Chapter Two. They know that the higher numbers in their bids are likely to be balanced out by the lower ones; and that if they give up the high ones while keeping the low ones, they are setting themselves

American Institute of Architects' Contracts

I resist using AIA contracts. They are written in legalese so dense that it is, I suspect, understood by few of the architects and fewer of the builders who sign them. Even their lawyers may not. While serving as an expert witness in a construction defects lawsuit, I pointed out a clause in the AIA contract allocating responsibility for writing up orders for extra work to certain of the defendants in the case. Their eight attorneys scrambled for their notepads. They seemed to have no idea the clause was in the contract.

The AIA, even as it incorporates overwhelming detail in its longer contracts, leaves out crucial provisions elsewhere. Some years back, I was asked to review a new AIA contract intended for small projects. To my amazement, I saw that the contract made no provision for changes, either extras or deletions, during construction. That's like producing a car without brakes. Inadequate change orders are one of the main sources of lawsuits arising out of construction projects.

up to take a financial bath. Unfortunately, for many commercial and public works projects, a bid submitted without the exactly prescribed breakdown may be disqualified and tossed out. If you really want to compete for that work, you may have no choice but to display your charges broken down into the MasterFormat categories or even something more complex.

Specifications should provide for *adequate time to build the project*. You don't want to have to rush a job. Haste makes waste in the form of errors, callbacks, insurance claims, damage to your reputation, and even lawsuits. Also, an overly demanding schedule may require you to hire too many new people at once in order to meet the schedule. New hires can turn out to be mistakes. Even when not, they need time to get up to speed with the way you do things. Until they do, productivity will drop off. An overly demanding schedule can force overtime work. Overtime, too, lowers productivity even while it increases wage rates. It tires workers unduly. Tired workers have accidents and get hurt; and injuries drive up workers' compensation insurance rates. A high workers' comp rate can seriously impair a construction company's financial performance or even put it out of business.

Injuries can also impair your spirit. They can drain away the satisfaction you can get from your work if your guys have been injured because of your bad management choices.

The specs should allow you to use your own *thoroughly developed, fair-minded, balanced, and comprehensive contract*, not mandate one designed to protect the designers (or owners) while exposing you to excess liabilities. Indeed, if a project requires use of an AIA contract, you should turn it away unless you thoroughly understand that contract. That's a major undertaking that will bring you to the conclusion that no builder in his right mind would sign such a contract in unqualified form. As Wayne DelPico writes, "It should come as no surprise that contracts written by architects tend to favor (them)." In fact, he adds, they "tend to favor the architect or owner to the point that the risks involved greatly outweigh the chance of success or profit and bidding would be a waste of time." Even if the specifications do allow you to use your own contract, they may nevertheless contain contractual stipulations. Your contract should adjust any such stipulations so that they are fair, especially any requirements for penalty payments from you to the owners for delivering their project allegedly behind schedule. You should not be penalized for delays caused by events, such as bad weather or labor shortages, beyond your control. And any damages should not exceed the real economic losses to the owners.

Likewise, you must systematically review the specs to determine that they do not burden you with unreasonable *warranties* such as this one from an actual project manual: A requirement that the builder guarantee, even after construction was complete, both the volume of water and the temperature of the water flowing from a nearby spring to a heat pump included in the project. Fortunately, builders who were bidding for the project caught the absurd requirement and complained. It was dropped.

Crucially, if the specs get into the matter of the extras and deletions that crop up in almost all construction jobs, they must provide for an *equitable change order process.*

They must also provide for *an acceptable payment schedule*—frequent payments made immediately after substantial (not total) completion of predetermined phases of work. For example, in my operation, when the first floor of a new structure is substantially framed, with only minor items such as a few blocks or sheets of subfloor not yet in place, whatever amount is stipulated in the contract for first floor framing is due within three days. I reject payment schedules that put me in the position of effectively making loans to clients—taking on obligations to pay for labor, material, subs, and services to build their project while they delay payment to me. I am, I explain to my clients if need be, a builder, not a banker.

Other builders accept a less rigorous payment schedule—and as a result, some of them find themselves stiffed now and again at the end of a project for tens or even hundreds of thousands of dollars. Whatever payment schedule you decide to live with, be wary of any that will cripple your cash flow. Avoid the experience of the roofer who took on work for a nearly insolvent city. He had to wait months for payment. As a result, he could not pay his suppliers. They in turn moved to close his account. He was on the verge of shutting down his business when, in the nick of time, the city paid a part of his bill.

It is not possible here to lay out just how every potential specification should be handled. But if you enter the commercial and public works arena, there are a final few issues beyond those encountered in residential work that may come up and that have especially significant financial consequences. Are *affirmative action requirements* in play? If so, can you fulfill them without taking on too many new and untested people? Do the specs mandate *prevailing wages*? If so, what are they in the area in which the project is located—union-level compensation or some other package? Are *bid and/or performance bonds* required? If so, and you don't intend to post them from your own cash reserves, can you obtain reliable quotes from a surety? And to what extent will that surety actually cover you if you fail to perform, and to what extent will you remain exposed?

Evaluating other bidders

Last, though not least, of the external factors you want to take into account when considering whether or not to competitively bid for a project are the other bidders:

- Are they so eager for work they are likely to bid at a level too low for long-term survival?
- Do they employ local qualified tradesmen and subs? Or do they import ultracheap labor, like the contractor working on a Tesla plant who was deploying Eastern European workers at five bucks an hour?
- Do they run a legal operation, carrying liability and workers' compensation insurance and paying employment taxes, or are they an under-the-table operation exploiting undocumented workers?
- Do they do good quality work? Or, as was the case with a contractor I checked out prior to accepting an invitation to bid against him, is their framing so sloppy that half of their shear nails miss their targets and you see a trail of shiners dangling along both sides of each stud?

Right job to bid?

Financial Factors
- Ability to estimate costs accurately
- Insurance availability
- Cash flow impact
- Adequacy of accounting

Physical Factors
- Safety
- Access
- Weather

Technical Factors
- Type of project
- Size of project
- Complexity of project
- Quality level required
- Managerial Factors
- Geographic location
- Supervision
- Suppliers
- Subcontractors

Professional Factors
- Development
- Learning
- Direction
- Relationships

In short, are the other bidders legitimate? Or, is the owner shopping around on Craigslist for some guy who works for cash under the table?

Sometimes you can't get answers to all those questions. But, it's worth the trouble to try. Even though the owners may say "quality" and "the relationship" are their first concerns, if they are going the competitive bid route, price is likely to be a high priority for them. You are wasting your time bidding unless you are bidding against other quality people.

Hitting winners

Lindsay Davenport, once the world's number one in women's tennis, was asked, "What's the key to success in your sport?" She answered, "Don't make mistakes. Hit winners." After retiring from the tennis circuit, Davenport began coaching a young player with awesome ability named Madison Keys. Keys had a problem. She ran out of patience. She went for winners too soon and made mistakes. Davenport began teaching her to keep the ball in play while "constructing" a better opportunity for a winner.

Likewise, builders often crash for lack of patience. They want that big win. They go for it prematurely. Paul Cook, in his fine book on bidding, says deciding to go after an "unsuitable project," one you are not ready for, is the worst kind of error. It is an error of leadership. You are taking your company up the wrong hill.

You won't escape the consequences of poor leadership decisions too many times. Ours is not a business with a lot of room for error. Klay Gould—who has accomplished the unusual feat of keeping a construction company going and growing through boom and bust for four decades—feels that if you take two unsuitable projects back to back, and they go south on you, "it's lights out." Klay is not sure that two is the right number. What is sure is that both Klay and I know of a number of companies that have gone out of business as a result of bidding and "winning" not *two*, but just a *single* unsuitable project.

A project may check out in all the various ways we have already reviewed. The client may be great. The designer may be well regarded and produce good-quality plans and specs. The lead sheet score may look promising. Yet the project may be a poor choice for a given company at a given time. You can get a good start on determining whether or not a project is right for you by considering the four groups of factors suggested in the list on the previous page.

To begin with, ask yourself if you can accurately estimate the direct costs of this job. Does it involve excavation for large retaining walls on a steep hillside, while your company is experienced only at trenching for spread footings on flat lots? If your answer is yes, either you find and carefully check references for excavators who can reliably bid and perform the work, or you walk away from the project. A lot of contractors have dug a hole for themselves when trying to dig a bigger hole in the ground than they have any business attempting.

By the time my company built this complex structure—on a steep site, during a winter of fierce rainstorms, for a tantrum-throwing client—we had acquired the skills needed to handle the challenges, and even enjoyed them. Ten years earlier the project might have overwhelmed us.

As you move down the list of factors, you come to those posing the greatest hazards of all—namely technical factors. Tom Schleifer, whose outstanding book is recommended in Resources, lays down the law regarding size: *Never take a job that is more than twice the size of one you have successfully completed.* To underscore his point, Schleifer tells the story of a company that violated his guideline, "leapfrogging" and more than doubling the size of projects in rapid succession. With the second leap, the company went bankrupt. All the assets, goodwill, and team building that had been accomplished over 40 years went down the drain. That is heartbreaking. Arrogance, "thinking you are above the rules," as construction consultant Judith Miller defines it, causes heartbreak. The rule against increasing project size too quickly is one we arrogantly break at our peril.

I would like to suggest a corollary to Schleifer's rule: *Do not increase total volume of work too rapidly.* Don't get beyond your capacity to staff your jobs reliably. Be sure that for every job you are bidding you will have available the project lead, tradesmen, and subs it will require. Don't assume you will be able to assemble a team of new people from scratch once the job is in the bag. In parallel with Schleifer's rule, my corollary also suggests you should no more than double your annual volume of work from one year to the next—and do that only rarely and when you are starting from a small base.

Over the longer term, doubling every five years—i.e., at an average rate of 15% a year—is a prudent limit. That would take you from $100,000 in annual revenue to $6.5 million across a 30-year career. It will take you to $26 million in 40 years. Builders do start out with a pickup and a box of tools and accomplish such growth over long careers. It's enough. Stay patient. Avoid losers. Hit winners.

Naïveté

A young general contractor had built several small projects designed by an experienced architect. He had come to trust the architect and enjoyed the architect's praise for his skillful production of details. Then the architect asked him to bid a project that was many times the size of the previous projects. The builder was flattered. He put together a bid and with some hesitance submitted his numbers. They seemed to add up but the bottom line seemed impossibly high to him. He had a hard time imagining that the project, any project, could cost so much. He felt embarrassed to be asking for such a huge sum.

In actuality, he was charging far too little. He had been unable to grasp from the scaled plans just how much more labor and material would be consumed by every single phase of the project than by his usual smaller jobs. After all, the job did not look any bigger on those scaled drawings than projects he had built before. They occupied the same size sheets of paper.

The architect knew the young builder's bid was woefully low. But he wanted to see the project built and the other bids that had come in were far beyond the owners' means. He encouraged the owners to accept the low bid. Partway into the project, the builder ran out of money and could not even pay for materials, much less other workers. With no other choice, he walked away from the job.

Danger zones

The rules of thumb Schleifer and I have arrived at are not mathematical theorems. They are far from absolute. They are simply guidelines based on observation. My experience with my own building operation does suggest, however, that they are about right. Both when I have doubled project size and when I have increased volume too aggressively, I have found my company and myself right up against our limits—with severe consequences. I will get to that in the next section.

Jumping up project size and/or volume too rapidly is especially dangerous if the projects you are taking on are of a complexity and quality level beyond what you have already handled successfully. You want to select projects that will stretch and expand your own and your company's skills, but to an achievable degree, not further.

Be especially sure, when you jump up a level in project size, complexity, and/or quality, that you have available a qualified, reliable person to run the job, whether that's yourself or a project lead well established with your company. Bringing an untried lead into your company to run a job that is already a stretch for you is hazardous. And your established leads should not be burdened with projects that are excessively larger, more complex, or more demanding of quality control than anything they have handled before. If the most demanding job a guy has handled is an interior remodel, he is not ready to manage construction of a house. He needs to first progress through an addition and then a bigger addition. Similarly, the steps from economy-grade to high-end work should be taken incrementally.

Projects should be bid not only with a lead available but together with suppliers and subcontractors that you will be able to count on when it comes time to build the project. Even a single flaky subcontractor can bring a job to a halt. Replacing him or her on an emergency basis may be so costly that the entire bid is thrown out of whack. What happens if your plumber decides to take off for the Bahamas right before you need him—as did the plumber of a certain East Coast builder? The other subs whose work must follow the plumbers must be rescheduled. Some may have prior commitments elsewhere when you need them at your project, so more delays occur, and then the owner (rightfully) gets agitated. He posts a scathing review on social media, and…the dominoes keep falling. Reliable subcontractors enable you to march your project steadily through to completion.

Bad suppliers can gum things up almost as severely as flaky subs. Hoping to lower the environmental impact of my projects, I briefly switched to a company touted for its sustainably harvested framing lumber. Foolishly, I chose the company on the basis of only a single reference, and that from a builder I like but who has never quite had his act together. On my first project with the supplier, they delivered my lumber load a week ahead of schedule and on a day when the job site was closed to allow for vacations. Not finding anyone at the job site, the driver dumped the lumber in my client's yard, crushing plants, blocking the sidewalk, and leaving the clients—since I was 3,000 miles away—to move the lumber to our staging area. (I docked the lumber company $500, knocking the same amount off my client's next bill, and took my business back to my company's tried-and-true supplier, who infallibly delivers just what we order and right on schedule.)

Sometimes, of course, you may want to bid a job that is beyond the reach, whether technically or geographically, of your usual subcontractors and/or suppliers. If that is the case, you should choose their substitutes with care. And you should limit the

How I say "no"

Personally, I prefer a direct approach to saying no. If the plans are lousy, if the owner wants my preconstruction services for free, if the project seems to me to be wasteful and/or damaging to a neighborhood, and I do not want to be associated with it, I will say no in as civil a way as I can manage. If the budget is so low the owners will get in trouble trying to build, I will say no and ask the owners if they would like me to explain the potential disaster they are heading toward. If the project is simply more than I think I can handle—if for any of these reasons, I do not want to bid the project, I say so as clearly and as respectfully as I am able.

Undoubtedly, prospective clients are sometimes offended. Surely the guy who wanted me to collude in a tax evasion scheme to help finance his remodel was offended when I told him, "No dice, you are a lawyer and you are asking me to help you break the law. You should be ashamed of yourself."

On the whole, however, I feel my candor works out well enough. People appreciate straight shooters. In our world, integrity sells. The schemers, the scam artists, and the slick salesmen have given our industry a bad name. Our fellow citizens are often uneasy about building contractors. "You just don't know whom to trust," a woman who needed extensive work done on her properties told me when she asked me to help her find a builder. Clarity and candor coupled with civility in saying either yes or no to invitations to bid projects is one way to bridge that distrust.

By the way, though disappointed that I would not participate in his tax evasion scheme, the lawyer contracted with me for his project.

Going for it

Mike Tolbert was rapidly expanding both size of projects and annual volume when he asked me to become his consultant. He knew he had a problem. "I tend to get out on the end of a limb, and I need help with that," he told me. I advised that he focus on growing in strength rather than size for a few years. I am not confident he will be able to rein himself in. "I see all those great jobs out there," he says, "and I think, 'why should we not be building them instead of someone else?' I want to go after them."

Mike does, however, have this saving grace. Though he overreaches on project size and volume, he avoids taking on complexity and quality requirements he cannot handle. He recently jumped abruptly from one- and two-story commercial build-outs to gutting and entirely remodeling a seven-story downtown apartment building. Reasoning that, though the job was larger than anything he had previously done, the work was straightforward and repetitive, and that he had the crew, subs, and project manager to handle it, he bid on the project and won it. As of this writing, it appears to be going well.

Even so, I feel that with his tendency toward headlong growth, Mike is likely skating on very thin ice. He reminds me of a company that once operated in my area. Led by a guy with a genius for marketing, it went from roughly $900,000 annual volume to $50 million (in 2016 dollars) in half a dozen years. Then it went "poof." It disappeared. (The genius resurfaced as the founder of an Internet company during the dot.com boom of the 1990s, momentarily enjoyed a net worth of over a billion dollars, no kidding, and lost all of that, too. But that is another story.) We'll see whether Mike avoids going "poof" himself.

number of new subs. My intuitive sense is that the number is two. All other suppliers and subs included in your bid should have been tried, tested, and found reliable—guys you know will help you avoid mistakes and construct winners.

Saying no

The warnings Schleifer and I suggest are fairly obvious. They have been posted by many industry savants. William Mitchell, who went broke en route to learning the lesson, says, "Know your limits.... Some jobs require more skill, more equipment, or more financial muscle than you possess." Iris Harrell says simply, "We try to get that big fish in the boat" when we should not.

So if the dangers are obvious, why are builders so often oblivious to them? Perhaps it's because the sort of folks who would even think of going into the construction business are prone to undue optimism. One construction attorney says he could give his builder clients a hundred reasons to not take on every one of the jobs for which they sign a contract and a thousand reasons to get out of the business altogether. But he just tries to help his clients avoid the worst potholes, because he recognizes that they are go-for-it guys, who believe that they can come out on top and that there is no stopping them.

Eventually, with experience, we may learn to keep our confidence in check. We may have even learn, as Fernando Pagés suggests in his *Fine Homebuilding* blog, to run regular "premortems" on our company in order to anticipate missteps before we take them. But even then, it is all too easy in construction to be seduced by the glamour project—that prominent structure that you think will make you proud. After all, as Paul Winans has pointed out, "Sometimes the most exciting projects are *not* those we make the most money on."

However, the siren song of that exciting project can lure us into serious trouble. "I have learned," a builder whom I have quoted before ruefully explained, "to be realistic, to not become so enamored of a project that I kid myself about what it will cost me to build it." Klay Gould seconds the conclusion. Because he believes that just a couple of badly chosen projects can take down even a large, mature construction business like his own, he would rather have his company do four $500,000 jobs than a single $2 million job. For him it's a "too many eggs in one basket" issue. Protecting his company's eggs is now more important to him than the bragging rights that come with having his sign hung on a high-visibility project designed by a prominent architectural firm. The jobs he has come to prefer are, he says, the reverse of glamorous. They involve standard stuff, "cookie-cutter" stuff. They are for school teachers, local businessmen. They aren't, Gould says, the sort of jobs commissioned by the "Twitterati" that you see featured in the Sunday supplements.

Construction of this curved retaining wall was a big step up for my then young company. We formed and poured it with guidance from a legendry local concrete artisan and learned much that enabled us to later take on progressively larger and more complex concrete work.

Its not just glamour projects that can lure us too far out on the limb. It's the glamour of the bigger, higher-profile business operation. That's the species of glamour that led me to once contract for a greater volume of work than I was equipped to handle. Plain and simple (and embarrassing), I was motivated by a desire for bragging rights at my builders' association's meetings. I was looking forward to strutting around and dropping mention of the millions of dollars' worth of work we had under construction even while other guys were struggling to stay busy through a tough recession. I did get some bragging done and enjoyed some admiring glances. But I also stressed my business systems to the point that I did serious damage—scarring professional relationships, disappointing and upsetting clients, and even briefly losing control over my financials. I got out alive largely because of the skill and determination of my project leads. They maintained control of their projects even as I came near to steering the company over a cliff.

Start-up builders, especially, must resist the temptation to grow their businesses too quickly, to make it across what construction industry consultant Judith Miller calls the "black hole" in one leap. There is a great divide between running a business that consists of yourself, your pickup and your dog, your tools, maybe a part-time helper and a few trusted subs; and a company that has moved to employing full-time field personnel and office staff. (By the way, that is not meant as a put-down of the pickup-truck-and-dog level of construction contracting. In fact, it might be the sweetest spot in our difficult industry. It can be peaceful).

There is an equally great divide, another black hole, between the company for which the owner still does the marketing, bidding, and estimating and has a hand in project management; and a company in which those roles have been largely delegated to employees. Miller has seen companies disappear down the black holes—and others successfully bridge them. As a result she has come to appreciate stability much more than rapid growth.

Risk or recklessness

I have several entrepreneurial friends who have spent their business lives swinging for the fences. They are always going for the huge financial hit. Always they exude confidence that this time up they are going to knock one out of the park.

Occasionally, one of them does make a pile of money. But inevitably he soon loses it on the next gamble. One, having blown through millions, now lives on an eight-hundred-a-month social security check in a small apartment he shares with an over-the-hill tennis pro. Another fled to a low-rent South American country with his family in hopes that his meager remaining assets could sustain them there for a few years. A third… Never mind, you get the picture.

My friends like to recite mantras like "No risk, no reward." And "Sometimes you just got to go for it." Without risk-takers, they assert, humanity would not ever move forward. But they never do point out, "Rome was not built in a day."

Watching them, and watching builders in my community flare brightly for a time and then flame out, it seems to me they have been seduced by a misunderstanding that permeates our culture. Risk-taking is seen as bold and brave. It is seen as good. And because "good things happen to good people," it follows that if you take risks, you should be and will be rewarded.

The problem is this belief blurs the distinction between risk and recklessness. It even aggrandizes failure as a necessary ingredient of success. It isn't. Too often it just precedes more failure.

If you are in construction, you don't have to worry about whether you are a properly brave risk-taker or not. You have stepped into one of the riskiest businesses on the planet. To achieve long-term financial success, you must emphasize risk management, not just risk-taking. You must manage your risks to avoid stepping over the boundary into recklessness.

That is why I distinguish bidding, with its strategic risk assessment, from estimating (a management task that can be routinized), and have devoted so many pages to the first phase of bidding—qualifying your projects and deciding systematically on where to place your bets.

She expresses admiration for a construction company owner who emphasizes that he does not reach for growth in volume until he is well organized and efficient at his existing level. Miller says she wishes more of her clients would do that and not try to "take on every project that they get called about. You really need to be able to say no at the front end of a project so you do not have to clean up the mess at the back end," she declares. She's right. Some builders have as hard a time with "no" as they have asking about budgets. Others have come up with ways of saying no that work for them. A builder who specializes in foundation and drainage work says he avoids badmouthing a project—even when he is saying no because the plans and specs are so bad he does not want to build from them. Instead he will tell the prospective client, "This is not really for us. There are other guys who will be glad to build this for you."

A custom homebuilder begs off with, "Ms. Johnson, we'd love to build your new home, but given our current obligations we could not start for a year." If Ms. Jones responds with, "That's okay, you are worth waiting for," he comes back with an admission that a year is probably optimistic. "Really it could be two years. Gosh, come to think of it, it might be three years!" Apparently that does it, and the client goes away. Whatever approach you develop to saying no, it is also important to learn *when* to say it, namely as early as possible. It may happen that you will commence a bid, and then have to withdraw. You might not realize that a project is beyond your capabilities until after you have agreed to bid the job and have begun to break down the plans, specs, and takeoff quantities of work.

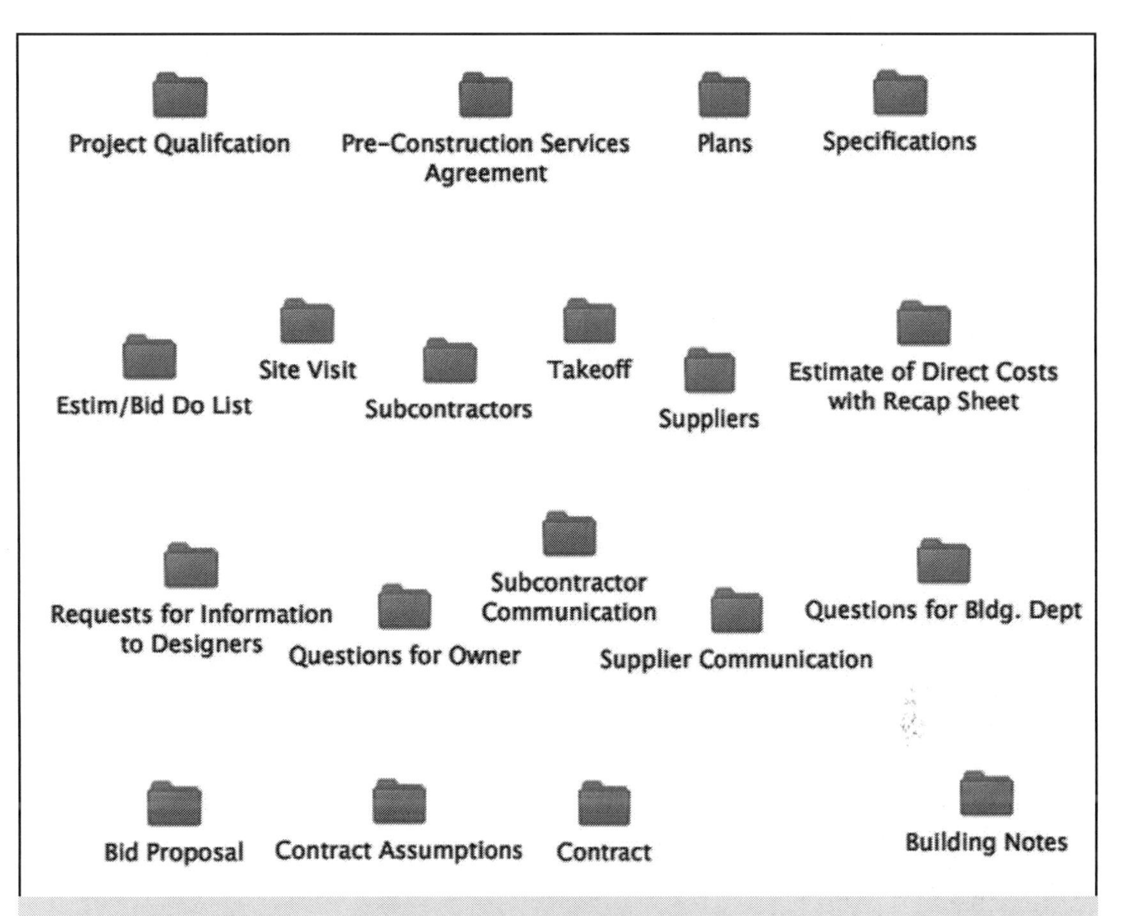

A basic set of subfolders (or tabs if you are using a three-ring binder for organizing your project paperwork) will be adequate for most projects. For larger jobs, you may want further breakdown, such as a subfolder for each trade.

But the deeper you get into bidding a job, the more of your time you will have used and the more other people—owners, designers, subcontractors, and suppliers—will be counting on you to complete your bid. Ask a sub to bid the work for his trade in a project, then fail to submit your bid, and you have just wasted a substantial amount of his labor. Ditto for a supplier you have asked for a quote. Tell a designer, who is limiting the number of bids for a project to three, that you will submit one of them, then fail do so, and you will undermine the client's trust for the designer. Likely, you will not hear from that designer again, nor from his or her friends in the profession.

In short, if you must say no, say it early. Qualify projects for a bid with care before saying yes. When you do say yes, move to nailing your numbers with the use of well-organized estimating procedures and systems.

From bidding to estimating

As a logical first step into creating a project estimate, set up a binder with divisions for each phase of your process from takeoff through initiation of construction. The binder might be an actual three-ring binder with tabs to mark phases of estimating. Or it might be a virtual "binder," a computer file with subfolders as shown on the previous page. It might even be—if your company has grown to the point where you judge the additional investment worthwhile—an advanced version of a virtual binder that is produced by a specialized construction project management program, stored in the "cloud," and available to everyone in your company involved in estimating, bidding, and managing a project.

In between the Project Qualification and Building Notes subfolders come a series of other folders for the documents that are produced during estimating and bidding. If you have moved beyond bidding for free and are being paid for your estimating, your Preconstruction Services Agreement will occupy a folder with that name. For competitive bids that folder, sadly, remains empty.

As the illustration of the subfolders suggests, you will begin your actual estimating by reviewing the plans and specification more carefully than was necessary to qualify the project. That review will determine just what you include in your Estimate and Bid Task List, contained in the next subfolder.

The Task List charts the path you will follow through all the work that needs to be done to compile an estimate and package it up in a bid proposal. At the start of your career, and if you elect to stay with smaller projects, a Task List may be only a single sheet in length, like the one displayed on the next page. You can create your own master Task List with word processing software, then make a copy for each new project, editing it as appropriate.

A table to track communication with subs can be easily inserted in the Task List using the table function in your software. The table enables you to see at a glance where you are in the process of gathering your needed sub bids. For each trade listed in the table you can enter the dates when you:

- Initially requested bids.
- Informed the subs when you will be at the project site to walk them through it.
- Answered their RFIs (Requests for Information).
- Notified them of changes in the plans or specs.
- Nudged them with a reminder that the deadline was approaching.
- Received their bids.

As you take on bigger jobs, the table for subcontractor communications may grow so large that you want to break it off into a subordinate list—especially if you are seeking bids from multiple subs and/or have moved to subcontracting virtually all of the on-site work.

Because subs can grow into such an important component of your business, we will discuss the management of subcontractor bids in more detail in Chapter Eight. Likewise, in later chapters we will go into greater detail about other steps in estimating and bidding suggested by the subfolder names. Here, we will move on to site inspection. Attentive site inspections, made together with your subcontractors, turn up issues that greatly impact costs of construction.

Site inspections

With web-based services displacing traditional means of doing work of all kinds, it was inevitable that someone would produce a virtual inspection application intended to replace actual site inspections. Now we have just such an app for roofers. It allows them to take an aerial look at a roofing job without ever actually going out to the site and, ostensibly, to figure the quantities of work for an estimate.

Maybe web-based site inspections work for roofers. For builders, there is still no substitute for making a drive to the site and noting the conditions

Estimate and Bid Task List   Project: Date:

I. Document Review
- Plans and Specs /Architectural
- Plans and Specs /Engineering

II. Subcontractors

TRADE	BID REQUEST	RFI's	CHANGES	DEADLINE	BID IN	COMMENTS
HAVAC						
PLUMB.						
ELECTRICAL						
ROOF						
DRYWALL						
TILE						
CABINETS						
TRIM						

III. Site Inspection
- Scheduled with owner & subs
- Subs walked through
- Access and other conditions noted

IV. Material Takeoff
- Concrete - takeoff done & quote requested
- Framing Lumber - takeoff done & quote requested
- Windows/Door – Schedule sent & quote requested
- Trim - takeoff done & quote requested
- Hardware - takeoff done & quote requested

V. Spread Sheet
- Labor
- Material
- Subs
- Services

VI. Recap Sheet
- Direct Costs
- Overhead
- Profit

VII. Present Bid (Request that all owners be present)

Expanded Site Inspection Checklist

Prepare
- Schedule inspection with owner
- Schedule site visit with subs
- Schedule site visit with consultants

Bring
- Overalls
- Raingear
- Boots
- Gloves
- Dust mask
- Hard hat
- Hand tools (shovel, pick bar)
- Flashlight
- Project binder/laptop
- Notepaper and pencils
- Camera
- Ladders
- Documents for subs
- Lunch

Off-site Factors
- Roadways
- Emergency medical
- Travel for crew
- Accommodations for crew
- Availability, quality and capacity of suppliers if they are new
- Availability of subs
- Availability of temporary labor
- Waste and recycling
- Distance to rental yards
- Licenses, taxes, and codes

Site Factors
- Parking
- Access
- Weather (during job!!!)
- Water
- Power
- Sanitary facilities
- Power lines
- Soil
- Grade
- Need for barricades and/or fences
- Tool and equipment security
- Site security
- Traffic control
- Erosion control
- Dust control
- Drainage system protection
- Tree and plant protection
- Driveway protection
- Other exterior protection
- Grubbing
- Streams
- Area for material drops
- Excavated earth stockpile area
- Waste and recycle storage area
- Operating space for off-hauling
- Opportunities to use heavy equipment (cranes, etc)
- Distractions (pets, etc).

Existing Building Factors
- Access
- Need for repeated mobilization
- Interior protection
- Special protection (computers, art)
- Hidden demolition and deconstruction
- Material layers
- Vulnerable existing pipes
- Vulnerable existing windows
- Other vulnerable items
- Plumb, level, and square
- Difficult-to-reach framing and framing cavities
- Space for HVAC, plumbing, & electrical runs
- Tie-ins, structural
- Tie-ins, finish
- Condition of finishes
- Matching issues
- Reuse possibilities
- Health and safety hazards
- Accuracy of as-built dimensions
- Space for temporary facilities for owner

Additional Design/Build and Cost Planning Issues (Limited sample)
- Existing structural
- Existing HVAC
- Existing plumbing
- Existing electrical
- Existing drainage
- Existing decay
- Existing code violations
- Window and door locations
- Existing exterior finish
- Existing interior finishes including hardware

My tried-and-true subcontractors were willing to service the reconstruction of this 120-year-old home, unusually distant from their shops, because they badly needed work during a severe recession. Had they not been willing, I would have had to hire too many unfamiliar subs to reliably estimate construction costs for the project and would have had to pass on it.

that will impact the costs of construction. If existing conditions at a site have been overlooked during the creation of the plans and specs, you may, nevertheless, have to shoulder any costs of construction that those conditions impose. Under standard-of-care laws, you can be held responsible for discovering the conditions for yourself and allowing for them in your bid. You need to go to the site and spot them.

Most items in the site inspection checklist illustrated here are self-explanatory. Some merit comment. A "Bring" section can provide helpful reminders, especially if you do site inspections only intermittently. The Bring section is long and not every item on it will be needed for every inspection. But loading up everything you *might* need is faster than figuring out each time what you *will* need—then arriving at the site to discover the one item you decided not to bring along is the one you can't do without.

Even before you get to the site, you may observe factors that will impact your estimate—particularly with respect to access. Roads deserve attention. Are they adequate to bring in any equipment or material you will need? Or, will you find yourself being told over the phone by your cabinet manufacturer, as I once was, that his tractor-trailer can not navigate the narrow street leading to your site, and that you will have to send a pickup and a couple of workers down the hill half a dozen times to collect the carcasses and doors?

Even if roadways to a project are clearly adequate for equipment and material delivery, it is worth finding out whether they will be clear when you need them. That's a real concern in congested metro areas. Are major repairs planned that will slow access to the job site? Can the roads handle the commute-hour rush? Or, will they be so congested that you and your crew will spend hours stalled in traffic every day? If so, you may need to include a travel allowance in your estimate for the project, or risk the possibly greater costs of demoralizing and even losing workers.

If you decide to bid for a job that is too far for a daily commute, another cost must be picked up—overnight accommodations that are affordable but also comfortable. A certain specialty contractor headquartered in Northern California regularly takes projects several hours away from his shop. He puts his crew up near the job, but generally in the cheapest dump he can find online. There his crew do not get a good night's sleep. During the day their productivity and attention to detail drop off. If you elect to bid out-of-town work, you'll get closer to nailing your numbers if you figure in decent lodgings for your crew.

If a project is beyond the range of the subs, suppliers, and services you usually contract with, you must determine whether they will go the extra miles or whether you will need to make other arrangements. Can your temp labor agency get workers to your more distant job site? Is there a rental yard within a reasonable distance? Will you be able to bring in dumpsters to handle waste and recycling from demolition and deconstruction? How about the subs? Will they take care of you on a job twice the usual distance from their shop and office? Can your concrete supplier make delivery to your more distant project?

If you are bidding in an unfamiliar jurisdiction, you will need to research local codes, license requirements, and taxes that will impact your estimate. They vary greatly jurisdiction to jurisdiction. A plumber based in Oakland, California learned that when he took a job piping a spec house on the other side of the San Francisco bay in the town of San Mateo. He completed his waste and drain system, installing supply lines using type M copper as usual, called for an inspection, and was then told that in San Mateo type M was not accepted. Type L was required. He was required to tear out and replace the entire supply system.

Likewise, you can get snared by unexpected license charges and/or taxes imposed on builders. But the reverse can occur as well. You may find they are lower than you usually experience and that you can tighten your estimate a bit. Something costing less than you expect? Yes, it happens. I bet it's happened for me at least two or three times in the 40 years since I got my license.

At the site itself, there is a crucial set of factors that is easily overlooked even though it can have severe cost impacts. That is access, including:

- Access onto the site.
- Access from one part of the site to another.
- Access from the site into existing buildings.
- Access within buildings.

If all access is easy and ample, wonderful. But if any access is difficult, make note of it and allow for extra cost in your estimate. In built-up urban areas, access to the site—including parking—can be critical. Will suppliers be able to readily park at the site and off-load directly onto a staging area, or will your crew have to haul materials from the suppliers' trucks to the staging areas? Will workers be able to park on or right at the site? Or will they have to hike in? When a project manager visits the site, will he be able to pull his truck right up to the job, or will he have to circle for half an hour looking for a parking space?

One estimator for a company that won a commercial project in a major metro area where the company did not often work forgot to check on parking. As construction began, his company learned that the only space available for workers' vehicles was a mile from the job site. Moving their tools onto and off site took an hour at the beginning and end of the day.

Even if parking is available, access *onto* the site bears watching. I once built a couple of new structures on a large site that was too steep for delivery trucks to come any closer than the top of the driveway leading down to the construction areas. All lumber and other material had to be dropped at the street, and then hand carried downhill for roughly 100 yards.

Fortunately, I had allowed for the labor under "material handling" in my estimate. Unfortunately, I was allowing for it in the summer when I was building my estimate, whereas we would end up building the project in the winter. When the heavy November-through-February rains hit during construction, the site became muddy and slippery. Access and material movement became difficult. My estimate proved inadequate.

The shortfall was not disastrous. Financial results on the job were acceptable even as my crew produced work pronounced by the pleased owner to be "bullet-proof." But slippage was enough to teach me that you have to estimate for the weather you will be building in, not for the weather outside your office window when you are punching numbers into your spreadsheet. I moved "weather" up close to the top of my site factors checklist and rewrote it as "weather during job!!!"

If there is a factor that compares to access in that it is both easily overlooked and can pile up costs, it is protection, including:

- Protection of entry drives.
- Protection of trees, of lawns, and of other plantings.
- Protection of neighbors from dust and noise.
- Protection of streams and drains from eroding soil and construction waste.
- Protection of existing building exteriors.

Even experienced builders who are attentive to protection of interior surfaces are occasionally prone to neglect aspects of exterior protection—and start-up builders more so, as I was early in my career. Another financially hazardous site condition, particularly deserving of attention if you are not experienced with it, is the presence of soil that must be excavated and removed from the site. The first few times you deal with excavation, you may be startled to learn—again, as was I—how heavy soil is, how greatly it can expand, and just how costly is the labor and machinery required to move it.

Even if the project plans and specs indicate grade and soil conditions, you would do well to check them to whatever extent is reasonably possible. If either is judged incorrectly, the error will throw off your excavation estimate. Grades that are steeper than shown in the plans will result in additional excavation for cuts. Some soils are far more difficult to remove than others. And, because some soils expand far more than others when they come out of the ground, an inaccurate assumption about soils can lead to serious errors in your estimate of the costs of both moving and off-hauling soil. Once soil is out of the ground, you will need a place to stockpile it for use as backfill or until off-hauling. Likewise, you will need a place to accumulate waste and to store building materials. You must be especially alert to those needs during your site inspection if you bid a project in a heavily built-up urban area where—in contrast to suburban or rural neighborhoods—open areas for handling and storing soil, waste, and construction material may well be sparse.

Of all site factors, inspection of existing construction is the most time-consuming. If buildings are to be remodeled or renovated, every system from foundation through mechanical to finish must be checked out. Sometimes system components are not apparent at a glance but just hinted at. For example, the presence of a three-story clay pipe vent stack concealed in walls that are to be removed may be hinted at only by the presence of a rusty old roof jack and cap. Similarly, there are often hidden layers—of plaster, of roofing, of flooring—beneath a visible layer of material. The hidden layers can double or triple the labor needed to expose the building frame, and can likewise multiply the quantity of material that needs to be removed from the building and hauled to the dump.

Along with demolition can come another cost—damage to vulnerable existing work. Vibration set up by demo can knock rust loose in old pipes, rendering them useless and requiring that they be replaced. Windows, especially the dual-pane glazing in modern windows, can fail due to nearby demolition. All such possibilities need to be noted and then mentioned in the assumptions and clarifications that are part of good final bid package. That is, along with giving them your price to build the job, you want to let owners know that their old pipes or new windows may fail due to the inevitable and necessary impacts of construction, and that your bid does not include the cost of repair or replacement.

During inspection of existing buildings, you must again focus on access. How easy will movement into, out of, and around the buildings be? How much labor will the movement of material and tools

For my first excavation job, my labor estimate was woefully low. I had no idea that soil could be so difficult to remove, weigh so much, or expand so greatly. By the time my company built this slab and retaining wall system, I had created records of productivity—cubic yards removed per man-hour—for excavation on a variety of sites. Relying on those records and my subs, I nailed my numbers. In Part II I will discuss effective ways of recording labor productivity and gathering reliable sub bids, two of the most critical estimating tasks.

therefore consume? Will there be room to set up all tools at the beginning of the day for all tasks and to store tools away quickly at night? Or will tools have to be moved back and forth to a vehicle or staging area repeatedly during the workday?

Two New England builders in the start-up phase of their careers were doing small residential renovation and repair projects and rapidly building a base of satisfied clients when they came to grips with the access issue. They realized that their projects were taking much longer than they had allowed for in their estimates. To figure out why, they started keeping a log of just how their time was spent each day. The answer emerged.

The builders saw that they had been accurately estimating their labor for actual production. But they had ignored access, often awkward because of the need to maneuver past their clients' stuff. Their estimates had not included the cost of what they came to call "remobilization," the moving of tools around their clients' homes and back and forth to their trucks as they shifted from task to task. To allow for that labor, they incorporated new lines in their estimates. Their labor charges went up by 50%. As a result, they lost a few clients; but others replaced them, and financial results got much better.

During building inspections you must also take a close look at possible routes for ducts and plumbing through existing and proposed structural work. Typically, at least for residential work, the designers leave mechanical layout to the builders. Meanwhile, they do specify structural elements. Sometimes the structure as designed leaves no room for HVAC ducts and waste lines. You are better off discovering that during the estimating phase than the building phase of a project.

From my own projects, I have learned to pay close attention as well to points of tie-in—both structural and finish—between old work and new. You need to find out to what extent the existing structure has gone out of plumb, level, and square. The worse the conditions, the more highly labor-intensive the tie-ins. In a building that is badly off, tying the frame, siding, and interior finish of an existing wall to that of an addition—with all the blending together of old and new siding and the shimming of out-of-plumb studs to allow a smooth transition of drywall from old studs to new—can require as much labor as the new wall itself.

If you are operating a company that produces its own designs as well as builds them, the site inspection demands ramp up steeply, far beyond the schematic list suggested in the sidebar. As mentioned earlier, the most thoroughly developed design/build checklist I have seen comes from Byggemeister, and roughly half of its 50 pages are devoted to site inspection issues.

Design/build, however, is a world unto itself. It is a subject largely beyond the scope of this book. Here, to bring this chapter to an end, I will return to the subfolders we have set up in our master estimate and bid folder. In the illustration, you may have noticed the folder designated "Estimate of Direct Costs with Recap Sheet." Recap sheet refers simply to an estimate summary that shows the cost of each trade involved in a project, the overhead and profit markups, and your selling price to the owner.

A last subfolder is named "Time Log." There you can keep a form for recording the hours spent on an estimate. If you have not kept a record in the past, you may find it sobering to learn how much time is required to estimate and bid a project from start to finish—from the clients' first call, to the actual number crunching, and on through the presentation of your bid. Once you realize how labor-intensive the process is, you may feel motivated to travel far enough down the path to skilled estimating and bidding so that you can charge for your services and move beyond bidding for free. So now, let's get on to the building of an estimating system that is well worth your clients paying for.

PART TWO

BUILD YOUR ESTIMATING SYSTEM

CHAPTER 4

Takeoff

The rubber hits the road

If you have taken all the steps suggested in Part I, you have come a far distance along the path to nailing your numbers. You have recognized that estimating and bidding are the heart of any construction business, have rejected the excuses for half-assed work, and are determined to go about bidding and estimating in a systematic way. You have set up a master three-ring binder or digital folders for organizing and storing all the documents you will create while estimating and bidding projects. Included in your binder are a thorough estimating Task List and a site inspection checklist.

Now, as a first step in actually building your estimating system, you are ready to refine your methods for performing "takeoffs." As mentioned in the previous chapter, takeoff has nothing to do with reduction, as the term might seem to imply. Instead it refers to quantifying all the items and assemblies in a project—so many yards of concrete, "x" number of studs, and so on. With takeoffs, the subject of this chapter, under your belt you will be able to move on in chapters five, six, and seven to the work of designing and fully developing the spreadsheet you will use to estimate the costs of projects.

COMING UP

- Takeoff guidelines and process
- Marking up the plans and specs
- Setting up your takeoff sheets
- Takeoff modification factors
- Efficiency
- Allowing for waste, laps, and expansion
- Speed tricks
- A takeoff walkthrough
- Auditing your takeoff
- Site inspections

Bear in mind that when actually doing takeoffs and estimates, as in construction itself, you want to consolidate tasks. Just as you would not go back and forth between layout and nailing when framing a wall, you want to avoid going back and forth from takeoff to costing. Instead, quantify all items of work in a project; then cost them all out.

Producing a takeoff requires intense concentration if you are to avoid violating the number one rule of estimating: Don't miss anything. Carl Hagstrom, a builder who has also earned his living doing takeoffs for other builders, advises us to achieve concentration by: Hanging up a "Do

Takeoff Guidelines

- Put up the "Do Not Disturb" sign, shut the door to the office, turn off the phone.
- Complete the takeoff in a single session.
- Mark up plans systematically, highlighting all items of work that must be quantified.
- Quantify work in logical order paralleling the order of construction.
- Record quantities on a takeoff sheet.
- Use one takeoff quantity to figure both material amounts and labor hours.
- Adjust material amounts for waste, expansion, and laps.
- Don't spend $100 worth of estimating time on a two-dollar item.
- Do not neglect the specifications.
- Make notes for contract assumptions and for construction while doing takeoff.

Not Disturb" sign on our office door. Closing the door. Turning off the phone. Ignoring e-mail. And bearing down on quantification.

Work only on the takeoff, Hagstrom urges; do not multitask. If at all possible, complete a takeoff in a single uninterrupted session. If you have to pause for more than a quick stretch or glass of water, leave yourself a note showing exactly where you paused.

To begin takeoff, first print out the plans and then systematically review them as suggested in Chapter Two. You can then move to marking up the plans—a process we will go into in more detail in the next section of this chapter. With markup completed, go about compiling your takeoffs in a logical order, more than likely the familiar order of construction. As Hagstrom puts it, "When you do a takeoff, you build starting with the footings and finishing with the interior trim work."

Record your quantities on well-designed takeoff sheets such as those illustrated later in this chapter. As you produce the takeoff, be a little generous in figuring quantities. In particular, do not shortchange yourself when applying factors for waste and for expansion of excavated soils—the specifics of which we will also look at shortly. A bit of generosity in figuring quantities will help make up for the inevitable small slippage that occurs no matter how hard you try to not miss anything.

Make efficient use of time. Bear in mind that even though takeoffs are sometimes referred to as "material takeoffs" (MTOs), often a single quantity for an item will serve as a basis for figuring both material and labor costs for that item. For example, for sheathing the roof of a building all you need to take off from the plans is the area of the roof in square feet. To figure the number of sheets you will need, simply divide the roof area by 32 (the number of square feet in a sheet of plywood) and add a waste factor. To figure labor cost, you will multiply the sheet quantity by an appropriate hours-per-sheet figure (hopefully one recorded in the kinds of labor productivity records we will discuss in Chapter Eleven), then multiply the result by an appropriate labor rate. (For example, 24 sheets × 1/3 hour per sheet × $45 an hour = $360).

As you do a takeoff, focus on the big stuff. Don't obsess over minor detail. For example, when taking off quantities for a new building, take time to make sure you have accurately quantified the total linear footage of all exterior walls. But don't burn time figuring the exact number of screws you will need to attach address numbers to an exterior wall. In short, to use an old industry adage, "Don't spend $100 of estimating time on a two-dollar item." Spend it on a $6,000 assembly.

To make efficient use of your time, minimize measurements. Take advantage of the fact that a single takeoff figure can support estimating of costs for several major components of a project. For example, your measurement of the linear footage of perimeter foundation can be used to determine quantities for footings, stem walls, mudsill, and exterior wall frame.

As you go forward with your takeoff, keep a record of assumptions and clarifications that you will want to include in the contract for construction you may eventually sign. Likewise, record good ideas for construction of the project. And make certain to not focus exclusively on the plans while neglecting the specifications. Tedious though all that dense print can be, you must pay attention to the quality levels it dictates. A single line—say, one specifying kiln-dried (KD) and Forest Stewardship Council (FSC) certified framing lumber instead of the usual green stuff— can impose large additional cost burdens.

Markup Sequence

1. Print out plans and specs.
2. Flatten plans.
3. Check that all pages are present.
4. Determine if plans are final and have been approved by building department.
5. Review the plans broadly.
6. Visualize construction of building.
7. Determine a pattern for going through the detail in the plans and specs.
8. Mark up plans using colored markers or pencils.
9. Mark up the specifications.

Marking up the plans and specs

In recent years software for doing takeoffs has been coming on the market. At this point, however, the sales figures reported by the principal vendor of takeoff software suggest it is used by only a tiny fraction of builders (and, frankly, I have my doubts about its reliability). So for the purposes of the following discussion, I will assume the use of paper plans—either plans that arrive rolled up in a tube or, more usual these days, printouts of PDFs sent to your computer by the project designers. And I will assume you will be manually marking up those plans—that is, highlighting all the items of work with colored pens or with a pencil rather than using software.

As the accompanying list, "Markup Sequence," suggests, before actually beginning markup double-check that all pages are present. If the plans came to you in a roll rather than as PDFs, flatten them out. You do not want them to be rolling themselves back up as you try to highlight structural hardware items or wall lengths. To the extent that your space allows, spread out the plans so that you can quickly access individual sheets rather than having to page back and forth through them.

Check whether the plans you are estimating from are final permit drawings that show all building department requirements. If they are not permit

Marking up with pencil

Though I originally learned to mark up using colored markers, I have come to prefer pencils with thick, dark lead. They are cheap and erasable, and with a pencil you don't waste time putting it down and picking it up, as you do with multiple colored markers.

Simple symbols written in pencil can be just as effective as colors at indicating differing items and concerns in the plans:

? A question mark can signal a problem in the plans that you need to ask the designer to resolve.

> An arrow (or an underline or a circle) can indicate standard work such as a 2x4 interior wall that you need to quantify.

!! Exclamation points signal something that is unusual and potentially costly.

A hash mark notes a detail that a sub needs to cover in his estimate and that you want to point out to him.

**** Backslashes or just some weird symbol you invent can signal other kinds of items you need quantify.

drawings, make a note in the folder of assumptions you are recording for the eventual construction contract. The note will remind you to include in the contract the date of the plans on which your contract price is based—and that, in turn, will let the clients know that any work subsequently added by the designers or the building department is not included in your price.

Rather than immediately diving into the detail of marking and quantifying all the items of work, take time to visualize how you or your crew will be putting the building together out on the job site. Take time to relate the engineering to the architectural drawings and the structural sections to the foundation and framing plans. For a remodel project, you may even want to make a tracing of the as-built plans and fit it over the new work. One builder who takes that step says that it helps him more clearly see the full progression of the project from demolition forward.

Once you understand how the project will be put together, you are ready to begin markup. It must be done systematically. Just ricochet around the plans and you will be like a carpenter who bounces back and forth between cabinet installation, trim work, and stair building. You won't be productive, and you will miss stuff.

For my own work, I have found it best to scan the plans slowly, as though I were reading difficult text. I begin at the upper left corner and move my eyes back and forth and gradually down each sheet, highlighting details such as foundation dimensions, wall thickness, and framing hardware as I go. If the specs are included on the plans, I read the specs on each sheet before moving to the next. If the specs are listed separately in a project manual, I grind through it from one end to the other after I have completed markup of the plans.

Builders prefer different ways of making marks on the plans. One professional estimator, who works with a variety of companies, uses an array of colored ink markers, as many as a dozen, accompanied by a color code indicating what each of the colors signifies. In his system, red signals unusual items that will need special attention during quantification and costing. Green signals high-cost items. Black signals details about which he has questions. Several colors are used to mark foundation walls, with each color signifying a wall of a different thickness.

Among builders who do the estimating only for their own company, many also rely on colored

TAKEOFF Project: Jacobs Addition Date: 11/23/17 Pg. 1

Assembly/ Item	Area	L	W	H	Install Amt	Unit	Waste/Laps/ Expansion	Total Amt	Unit	Comment
FOUNDATION										
Excavate										
Form Footing 2x10s										
Form Stem 2x8s										
Stakes 1/ 2',3', 4' @ 3' o.c.										
Snap ties w/ wedges										
Footing Rebar 2/5/8"										
Stem Rebar 4/5/8"										
Vert. Rebar 5/8"x4' @ 2' o.c.										
ABs 3/4 x12 w/plate wash. & nuts @ 4' o.c.										

Takeoff sheets can be formatted in a variety of ways. All, however, feature a vertical left-hand column listing items and assemblies of work. All feature a series of column labels across the top. Some takeoff sheets provide space only for dimensions and quantities of items shown on the plans.

Here I allow for the quantity shown on the plans in the column labeled "Install Amount." Two additional columns provide space for the waste, laps, and expansion associated with many items of work and for a Total Amount, which includes installed amount plus waste, laps, or expansion. Material cost can be calculated on the basis of the Total Amount, while labor is more likely to be figured on the basis of the Installed Amount

A "Comments" column provides space for noting any unusual requirements indicated in the specifications or other considerations that have caught the attention of the estimator. The comments column can also be used to call out quality levels, if they are not adequately covered in the description column.

markers, but fewer of them. One builder uses yellow to highlight all items of work in a project. After he has quantified an item, he checks it off with blue. Another simply highlights with yellow and calls it a day.

I once preferred colors myself. In fact, in my book *Running a Successful Construction Company* I suggested a five-color system. Since writing it, I have realized that using colored ink for markup has two drawbacks. You cannot erase it. If you make an error, you have to cross it out with still another color. Make too many errors, and you have created clutter, which is distracting and gets in the way of efficient takeoff.

I have come around to thinking that sharp carpenters' pencils are the best markup tools. They are cheap. They make marks that are prominent yet erasable. Using a series of symbols such as those illustrated on page 90, you can distinguish the meaning of your marks as readily as you can using different colors. And with pencils, unlike with ink makers, you don't spend time putting down one marker and groping for the next. If you begin take-off with a small pile of sharp carpenter's pencils, you will never lift your head from the plans during markup other than to set aside a dull pencil and select another. That's the kind of focus you need during markup. It limits your oversights and ensures that you pick up every item that needs to be listed in your takeoff sheets.

Formatting your takeoff sheet

At a glance, takeoff sheets look identical to the spreadsheets used for estimating the costs of work recorded during takeoff. Like an estimating spreadsheet, a takeoff sheet is simply a page or series of pages, paper or digital, divided into a grid of labeled columns and rows. However, an estimating spreadsheet is used not only to record numbers, but very extensively to multiply, total, and audit them. A takeoff sheet (sometimes also referred to as a takeoff "matrix," though I will try to avoid using that fancier term) is used largely to record information that will later be forwarded into the estimating spreadsheet.

Adherence to a few basic practices will help ensure that the takeoff record is complete and reliable:

- Fill out the sheet methodically in the order of construction, checking off each marked item or assembly on the plans and specs as you quantify it and enter a quantity on your takeoff sheet.
- Leave blank lines between items to make reading of the sheet easier and to provide space for additional information.
- Number each sheet so that you will notice if one has been misplaced or deleted.
- Make necessary comments as close as you can to the line indicating the quantity of an item—either in a comments column or on a blank line below the quantity.
- If you need to do written calculations in order to figure quantities, do them on a separate worksheet that you can insert right behind the related takeoff sheet. Give the worksheet an appropriate number, such as "4a" for a worksheet backing up Takeoff Sheet 4.
- If you are filling out your takeoff sheet by hand, take care to form your numbers clearly. You do not want to be squinting at them, trying to figure out whether those two digits are "78" or "16" when you are costing out the work in the project.

That may seem like a lot of guidelines. But it will be worth your while to get in the habit of habitually working in such an organized way. You will then compile takeoffs efficiently and avoid missing items. Just like framing layout, takeoff requires efficient, precise movement.

Your choice of format, i.e., the layout or arrangement, for your takeoff sheet will depend on your choice of business model. If you are using the developer model (sometimes called the broker model) and subcontracting virtually all the trades—doing nothing with your own forces other than coordinating the subs and taking care of site maintenance along with a bit of pickup and touch-up work—you may need only a stripped-down takeoff sheet. If you operate a traditional company, as I have done, with your own forces doing all the carpentry from foundation through finish, you may choose to format your sheet like the one shown in the illustrated takeoff sheet.

For ease of reference the takeoff sheet should be organized in the same sequence as your estimating spreadsheet, likely the natural order of construction. The takeoff sheet will probably be much shorter, perhaps listing only a couple of dozen quantities while the spreadsheet may list hundreds of items. After all, the takeoff sheet lists only quantities of work that *are* in a project. The estimating spreadsheet, as we will see in detail in coming chapters, does double duty as a checklist to remind you to pick up all items that *might* be in the job. But if the major divisions of the takeoff sheet are in the same order as those of the estimating spreadsheet, you will be able to move through both documents in parallel as you estimate costs, enjoying efficiency and keeping track of just where you are in your work.

Though in the illustration I did not do so, because of space constraints it is a good idea to format your takeoff sheet with a lot of open lines for the sake of readability. When you are costing out work, your eyes are moving back and forth from the takeoff sheet to the spreadsheet. If information on the takeoff sheet is crammed together, you will find it more difficult to locate quantities. You will also be more likely to misread quantities. Intending to pick up a quantity of anchor bolts, for example, you instead pick up the quantity for stakes, just one cell higher up the sheet, and as a result may find yourself with dozens of unneeded anchor bolts (ABs) after your foundation is formed.

Efficiency

To accomplish takeoff efficiently, it's necessary to quantify items to a useful and appropriate level of detail. Producing too little detail during takeoff will lead to inefficiency. During estimating of costs, you will find yourself being forced to pause and return to counting items, figuring areas, or measuring linear footage. Producing too much detail during takeoff, on the other hand, is also inefficient. It wastes time twice, first during takeoff, when you are doing a lot of unnecessary calculation, and then during costing, when you grope around in the excess detail looking for the numbers you really need.

During takeoff you need calculate only the quantities necessary for costing a project. You do not need to produce a complete list of all the parts of the building such as will be necessary for submitting orders to your suppliers during construction. For example:

- For wall frames you need record only the height of the walls, their thickness, their total length, the total length of headers, and possibly the number of headers. You do not need an exact count of the sticks needed for sills and plates or studs, or the exact length of each header.
- For exterior wall sheathing and cladding, you need record on your takeoff sheet only the total square feet of surface. You do not need to know the number of sheets of sheathing, rolls of wall wrap, or pieces of siding.
- For interior base, you need record during takeoff only the length of exterior and interior walls (and possibly a count of the number of pieces that will have to be installed). You do not need to know the exact length of each of the pieces of baseboard that will be installed during construction.

Hold detailed takeoff until construction

Shawn McCadden, a builder who conducts estimating workshops in New England, says, "I don't give lengths of lumber to my lead carpenters. That is their job."

I agree. I leave it to my lead carpenters to produce the detailed list of material they will need for their projects. That produces two benefits. First, it saves estimating time—most notably the time spent creating long lists of items for projects I might never build. Second, by figuring out exactly what they need to order from suppliers, leads become thoroughly acquainted with just what it is that they are going to build. As a result, they are likely to run their jobs more effectively.

Of course, for those benefits to materialize, you need a lead capable of doing detailed takeoffs efficiently and accurately. If he cannot, you will have to teach him how. If you see, from timecards and invoices, that he is making too many trips to the supplier to pick up missing items, or sending too much stuff back for restocking, you know that he is missing the mark with his takeoffs. He needs more training or needs to be replaced.

If you are certain you will be signing a contract for construction of a project, going all the way to a

Wear out your calcs: When you are creating a takeoff, you want to take as few measurements as possible. For example, to determine quantities of redwood 2x4s and metal flashing needed for the water table shown here at the junction of the stucco and the siding, I began with the perimeter measurement of the foundation and subtracted the width of the bay rather than remeasuring all the way around the house.

stick-by-stick takeoff may be worthwhile. Otherwise it is too likely to be a waste of time. You run unnecessary risk of spending time on a highly itemized takeoff that you will never use.

Particularly important for achieving takeoff efficiency—you must, as Carl Hagstrom puts it, "wear out your calcs." For takeoff, an individual calculation (or measurement), such as foundation length, is like a jig on a job site. You can use it over and over.

Thus, to figure material and labor quantities for your foundation, you will need to calculate the total length of the foundation. That length measurement can also be used to give you the quantity of mudsill you will install (and, modified for waste, it gives you the amount of mudsill material you will need to purchase). Divide the measurement by the appropriate number and you have the number of anchor bolts you need. For example, if you have 100' of foundation and are spacing bolts every two feet you will need 50 ABs (100 ÷ 2 = 50), plus one more for the end of each run.

Multiply that same foundation measurement (your "calc") by three, adjust the measurement upward for any bays, and you have the total length of the sill and plates for a one-story exterior wall frame. Multiply the length of the sills by the height of the exterior wall surface, adjust for any major openings in the walls and for waste, and you have the square footage of sheathing, wall wrap, and exterior cladding you will need.

The measurement originally taken to determine foundation length can even contribute to calculating quantities of interior material. Thus, to calculate quantity of baseboard, start with the length of your exterior walls, i.e., the length of the foundation adjusted for bays. Reduce that figure by the length of walls where you will not need baseboard such as walls with base cabinets. Now you have the quantity of baseboard you need for exterior walls. Similarly, you can begin with the length of interior walls to figure the quantity of baseboard needed for those walls. Just make sure to take into account the fact that you will need to double the length of the walls to allow for baseboard on both sides.

When you get to the roof, wear out your calcs again. Measure for your fascia. Then use those measurements again when figuring area of roof sheathing and of roofing material. In short, from the foundation up, stretch the use of your measurements to the max, squeezing the greatest possible utility out of each one.

Just how you will make your measurements is your call—a matter of experience and preference. Soon perhaps, we may have inexpensive and reliable software available that enables us to take all the measurements necessary for takeoff directly and quickly from plans that arrive in PDF format. In the meanwhile, if you are compiling your takeoff from printed plans, you may want to use an architectural scale wheel. (To see the variety of scale wheels available, just Google "architectural scale wheels.") The most sophisticated digital versions will take the distances measured by rolling the wheel along the plans and will then transfer the measurements directly into an electronic takeoff sheet. The simplest mechanical versions simply give you the distance measured, and you must then transfer the result to your takeoff sheet.

If you operate a compact company and do relatively few estimates, a mechanical wheel or even just an architectural ruler is all you need. (In fact, I'll confess that I often make do with my carpenters' tape measure.) You don't need to take many measurements to do a complete takeoff if you stretch the use of those measurements and "wear out your calcs." Dealing with the complexity of a fancier instrument may cost you more time than it saves if you do only a limited number of takeoffs. If you decide to grow your company, more elaborate instrumentation may prove cost-effective.

Modification factors

When you are taking measurements for takeoff, as always in estimating, the goal is to nail your numbers—not perfectly, but closely. You need to be on target. You can't expect to, nor do you need to, hit the center of the bull's-eye. There's no point in obsessing over quantities to the point that you invest 20 bucks worth of takeoff time to get two dollars closer on a $1,000 item. That is especially true because any obsessive precision will be erased by the necessary, inevitable imprecision that creeps in when you modify your takeoff to account for expansion, laps, and, especially, waste.

Every construction project results in some degree of waste. Just how much depends on the character of your operation. Some construction companies are profligate. At jobs where I worked as a union framer, waste occurred so regularly the guys even had a phrase for it, something like "throw that s—— over the side," meaning toss the 6x6 you just miscut over the side of the building before the super sees it. At my own projects, I try to minimize waste—in part from environmental concern, in part because I believe frugality is the surest way to financial independence for a builder. Even so, as a job progresses, my crew's pile of occasional miscuts, necessary off-cuts, and leftovers grows faster than we can make use of it for blocking and miscellaneous items. Waste happens. If you ignore it in your takeoffs, you will impair the accuracy of your estimates.

The modification factors for waste, along with those for expansion and laps suggested in the table on the following page, are derived from my own experience and from factors suggested in training manuals for professional estimators. The minimum factors provided in the table are a bit to the generous side, and the maximum factors more so, especially for soil expansion. Here's why: The highest published figure I have seen for clay expansion is about 35%. But I work in an area where builders encounter extremely expansive clay. On my jobs, I have seen it nearly double in volume upon excavation. Five yards comes out of the ground. By the time it gets to a 10-yard dump truck for off-hauling, it fills the truck close to the rim of the bed. For that reason, I have indicated 50% as a standard "maximum" expansion factor for clay but emphasized in a note that it can run even higher.

For expansion of other types of soil, I have also indicated that you can encounter increases higher than the maximums that you may find recommended elsewhere. I am inclined to think that for light-frame builders, the excavation modification factors do need to be high. We do relatively small volumes of excavation compared with large-scale commercial and industrial builders. We use small equipment such as power shovels or even excavate by hand. As a result, our excavated soil gets chopped up more than the soil removed in huge and still-coherent chunks by heavy equipment and is likely to expand by a larger percentage.

Takeoff Modification Factors

ITEM	MINIMUM	MAXIMUM	COMMENTS
SOIL EXPANSION			
Average	20%	35%	Per Paul Cook
Clay Dry	25%	50%	Or More!!!!
Clay Wet	25%	50%	Or More!!!!
Clay/Gravel Mix	15%	40%	Or More!!!!
Gravel Dry	15%	29%	
Gravel Wet	18%	25%	
Sand	15%	20%	
COMPACTION	10%	20%	Very Approximate
CONCRETE			
Forms	10%	20%	
Rebar	10%	20%	Laps. cut-offs
Welded Wire Fabric	10%	40%	Laps
Concrete	10%	20%	Spillage. leftovers, uneven base
LUMBER			
Studs, plates	5%	25%	
Solid lumber, joists	10%	20%	
Plywood	10%	25%	
Engineered Lumber	5%	10%	TJIs, glulams, PSLs, etc.
Trim	10%	20%	
HARDWARE			
Rough	2%	10%	Lost, leftovers
Finish	1%	5%	Lost, walk-offs

Compaction, as you can see in the table, does not exactly mirror expansion. For example, let's say you have excavated 10 yards of clay for a foundation trench. It expands by 50%, giving you 15 c.y. of excavated soil. You send 10 yards to the landfill and plan on using the remaining five yards for backfilling four c.y. of vacant trench around your installed foundation. Why five yards to fill four yards? Answer: You are assuming 20% compaction because you know you will not be able to compact the backfill soil as tightly as it was compressed when undisturbed in the ground.

To develop reliable expansion and compaction factors for your own projects, you will need to keep records (using forms such as those illustrated in Section III), for the soils you actually encounter in your area. Likewise, for concrete work—while the range in the table may prove a useful starting point—you will need to develop records to determine waste modification factors that fit your operation. Hell-for-leather production carpenters can waste a great deal of lumber and reinforcing steel or wire. Methodical carpenters committed to careful use of resources will waste a lot less.

However, even frugal carpenters can have waste forced upon them by the availability of material. If they must order welded wire fabric in 100 s.f. rolls for a slab that requires 240 s.f. including laps, they are going to have 60 s.f. left over. And if you take the leftovers home to your "shop" rather than immediately recycling them, they will likely sit there for years. Now and then you will move the leftover wire to make room for other leftover stuff you think you might need one day (but probably won't). End result, you are saddled not only with leftover wire but also with wasted space and time. (Alas, I speak from experience.)

Waste of concrete itself can also be forced by supply contingencies. You always order a little more than you think you will need to complete a pour. You don't want to run short during the pour, and so you end up sending a yard back to the batch plant. Additionally, waste of concrete can result from sloppy craftsmanship. A wobbly form, wider than need be at points, or an uneven slab base, dipping here and there to excessive depth, can absorb significant amounts of concrete on pour day.

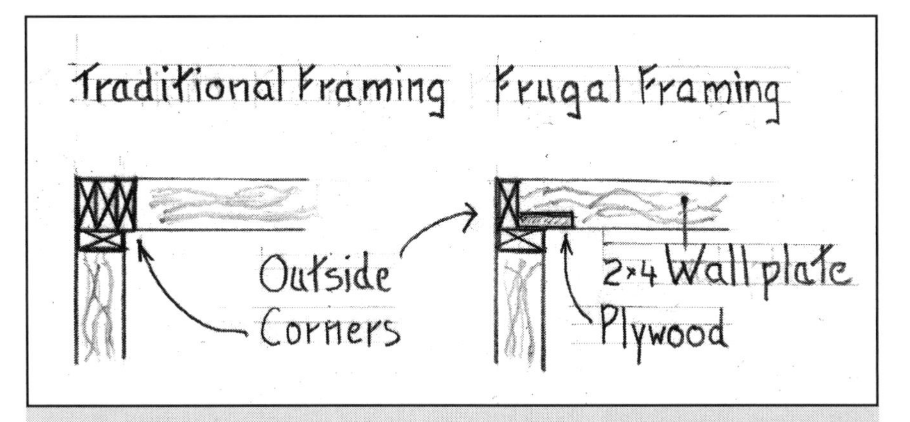

I have always regretted the huge waste of plywood scrap that piles up on framing jobs. Finally, when I was building the house whose design and construction is chronicled in *Crafting the Considerate House*, I hit on a use for the scrap. I had my apprentice spend an hour ripping it into strips, which we then installed as drywall nailers at inside wall corners and across top plates. That saved us dump costs and a considerable amount of lumber, beginning with two studs at every corner, as you can see in the drawing.

The plywood nailers have proved effective. Ten years into its life the house had been through a dozen or so mild earthquakes. There was not a single crack at any of the drywall corners.

During framing, the skill and care with which you and/or your crew work obviously influences the degree of waste you experience. A capable crew wastes less. However, no crew can prevent the waste created by thoughtless design requirements. If a breakfast nook requires 10'3" joists, you will be forced to buy 12' joists, cut them to length, and toss the cutoffs unless you can utilize them for blocking. Likewise, with plywood and OSB (Oriented Strand Board) for subfloor and for wall and roof sheathing, if your building is not laid out on a four-foot by eight-foot grid, you will end up with cutoffs piling up until they are hauled off to the dump. (Or at least that is what will happen if you use typical framing practices. If you want to do better, utilize the cutoffs for drywall backing as in the frugal framing approach illustrated here.)

At the end of the modification factor table you will see two lines for hardware. Requirements for rough hardware have been ramping up rapidly for decades. When I started out as an apprentice carpenter, about the only hardware we saw on our job sites were foundation form ties, 8d and 16d nails loose

Takeoff Speed Tricks

Forms, Plywood	Double area of one side of foundation wall	Divide by 32 and add waste factor to get number of sheets
Forms, 1x Lumber	Double area of one side of wall for square feet	Same if using 2x lumber
Concrete	Wall 8" x 4' x 10' takes one cubic yard	For other widths and heights adjust proportionately
Joists @ 16" o.c.	3 joists per 4 feet of floor width plus 1 for end of run	Add for doublers at concentrated loads (stairs, partition walls, etc.)
Joists @ 2' o.c.	Divide floor width by 2 and add 1 joist for end of run	Add for doublers at concentrated loads
Wall sill & top plates/ conventional framing	Multiply total wall length by 3	
Wall sill & top plates/ frugal framing	Multiply total wall length by 2	
Studs, conventional framing	1 to 1.5 studs per foot	Somehow, there are never enough studs!
Studs, frugal framing	Divide total wall length by 2 and add 1 for end of run	Add 2 for each door and window
Wall Sheathing (sheets of 4x8 ply)	Multiply length x width and divide by 32	Do not deduct for openings less than full sheet in size

in their boxes, rolls of building felt and leatherback window flashing, and perhaps a few lag bolts along with a few hangers and clips from a tiny new outfit named Simpson Manufacturing Company.

Now nails come in a variety of strips and coils. Building wraps and flashing material comes in an endless and ever-changing array of forms including adhesive and liquid materials. Simpson now offers such a huge variety of hardware that a catalogue the size of this book is required to display it.

Even so, and oddly, I have never seen any publication or heard any estimating expert mention a waste factor for hardware. Perhaps that is because hardware items are largely counted during estimating, and it is assumed that estimators will get the count right. However, waste there will be—and costly waste, because hardware is costly. Nails and connectors will get misplaced, spilled, and lost. Partial rolls of building wrap and flashing will be left over and discarded. Partial tubes of sealants, foams, and adhesives will dry out and get tossed.

Even finish hardware will go to waste, though I have not allowed for that in the modification table. Occasionally a cabinet knob is lost. A passage lock is somehow damaged. A window screen is accidentally torn. Ten percent of a box of finish nails is left over and discarded or given away. And sometimes finish hardware just ends up going home with one of the many workers who pass through your job sites.

You may have noticed in a previous section that the illustration of a takeoff sheet includes a separate column for installed amounts and that waste and like factors are included under "Total Amount" but not under "Installed Amount." There's a reason for that: Waste, expansion, and laps always impact the cost estimate for material. They may or may not impact the cost estimate for labor. For example, if you are estimating the costs of an exterior wall frame, you will want to add for waste when figuring the cost of materials for the wall. You will not, however, want to include the waste factor in your labor cost estimate, as it will be based on hours of labor per foot of installed wall.

That's not to say that there is no such thing as wasted labor. Even highly focused and experienced tradespeople waste a little motion here and a bit of time there. But that normal waste will be included in your labor productivity figures from past jobs that you will use to estimate new jobs. If you also provide for labor waste in your takeoff for installed work, you will, in effect, be doubling up the waste factor. You will be moving a little further from the bull's-eye you are aiming for, if never quite hitting, as you attempt to nail your numbers.

Speed

If the first priority during takeoff is accuracy, then the second is speed. In addition to "wearing out your calcs" you can use a variety of speed tricks to accelerate takeoff. For concrete formwork, if you are using plywood, you can figure the quantity needed with three simple steps. Measure the total length of the wall. That will likely be the figure you will use to estimate your labor costs—or at least you will if you have created labor productivity records of man-hours per linear foot spent forming concrete walls as I recommend and which we'll go into in Chapter Eleven. To figure material:

1) Multiply the length by the height of the wall to get the area of one side of the wall.
2) Double the area to get a quantity of material for both sides of your form.
3) To figure the sheets of plywood needed, divide the area by 32 (the area of one sheet of plywood), and add an appropriate waste factor (L × W ÷ 32 + waste = number of sheets).

For figuring volumes of concrete, first note that a wall that is 4' high by 8" (i.e., .66') thick will require a yard of concrete every 10'. Here's the math: 4' × .66' × 10' ÷ 27 c.f. = .97 c.y., which rounds off to 1 c.y.

Once you know that, to get the cubic yards of concrete for any 4' × 8" wall, you simply measure the total length of the wall and divide by 10. The result, plus an appropriate margin for waste, is the total cubic yardage you will need to order for the wall. For example, a 4' high wall that is 80' long will require 8 c.y. of concrete (80 ÷ 10 = 8) plus a yard for waste.

If a concrete wall is taller or wider or narrower, you can readily adjust from a cubic yard per 10' of 4' × 8" wall to obtain cubic yards of concrete needed for the wall at hand. For example, if a wall is 8' rather than 4' high, it will require double the yards of concrete, or 2 c.y. rather than a single cubic yard for every 10' of length. Likewise, if thickness increases or decreases, concrete increases or decreases proportionately. If a 4' wall is 12" rather than 8" thick, it will require 50% more concrete, and so on.

For masonry blocks or bricks of different types, a similar base reference can prescribe a certain quantity per 1' of wall. The quantity can be adjusted proportionately for increases or decreases in wall thickness or height. Whether for concrete or masonry, once you have the standard trick committed to memory, you can very quickly figure quantities for any wall.

Please note, however—very important—that labor for concrete work does not increase in such strict proportion as material. An 8' wall will require more than twice the labor of a 4' wall because the need for bracing will increase more than

If you are forming your wall with solid lumber, you may prefer to use board feet (b.f.) to figure your material quantity. A board foot is the amount of material contained in a board that measures 1" x 12" x 12."

A 1x12 that is 10 feet long contains 10 board feet. A 2x12x10 contains 20 b.f.

A solid lumber form for a wall that is 4' high will require 4 b.f. per foot per side if it is formed from 1x material (1' x 1' x 4' = 4 b.f.). It will require 8 board feet if formed from 2x material (2" x 1' x 4' = 8 b.f.).

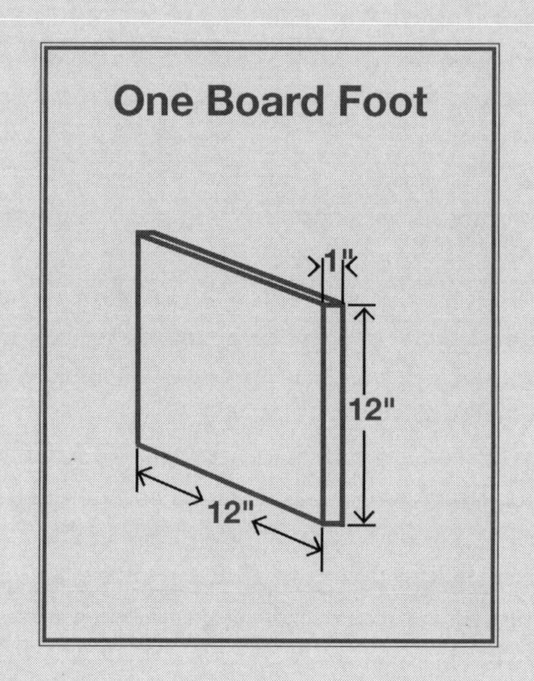

Board feet calcs may be useful for professional estimators figuring costs for large commercial and industrial jobs where huge amounts of concrete formwork are built. However, board feet calculations are not much used by light-frame builders, who typically came up through the trades as carpenters and think in linear feet, not board feet.

Also, with the increasing prevalence of engineered lumber from TJIs to paralams, we are using less and less solid lumber on jobs; and engineered lumber is priced by the linear foot, not by board feet. A senior sales rep at my own lumberyard tells me he no longer sees requests for quotes submitted in board feet, but that these days, lumber is always quantified in linear feet.

proportionately and because carpenters will not be able to work from the ground but must clamber up and down staging to work at the higher reaches of the 8' wall. If you merely double the labor needed for installing a 4' wall for a wall that is 8' high, your estimate will be low. The only way to reliably estimate how long it will take your crew to build walls of different heights is to keep and rely on accurate records of their productivity for construction of those varying walls.

When you move from taking off the concrete work in a project to taking off framing, you can likewise accelerate your work with speed tricks. Thus, for floor joists: For 16" centers, divide the width of the floor by 4, multiply the result by 3, and add a joist for the end of your run.

Here's an example for a 24' wide floor: 24 ÷ 4 = 6. 6 × 3 = 18. 18 + 1 = 19. So, you will need 19 joists for your 24' wide floor.

For 2' centers, simply divide the length by 2 and add a joist for the end of the run: (24' ÷ 2 + 1 = 13 joists). For both 16" o.c. and 2' o.c. floor frames, add doublers required to support wall partitions, stair carriages, or other concentrated loads.

Figuring wall plates can likewise be accelerated with a simple speed trick, as suggested in the sidebar.

When it comes to studs, however, we suddenly enter a zone of murk and mystery. Early in my career, when all wall framing was done at 16" o.c., I was taught to figure a stud per foot rather than a stud every 16". The extra studs included in the stud-per-foot calculation were assumed to be adequate for kings, trimmers, ponies, sills at door and window openings, and corners. However, in my experience, use of the widely accepted stud-per-foot rule for takeoff resulted in a shortfall of studs on project after project.

Other builders told me they also experienced the shortfall. In fact, it seemed that no matter how many studs we ordered for 16" o.c. framing, there were never quite enough. The mysteriously missing studs, we concluded, must have been sucked into a fifth dimension along with such perpetually disappearing items as tape measures, nail punches, and socks. Eventually, I came across published recommendations to allow for one and a half studs per foot for 16" o.c. framing. Maybe that would do the trick; I never tried it. Instead I moved to frugal framing (a full illustration appeared in Chapter Three), with the following astonishing results.

When I built my first frugal frame, the missing stud problem evaporated. Allowing for one stud every two feet, adding one for the end of each run, two more for each door and window opening, and using scrap plywood rather than studs for drywall corner support, I found myself with exactly the right number of studs during construction. And I mean exactly, dead on the money. That was an exciting moment. The monster from the fifth dimension was put down at last!

There are other speed tricks beyond the basics we have discussed so far. We will encounter a few of those additional possibilities in the next sections as we work through sample takeoffs.

A takeoff walkthrough

To keep things simple and digestible here, I am going to limit myself to doing an abbreviated takeoff for a single-story structure that is 40' long and 20' wide with 9' high ceilings—a simple box of a building, in other words. A more complex, light-frame building would usually be just a collection of such simple shapes—one box for a central area, another for a garage, a third for an additional story or wing, and so on. Taking off the quantities of work for the more complex shape would require the same steps as this simple sample building, the difference being only that the steps would be repeated more often.

For the foundation takeoff, I have determined, first of all, that the lengths of its four sides total 120' (40' + 40' + 20' + 20' = 120'). I will use that figure repeatedly in the takeoffs that follow. For excavation, I simply multiply the 120' by the width and depth (shown under W as 1.5' and under H as 2.5' in the takeoff sheet) of the trench. Then I add 50% for expansion because the soil is predominantly clay and divide by 27 (because there are 27 cubic feet in a cubic yard). Thus, I calculate that after excavation I will be dealing with 26 c.y. of soil: (120 × 1.5 × 2.5 ÷ 27 = 17. And 17 × 150% = 25.5, or 26 rounded off).

To figure the total length of the form boards for the foundation's footing, I double the 120 feet of total foundation length because I will need one form board for each side of the footing (120 × 2 = 240). For waste, I add a minimal 10%, bringing my total to 264 l.f. of 2 × 10—not unduly optimistic because the shape of the building is so simple; and I should be able to use 20' boards at full length to form the footing and thereby experience little waste.

To figure lumber for the stem, I simply multiply the linear footage of lumber needed for the footing by five (264 × 5 = 1320). Why? Because I am planning to form the stem with 2x8s that I will then reuse for the floor frame. And since 2x8s are actually, these days, about 7⅜" (or 7.375 in decimal form) wide, five of them stacked edge to edge will be just over 36" or 3' high (5 × 7.375' = 36.875"). By ripping .875" (that's ⅞") off the top board I will form my stem at exactly 3' high.

Both the stem and the footing forms will be held in place with stakes set about every three feet. I could probably get by with wider spacing. However, a buddy of mine once worked on the construction of a

TAKEOFF	Project: Jacobs Addition						Date: 11/23/17			Pg. 1
Assembly/ Item	**Area**	**L**	**W**	**H**	**Install Amt**	**Unit**	**Waste/Laps/ Expansion**	**Total Amt**	**Unit**	**Comment**
FOUNDATION		40	20		120	lf				
Excavate			1.5	2.5		lf	50%	26	cy	Heavy Clay
Form Footing 2x10s					240	lf	10%	264		
Form Stem 2x8s					1320	lf	10%	1452		
Stakes 1/ 2',3', 4' @ 3' o.c.					80/2',3',4'		10%	88/2'3'4'		Steel Stakes OH?
Snap ties w/ wedges								2	bxs	1 bx ties, 1 bx wedges
Footing Rebar 2/5/8"					240	lf	20%	288	lf	
Stem Rebar 4/5/8"					480	lf	20%	576	lf	
Vert. Rebar 5/8"x4' @ 2' o.c.					240	lf	10%	264	lf	Tot Rebar 1128' - 56 sticks 5/8x20
ABs 3/4 x12 w/plate wash. & nuts @ 4' o.c.					34		10%	38	#	38 ABs w/plate wash & nut

SAMPLE FOUNDATION TAKEOFF

L/W/H = Length, Width, Height (or Depth) l.f. = linear feet c.y. = cubic yards # = number

parking garage where the inadequately supported forms blew out, releasing hundreds of yards of concrete onto the work site. Ever since he told me about the blowout, I have been conservative in bracing my own forms—and as a result, I've never had one sag out of line, much less collapse.

For this foundation, I am assuming use of 2' stakes for the footing and 3' and 4' stakes for the stem. To get the total numbers of stakes, I once again start with 120', double it to provide for stakes on both sides of the form, and then divide by 3 to provide for stakes at the 3' interval. I add 10% for the end of runs and for the few stakes that will inevitably be wasted during

TAKEOFF	Project: Jacobs Addition					Date: 11/23/17			Pg. 1
Assembly/ Item	Area	L	W	H	Install Amt	Unit	Waste/Laps/ Expansion	Total Amt	Unit	Comment
FRAMING:										
Mudsill					120	lf	10%	132	lf	3x6 PT
Floor joists & rims w/blkng					540	lf	15%	620	lf	25 jsts & 40' blks - Reuse for m 2x8s - See work sheet
Subfloor	800	40	20		800	sq ft	5%	840	sq ft	3/4 t&g ply
Ext wall sill & plates					360	lf	10%	396	lf	2x6x20 DF SB
Ext wall studs					64	#	20%	77	#	2x6x9 KD DF
Ext Wall headers					60	lf	20%	72	lf	4x10 12 openings
Int Wall Sill & Plates		84			252	lf	15%	290	lf	2x4 DF SB
Int Wall Studs					50	#	10%	55	#	2x4x9 KD DF

SAMPLE FRAMING TAKEOFF

L/W/H = Length, Width, Height (or Depth) l.f. = linear feet c.y. = cubic yards # = number

construction. That brings me to a total of 88 for each length of stake (120 × 2 ÷ 3 × 1.1 = 88).

I plan on using steel stakes and I probably have most of what I need on hand (OH) in my storage shed. Even so, I put a question mark after the OH in the comments column. The question mark will remind me, when I am extending my takeoffs into my estimating spreadsheet and costing out the project, to take a look in the shed and determine whether I will have to purchase additional stakes for the job.

For figuring the quantity of ⅝" rebar needed, the 120' foundation length again serves as a starting point. For the footing rebar, I double the 120' for two bars all the way around, add 20% for waste and laps, and get a total of 288' (120' × 2 × 1.2 = 288').

For the horizontal stem rebar, called out as four bars all the way around, I quadruple 120', again add 20% for waste and laps, and arrive at 576' (120' × 4 × 1.2 = 576').

For vertical rebar, I figure I will need a 4' stick at 2' o.c. all around the 120' perimeter and that waste will run to about 10%. That gives me 264' of rebar for the verticals (120 ÷ 2 × 4 × 1.1 = 264). Adding together my rebar totals for footing, stem, and verticals, I get 1,128 l.f. In my comments column, because my lumberyard will be happier quoting by the stick than by the foot, I note that I will need 56 sticks of the ⅝" × 20' rebar (1128 ÷ 20 = 56.4; and I round off to 56 rather than up to 57 because I have already allowed 10% extra).

To figure the number of anchor bolts (ABs) with nuts and washers, I divide the familiar 120' by the spacing of the ABs (4'), add one for the end of each run, add 10% for misplaced, bent, or cross-threaded bolts and arrive at a rounded-up total of 38 (120' ÷ 4' = 30; 30 + 4 = 34; 34 × 110% = 38 rounded up). For any specialty foundation hardware, such as anchors for hold downs or strong walls, I will need to make a count. I have not shown the results here. But in actual practice, to continue my takeoff in the natural order of construction, I would make the count right after calculating the number of ABs.

Moving to the framing takeoff, I again "work my calcs," drawing on the measurements used for the foundation takeoff. Thus, for the amount of installed mudsill, I bring 120' over from the foundation takeoff because, of course, the perimeter of the foundation and the total length of the mudsill are the same. As you can see in the sample takeoff sheet, I have specified a 3x6 pressure-treated mudsill. In the earthquake country where I build, I want a strong connection between my frame and foundation. A 3x6 will handle the anchor bolts, plus all the nailing for shear panels required for seismic strengthening, with less risk of cracking than a smaller sill.

For the floor frame, the calculation is more complex. Setting up a separate worksheet (not illustrated), I calculate as follows:

1) For each joist, I will use a 20' board because the building is 20' in width.
2) To get the number of joists, because I am using frugal faming and installing joists at 2' o.c., I first divide the 40' length of the building by 2. That gives me 20 joists.
3) I add one more joist for the end of the run.
4) For rims, I start with the 40' length of the building and multiply by 2 for its two sides. That gives me 80'. Dividing 80 by 20, I see that I need four additional 20' sticks for rims.
5) For mid-span blocking along the 40' length, I figure on two more 20-footers.
6) I total up all the calculations: 20 (joists) + 1 (joist for end of run) + 4 (for rims) + 2 (for blocking) = 27 sticks.
7) Multiplying 27 × 20 (because I have determined all the sticks for the floor frame will be 20' in length) I come to an installed linear footage of 540 feet.
8) Adding 15% for waste, I come to a total amount of 620'.
9) In the comments column, because I will base my estimate of labor cost on the number of pieces of material installed, I note that I will be installing 25 joists each 20 feet long and cutting up 40 feet for blocking.

I could forgo making a note of the number of pieces in the floor frame and instead assume that when I figure my labor cost, I will base that calculation on the square feet of floor area, 800 s.f. in this case (40 × 20 = 800). In fact, some estimators do just that. But in my view, using square footage to figure labor for floor framing does not nail the cost down closely enough. Here's why: An 800 s.f. floor frame might require the installation of 25 joists and rims, plus blocking, as does the simple building we are visualizing for our takeoff here. With a more complex configuration, however, it might require many more pieces. And a frame with the greater number of pieces will require a greater amount of labor, even if it uses the same amount of material.

To obtain a quote for the joist material to be used in the project, I could use 620'—the linear footage including waste that I have recorded in the Total Amount column. For this project, however, when I do move from takeoff to costing out the job, I will not, in fact, be pricing material for the joists. Why? Because, as noted in the comments column of the foundation takeoff, I plan on reusing the 2x8 form lumber for joists. I have learned that if I purchase straight lumber stock of good quality, use it for forms without form release oil, then cure the concrete fully before stripping the forms, the form boards come away clean and in sound condition and can be reused for framing. (Frugal framing = environmental consideration + money in the bank!)

From my foundation takeoff, I can see that I will have 1,380 l.f. of 2x8 material available after the forms are stripped. That will easily provide the 620 l.f. needed for the floor frame while also providing material for a mid-span girder and miscellaneous other framing (not shown in this abbreviated sample takeoff). To remind myself to count on using the form lumber for joists, I boldface a note in the comments column. I will also put a reminder into my Building Notes folder—the list of ideas for construction, discussed at the end of Part I, that come up during estimating.

As I move to takeoff for the wall frame, I make use yet another time of the 120 l.f. figure. For the sill and two plates, I triple it and add 10% for waste.

For studs, since my building is to be framed at 2' o.c., I divide 120 by 2 and add a stud for the end of each run of wall to get a total of 64 installed studs. To that I add 20% for waste. With frugal framing a much lower waste factor can usually be used. However, here we are taking off quantities for a building with 9' high walls. Because studs do not come in 9' but only in 10' lengths, I will have to cut them to length, wasting 10% of each stick because the short leftover lengths will be of no use for anything other than firewood. To allow for a bit of additional waste such as a bad stick or a miscut, I have bumped the waste for the studs up to 20%. Though I have not included the quantification here in my simplified sample, in actual practice I would also add a pair of studs for each door and window opening.

More takeoff savvy

The more experienced with takeoff you become, the sharper you get at it. In particular, you come to understand more precisely just how far you need to go with takeoff to nail your cost numbers and how much of the more detailed takeoff you can leave till the project is underway. During the first years of my career I tended to go into excessive detail. For example, I would worry over the headers, figuring the exact length of each one.

Eventually, I realized that for takeoff purposes all I needed to do was count the number of openings in the exterior walls, multiply the total by a rough average of the widths of the openings, repeat the procedure for the interior walls, and call it a day. The count could then be used to figure labor costs for framing openings. And the approximate total length of all the openings was all that was needed to get a material quote. If the length was off a few feet and the material quote I got from my supplier therefore off a few dollars, that was not enough to sweat over during takeoff.

Similarly, with experience you become more judicious and efficient at the many takeoff tasks beyond those illustrated in the brief walk-through we have just completed. You learn just how much detail you need to develop and just where you can make use of speed tricks for tasks such as the following: Batterboard installation. Bracing. Scaffolding. Blocking. Wall sheathing. Roof framing and sheathing. Windows. Doors. Trim. Stairways. Finish hardware. While I won't walk through such additional takeoff lore in detail, here's a bit of extra savvy that you may find especially helpful:

- Roof trusses for gable end roofs often parallel floor framing. If so, and if the roof trusses are spaced on the same centers as the joists, you will have the same number of trusses as joists, minus the number for any additional joists allowed for concentrated loads.
- For doors and windows, if the designer is not providing a complete and accurate schedule, you must create one that describes each variation of doors and windows in detail—manufacturer, model, dimensions, hardware including screens—and the number of times it is used. Accurate window takeoff is essential for getting accurate supplier quotes, and mistakes are costly.
- Once you do have an accurate and complete door and window schedule in hand, interior trim can be easily tallied using speed tricks. To figure trim quantities for interior doors: (1) Count the number of hinged doors. (2) Multiply the number of doors by two (assuming that identical trim goes on both sides).

(3) Multiply by 20, since 20' of trim will provide for two pieces of vertical casing and one piece for horizontal casing with a waste factor built in. Finally, to complete the interior door casing takeoff figure, add material for the interior side of each exterior door and for any special doors such as sliding closet doors.

- For windows, use a like procedure. Count the windows and multiply by appropriate lengths to obtain necessary quantities of sill, apron, jamb, and casing material. For example, say window variety A needs 4' of material for sills, B needs 3', and there are 6 of the A and 4 of the B windows. Including 10% for waste, you then need to provide for 40' of sill material (6 × 4' = 24'; 4 × 3' = 12'; 24' + 12' = 36'; 36' × 1.1 = 40' rounded up). Alternately—and often close enough for estimating—just as for headers, assume an average window width and multiply it by the total number of windows, then add a waste factor to arrive at a total.

As with rough work, you should amplify your finish takeoff with any additional numbers and comments you will need to cost out labor.

Just what amplification is necessary will depend on how you have developed your labor productivity records (again, that's a crucial subject we will explore in Chapter Eleven). For example, if your labor productivity records for baseboard show labor hours needed per foot (or perhaps 10 feet), you can figure labor directly from the total baseboard length you worked up for material. On the other hand, if your baseboard labor productivity record is based on the more accurate per-piece rate, you will want to take the additional step during takeoff of counting the number of pieces of baseboard.

For door casing, you will likely develop per-door labor productivity records; and for windows a per-window labor figure for various combinations of trim—sills and aprons only; sills, aprons and casing; and so on. If that's the case, the count of doors and windows already in your takeoff will be all that you need to figure labor costs for door and window trim.

Unfortunately, for two critical takeoffs—namely hardware, both rough and finish—there are no speed tricks (that I know of) other than reusing a count you have already made for another item, such as a count of joists that will be installed in joist hangers that can be used to figure the number of joist hangers. As we have discussed, buildings now consume huge amounts of hardware from bolts through adhesives for the rough work, and much finish hardware as well. To take off the quantities, you have no choice but to list all the hardware items indicated in the plans and specs, determine the number of times each item is required, and record the results in your takeoff sheet. In my abbreviated sample takeoff, I did the takeoff for the one item of hardware that came up, namely anchor bolts, along with the rest of the foundation takeoff, to adhere to the natural order of construction. However, given the variety of hardware now required, it would be reasonable during a full takeoff to instead create a separate sheet for rough hardware and another for finish hardware.

All the counting is tedious. Therefore, the ability of the estimating software that is coming into use to count for you is highly appealing. For example, if you show the software a particular light fixture and ask it to count how many times the fixture occurs, it will count each iteration of the fixture and tell you it is required, say, 56 times—which will cause you to say, "Wow!"

Unfortunately, however, all repetitions of hardware are not necessarily called out in plans, particularly for rough work. Often, for example, a strap or clip is shown once and labeled "typical." It is up to you to spot all the other places it will occur, and the software won't do that. It counts only what is visible, not what is invisible. You are responsible for seeing the invisible. For now, estimators are stuck with a lot of counting.

Clearly, takeoff work is not mere clerical work. It is not "back-office" work, as one builder dismissively referred to it. It is skilled work. It involves knowledgeable visualization of what is actually going to happen out on the job site. And it requires your concentrated attention so that you don't overlook work required by the project.

Takeoff errors arise from many sources: Mishearing the answer to a question you have asked the

Auditing your takeoff

A number of construction industry manuals emphasize that once an estimator has completed a takeoff, it must be checked by a second skilled estimator. That's good advice—for a company that actually has more than one estimator on board. But what about the typical light-frame operation where the owner is the estimator?

One builder friend of mine suggests having your spouse listen while you carefully explain to her how you arrived at your numbers. His implication is that she or he will catch your missteps.

When I heard that I thought, "Are you kidding me, man?" My wife is steadfast and loving. But I fear that if I asked her to listen to me drone on about quantification of mudsills, rim joists, frugal wall framing, and waste factors, I would bring down the curtain on our marriage.

Fortunately, there are ways of auditing your takeoff other than torturing your spouse. First, put it aside overnight (or longer). When you come back to it, first check major calculations that you have used over and over in the takeoff, such as a measurement of foundation perimeter that is used to figure forming material, mudsills, exterior wall sills and plates, and more.

Next check your quantification of major items, such as lumber for wall plates and studs, to make sure they too are correct.

Finally, go over the rest of your takeoff, line by line, checking your math and reviewing your comments.

Unless you are a lot better at takeoff than I am, you are likely to find a mistake or two (I found a couple in the first pass of the foundation and framing sample takeoffs illustrated above). But if you have created a takeoff that includes just those numbers you need for costing out the job, and not all the stick-by-stick detail you will need for construction, your takeoff will not be very long. A brief but careful audit will help insure you've nailed your numbers.

designer. Misreading the plans or specs. Mismeasuring. Miscalculating. Or outright forgetting something, such as batterboards or wall bracing, that is not shown in the plans and specs but that you must nevertheless envision.

However, you can nail your numbers closely, minimizing the impact of errors and lapses, by systematically using the procedures we have discussed here. Add a careful audit such as the one suggested in the sidebar, and you should be able to produce accurate takeoffs. With a takeoff complete and audited you are ready to move on to estimating the costs of your project. That is, you are ready to take that step if you have created a well-designed and comprehensive estimating spreadsheet, the subject of our next chapter.

CHAPTER 5

Design Your Spreadsheet

The wheel of the construction business

The first time I heard the term "spreadsheet," I was only a year or so into my career as a general contractor. Though I proudly called myself a "builder," my "construction company" consisted only of a pack of business cards, the 10-year-old GMC pickup I had purchased for $2,200, a wooden box of hand tools, and a few power tools. I hadn't yet taken a business course or attended a seminar about the construction business and was not familiar with the concepts discussed in this book.

Fortunately, I met an experienced general contractor who invited me to a meeting of his builders' association; and there I heard a presentation on bidding by the owner of a respected structural repair company. Peppered with terms like "overhead," "margin," and "spreadsheet," her talk baffled and even intimidated me. But it also made clear that if I were going to operate in the world of the seasoned pros in the audience, I had to somehow acquire a business education. I began that very evening by asking the presenter if she could explain a few of her terms to me.

COMING UP

- The indispensable tool of estimating
- Creating your spreadsheet
- Pros and cons of MasterFormat
- Alternative formats
- Final touches
- The recap sheet

A "spreadsheet," I was relieved to learn, and as illustrations on the following pages show, is really nothing more than a page (or series of pages), either paper or on a computer screen. In either case the page, or "sheet" as it is called, is divided by vertical and horizontal lines into small boxes called "cells." You can enter, i.e., "spread," numbers into the cells. And once you have spread numbers across the sheet, you can total them up. If you are trying to figure the cost of a construction job, you enter costs of different items across the sheet (in an organized way, as we will discuss shortly) and then total them to arrive at the costs of your project.

In Chapter Four, we got a preliminary look at spreadsheets when we developed takeoff sheets for recording the quantities of work in a project. Here we will move up a step to the estimating spreadsheets (which I will refer to as just "spreadsheets" from here on out) that we use to build on the information developed during takeoff and to arrive at the costs of a project and a contract price. Simple though it may seem—and actually is, once you understand its basic characteristics and function—the spreadsheet is to business, including the construction business, as the wheel is to transport and industry. Without it, skilled estimating and bidding, not to mention accounting, would be far more difficult, just as moving goods without wheeled vehicles would be far more difficult. Without being able to enter costs for a possible project item by item, then total and evaluate the costs, you would be limited to guesstimating the price of a job. You would be limited to tilting back your head, eyeballing a project, and saying, "Well I reckon this one will run you, Mr. Client, about 60K"—and likely be off by a whopping percentage.

A well-designed spreadsheet enables you to move beyond guesstimating to skilled estimating and, therefore, to astute bidding as well. And it is doubly valuable because it does double duty as a checklist that helps you avoid overlooking items in a project.

Additionally, it enables you to array the costs so that you can more readily see opportunities to "value engineer" a project—to trim costs while retaining quality. It will even support creation of both a schedule for the construction of the project and a schedule of payments. By developing skill

The Power of Estimating Spreadsheets

A well-designed estimating spreadsheet:

- Acts as comprehensive checklist that helps you to avoid the cardinal sin of estimating: Missing something.
- Enables efficient costing of all work in a project, both by item and by items grouped into assemblies.
- Totals costs for items, assemblies, and trades by division of work, such as Concrete or Framing, so that you can see clearly where the costs of a project are most concentrated and where you can most readily trim costs to fit a budget.
- Enables you to show clients in detail the sources of the costs in their project, thus giving them confidence that your bid is rooted in reality and is not simply the highest number you can get them to pay.
- Displays not only the dollar cost of divisions of work in a project but the labor hours required as well, and thereby gives you a basis for generating schedules for your projects.
- Enables you to create a payment schedule for phases of work by showing you the projected costs for each phase.
- Helps you move beyond bidding for free. A clear, comprehensive, and attractively formatted spreadsheet makes vivid to clients the amount of work a reliable and accessible estimate requires and helps them appreciate that a good estimate is a product worth paying for.

at creating and using a well-designed spreadsheet, you also enhance your own value and earning power. Once you can use it to produce reliable numbers for a client, you are in a much better position to move beyond bidding for free and up to being paid for your preconstruction services.

Creating a basic spreadsheet

In its most fundamental form, as you can see in the illustration on the opposite page, a spreadsheet has three basic components:

- A *header* where you can name the project, typically with the client's name or the project address, indicate the date the estimate was created, and list other vital general information.
- *Rows* listing items and assemblies. It is the rows that act as a checklist to help you catch all items in a project. So your spreadsheet should include a comprehensive list of items.
- *Columns* for the quantities and costs of each item or assembly listed in the rows.

Only a couple of decades ago, builders set up their estimating spreadsheets on preprinted forms with blank rows and columns and filled them out in pencil. If you are just making your way into the use of spreadsheets, you may find it best to create your first spreadsheets on such forms. They are still available at office supply vendors. Alternately, if you have become skilled in the use of a word processor, you can use the table function of your processor to create a spreadsheet form and print out copies as needed.

When you create an estimate on a printed form, you will need to fill in the spreadsheet a cell at a time and total your numbers with a calculator. After doing a dozen estimates and developing confidence in your skills, you may find yourself eager to jump up to an electronic spreadsheet, such as Excel, the primary subject of Part V, so that you can automate the math. If you are already skilled in the use of an electronic spreadsheet, you may be tempted to skip both the paper and the word-processed versions and create your first estimating spreadsheet directly in a computer spreadsheet program. But that's a big, big step. I strongly advise not taking it before you have a grip on estimating fundamentals.

To begin creating your spreadsheet, either handwriten or word processed, set up a header. Do make sure to provide a place in the header for the number of each page so that when your estimate is complete you can make certain all pages are present. As with the pages of a takeoff sheet, spreadsheet pages can get misplaced, with the result that the costs recorded on them get left out of your bid (Ouch!).

With the header complete, you can move to labeling the rows for the divisions and for the items and assemblies of work in a project. In the illustration of the word-processed spreadsheet, as you can see the first division is for GENERAL REQUIREMENTS. Below the division name, and slightly indented to show that they are part of GENERAL REQUIREMENTS, are five items from Project Management through Project Closeout. You do want to distinguish divisions from items and assemblies. My style, as illustrated, is to use all capital letters for divisions and indents and smaller letters for items. You may prefer a different design.

With the rows labeled you can proceed to labeling your columns. In the illustration of the word-processed spreadsheet, the first column is labeled "DIVISION Item/Assembly" to indicate that the rows for divisions, items, and assemblies are listed below. To the right of that first column are columns for the quantities that will be brought over from the takeoff sheet and the units (such as square feet) that they are expressed in. I have labeled the quantity column "Quantity Installed" because that is the quantity that will be used to figure "Labor Hours" and "Labor Costs" in the two columns next to Units.

To the right of the labor columns come three more columns:

- Material Cost, which includes cost for waste, laps, and expansion, i.e., the "Total Amount" figure from the takeoff sheet.
- Sub Cost for subcontractor charges.
- Service Cost for such items as sani-cans and scaffolding.

Project:	Estimator:	Date:	Page:

DIVISION Item/Assembly	Quantity Installed	Units	Labor Hrs.	Labor Cost	Material Cost	Sub Cost	Service Cost
GEN. REQUIRE.							
Project Manag.							
Project Setup							
Daily Cleanup							
Sani-Can							
Project Closeout							
SITE WORK							
Drainage							
DEMOLITION							
Remove (E) Deck							
CONCRETE							
Piers							
FRAME							
FINISH							
Decking							
TOTALS							

You can readily create a spreadsheet by using the table function of your word processing software. While this word-processed spreadsheet is abbreviated for purposes of illustration, you can extend your table for as many pages as you like to create a comprehensive spreadsheet/checklist. To enjoy a spreadsheet that automatically totals numbers, however, you will have to learn to use a computer program such as Excel.

Electronic spreadsheets are awesomely efficient and powerful tools. We will go into them in detail in Part V of this book.

However, it does take time to build and master an electronic spreadsheet. You do not want to take on that challenging task even as you are learning to create, format, and use spreadsheets. To return to the wheel analogy used in the main text, that would be like moving directly from your first hand-drawn wheeled cart to an Indy 500 racer. You are likely to drive into a wall.

Creating and using a spreadsheet created with your word processor is an excellent interim step. You will be using your computer to build your spreadsheet, if not to crunch numbers, and that will pave the way for the big move to electronic spreadsheets.

All numbers placed in those columns should come from quotes supplied by outside firms such as a lumberyard, plumbing subcontractor, or scaffolder.

Separation of subcontractors and services is not essential; some estimators forgo it. I make the separation because the relationships of subs and service providers to a project are quite different. Subs provide and install work that becomes a permanent part of the project. Service providers, on the other hand, supply items such as scaffolding, sani-cans, rented equipment, and special inspections. While necessary for the construction of a project, they do not become a permanent part of it.

A final column in the illustrated spreadsheet provides space for comments. Though not needed for every row, item, and assembly, well-chosen comments can serve to flag a decision that underlies a tricky or complex cost in the spreadsheet and to refer to a worksheet where the math that led to that decision is recorded. For example, in one of my estimates, a comment on a row for excavation might read "See Wksht.#1/Excavation costs." There I will have compared the cost of having my crew excavate trenches with rented electric spades to the cost of hiring a sub with a Bobcat. The worksheet will show that the Bobcat is the way to go. And that will explain why the cost for excavation is largely in the sub costs column with only minor amounts for incidental work by the crew in the labor hours and costs column.

Beyond the basic spreadsheet

As you advance in your career, you may want to expand your spreadsheet to accommodate additional information. To begin with, you may decide to place an additional narrow column to the left of the divisions/assemblies/items column and use it to record numerical cost codes, such as 01 for General Requirements, 02 for Site Work, and so on. Such cost codes can serve to connect your estimating spreadsheet to an accounting spreadsheet. With the connection established, you can more efficiently "job cost"—that is, track actual costs against estimated costs as construction of a project proceeds.

Some builders elect to include in their spreadsheets not just one but a series of columns for labor hours. The multiple columns serve to record hours for crews of different sizes and compositions. For example, an initial labor hours column, labeled "1-crew," is for a one-person crew, a lead carpenter who will work alone doing a small project such as a bathroom remodel. A second column is for a two-person crew ("2-crew"), a lead plus an apprentice. Others are for three- and four-person crews ("4-crew").

Along with multiple labor hour columns, you may wish to set up your spreadsheet to include a percentages column. There you can display the percentage of total project cost consumed by a particular division, such as 10% for Concrete or 15% for Framing, or by a major assembly such as wall framing or windows. You may find that figuring more than a few selected percentages when you are still using word-processed spreadsheets will be too time-consuming. When you move to electronic spreadsheets, however, you can quite easily embed formulas in the spreadsheet so that it will automatically generate the percentages for all divisions of work. They are very useful for determining whether the costs you are placing in the spreadsheet are reasonable. For example, if you are estimating for a kitchen remodel, and your plumbing sub's charges typically run about 9% of total project direct costs, but for this project he is at 16%, you will know to ask him for an explanation.

Along with adding columns, you may want to enhance your spreadsheet with special graphics that make it easier for your eye to find its way to the appropriate columns and rows when you are producing an estimate. Toward the end of this chapter on page 116 I have included an illustration of an abbreviated electronic spreadsheet. The illustration gives you a sneak preview of the electronic spreadsheet I discuss in detail at the end of the book. But it also serves to illustrate the use of graphic elements including:

- Boldface (**FRAMING**) to distinguish the division heading from the subordinate assemblies and items.
- All Capital letters (WALLS) to distinguish assemblies.
- Colors plus boldface to set off the Labor Hours and Material Costs columns. (With Word or Excel, the color formatting can easily be extended down the columns so that numbers placed in the columns are also colored).
- Spaces between assemblies to break up the spreadsheet so that it is easier to accurately place and find information.

When it comes to rows, as you more fully develop your spreadsheet, you may want not only two but three levels of indents. That, too, is illustrated in the electronic spreadsheet toward the end of the chapter. There you will see one level for divisions, a second for assemblies, and a third for items included within an assembly. For example, below "FRAME," for the division of work, you will see an indent for "WALLS." Then you will see an additional indent for the list of items in the wall frame—sill, plates, studs, headers, sheathing, and so on.

Three levels of indents can be helpful because often you will find it advantageous as you work through a spreadsheet to estimate work in the project partially by assembly and partially by item. With wall framing, for example, you may estimate labor by assembly, allowing a certain number of man-hours per linear or square foot of wall. But for material, you may choose to put in your numbers item by item rather than trying to capture the cost of the diverse materials used for sills, plates, studs, headers, and sheathing in a per-foot number.

Beyond choosing your level of indents, you will want to decide just how you go about organizing your rows. Surprisingly, that is a subject so controversial that it deserves a section of its own.

MasterFormat

You might think that a subject as straightforward as labeling and organizing rows for an estimating spreadsheet would be exempt from our human tendency to develop fiercely held and opposing opinions. It is not exempt.

The argument begins with the presence of an 800-pound gorilla in the world of estimating, the "MasterFormat," developed by the Construction Specifications Institute. CSI's mission is maintenance of a common and consistent language for construction so that designers, builders, and material suppliers can communicate more clearly. And its MasterFormat does offer benefits, even to light-frame builders, at least those who have moved beyond the start-up stage, and especially to those involved in commercial work. If your spreadsheet is organized with MasterFormat, it will be in sync with the numerous construction industry references—checklists, cost data books, and material catalogues—that rely on MasterFormat.

Additionally, if you plan to grow your company to the point where you will hire professional estimators who have been trained in academic construction management programs, the likelihood is that you will be bringing in folks accustomed to thinking in "MasterFormatese," or some variation of it. They may have less trouble getting up to speed in your company if it is using MasterFormat rather than a format you have invented and that they have never seen before. Finally, for certain projects, you may

MasterFormat Divisions

In 2004 the Construction Specifications Institute expanded MasterFormat from 16 to 48 divisions (with some divisions reserved for future use). Here the older divisions, those from a pre-2004 edition of MasterFormat, are italicized. Stars indicate divisions whose names have been changed.

00 Procurement and Contracting Requirements
*01 General Requirements**
02 Existing Conditions*
03 Concrete
04 Masonry
05 Metals
06 Wood, Plastics, Composites
07 Thermal and Moisture Protection
*08 Openings**
09 Finishes
10 Specialties
11 Equipment
12 Furnishings
13 Special Construction
14 Conveying Equipment
21 Fire Suppression
*22 Plumbing**
*23 HVAC**
25 Integrated Automation
26 Electrical
27 Communications
28 Electronic Safety and Security
31 Earthwork
32 Exterior Improvements
33 Utilities
34 Transportation
35 Waterway and Marine Construction
40 Process Integration
41 Material Process and Handling Equipment
42 Process Heating, Cooling, and Drying Equipment
43 Process Gas and Liquid Handling, Purification and Storage Equipment
44 Pollution and Waste Control Equipment
45 Industry Specific Manufacturing Equipment
46 Water and Wastewater Equipment
48 Electrical Power Generation

be required to submit your bid with your charges sorted into MasterFormat divisions. If you have estimated the project using another format, conversion to MasterFormat will be, at best, a nuisance and, at worst, a potential source of errors.

MasterFormat is deeply embedded in the construction industry and enjoys staunch support from its advocates. The chief estimator for a mid-sized commercial and institutional builder located in Northern California, speaking to a group of residential builders, chastised them for resisting MasterFormat. Get with it, he told them. It's not your job "to reinvent the wheel"; get your people "CSI literate."

But others take an opposite stance. Even the Associated General Contractors notes that CSI's work largely supports the "nonresidential" side of the industry. Carl Hagstrom, the professional estimator whose insights contributed to the preceding chapter on take-off, says, "for residential work, CSI never came close to being an appropriate system." Another builder says simply, "I hate it."

Builders have good reasons for resisting MasterFormat, not least its increasing complexity. In 2004 it was expanded from 16 to 36 active divisions with another dozen set aside for future use. Many of the divisions are irrelevant for light-frame builders and even for sizeable commercial construction companies. If that's your world, you might bid a project that requires "conveying equipment" (Division 14), say a dumbwaiter in a restaurant or an elevator in an office building. But are you likely to get involved with any of the work from waterway construction through electrical power generation covered in Divisions 35 through 48?

Of course, if you are attracted to MasterFormat as a way of organizing your spreadsheet, you can elect to exclude divisions that are of no use to you. But even then, you will run into what I think is the larger problem with MasterFormat. As it breaks down its divisions into phases of work, it regularly departs from the natural order of construction. For example, Division 5, for metals, includes both "structural metal framing," which is installed early in a job, and "decorative metal," typically installed during finish work. Division 7 is a particular mess. It places "Thermal Protection" before various roofing and flashing tasks, thereby placing insulation before the work of drying in a structure.

There are numerous other such departures from the natural order of building in MasterFormat. It is difficult to discern the logic that governs its organization. It appears that work is grouped sometimes by function, as in "Fire Suppression," and sometimes by type of material, as in "Wood, Plastics, Composites." Perhaps for some estimators—maybe those who have been academically trained rather than coming up through the trades—MasterFormat works.

I feel, however, that if you know how to build, there are advantages to estimating in parallel with the natural order of construction rather than in accordance with whatever logic underlies MasterFormat. When you estimate in the natural order of construction, you are building the project in your imagination. You are visualizing installation of items and assemblies just as your company will actually do it, one step at a time, out in the field. If you are experienced at building, then estimating in parallel with the usual flow of work in the field is like retelling a story you have told many times before. You realize when you have skipped a part. You backtrack and put it in. When you are estimating the installation of a concrete slab. you include your membrane right after the drain rock and before the sand and rebar. You don't, as with MasterFormat, have to jump over the membrane and pick it up in an entirely separate division called "Moisture Protection" (which you will find mixed up with "Thermal Protection" and "Fire and Smoke Protection").

All that is not to say, however, that MasterFormat won't be of any use to you, even if you choose to set up your spreadsheet differently—perhaps using either what I call the "traditional format" or the "developer format," both of which are discussed in the next section. Its list of divisions and the breakdown within the divisions can be mined for row labels for your spreadsheet. For example, "Administrative Requirements" is a good label for such items as meetings with the owners during construction. "Closeout

Requirements" can cover completion of an end-of-project punch list and final project cleanup. Additionally, MasterFormat can suggest many items and assemblies that you can include in your spreadsheet to make sure it is comprehensive enough to serve as an effective checklist. (If you want to mine MasterFormat for ideas for your spreadsheet, you can do so at low cost by purchasing a slightly out-of-date copy of one of RSMeans cost books, which are organized in accordance with MasterFormat. See Resources at the end of the book).

Traditional and developer formats

The "developer format" and the "traditional format," in contrast to MasterFormat, are not products of a big institution like CSI. Rather they are my own creations, composites of other light-frame builder's spreadsheet formats that I have had a chance to review. The developer format works for builders who have adopted the business model long used by developers—namely subcontracting virtually all construction from demolition through structural work to landscaping, with their own crews handling only supervision and utility work.

The developer business model was once an object of scorn. Builders who used it were derided as mere "paper contractors." Insurance companies refused to issue them policies, assuming that builders who subbed out so much work did so because they did not themselves know how to build. But over the years the developer model has become increasingly popular and accepted. Now

Alternatives to MasterFormat

Developer Format Divisions	Traditional Format Divisions
General Requirements	*Crew*
Deconstruct. & Demo	General Requiremts
Earthwork	Deconstruct & Demo
Other Site Work	Site Work
Concrete & Masonry	Concrete & Masonry
Building Drainage	Building Drainage
Rough Carpentry	Rough Carpentry
Structural Steel	Ext Drs & Windows
Ext. Drs. & Windows	Bldg Wrap & Flash
Plumbing	Siding & Ext Fin Carp
HVAC	Int Drs & Int Fin Carp
Water Management	
Roofing	*Subs*
Electrical	Structural Steel
Security	Plumbing
Media & Communica.	HVAC
Exterior Cladding	Roofing
Thermal Boundary	Electrical
Drywall	Media & Communica.
Finish Carpentry	Security
Cabinets	Stucco
Countertops	Thermal Boundary
Tile	Drywall/Plaster
Finish Flooring	Tile
Paint	Finish Flooring
Appliances & Equip	Cabinets
Specialties	Countertops
Landscaping	Paint
	Appliances & Equip
	Specialties
	Landscaping

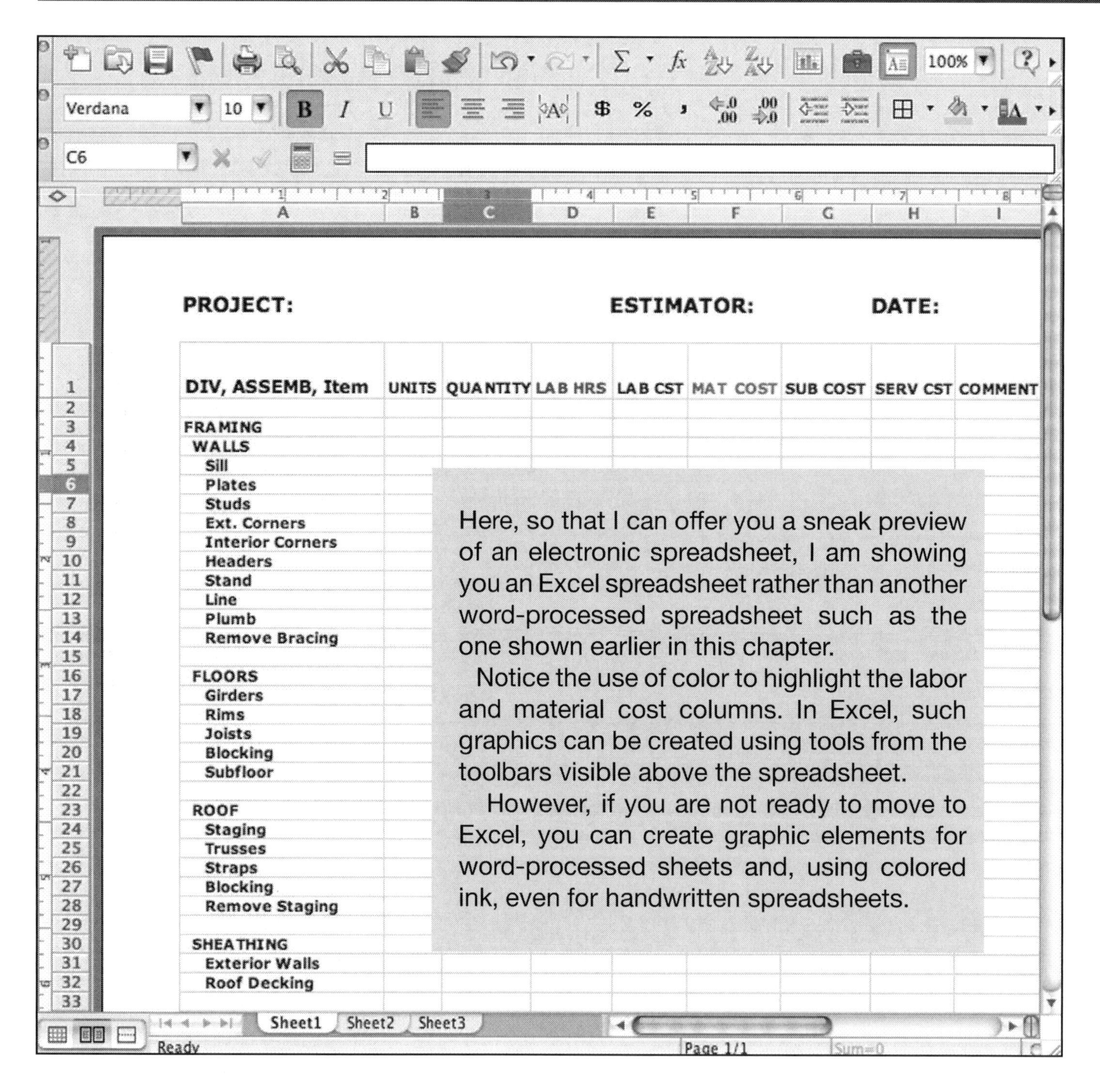

Here, so that I can offer you a sneak preview of an electronic spreadsheet, I am showing you an Excel spreadsheet rather than another word-processed spreadsheet such as the one shown earlier in this chapter.

Notice the use of color to highlight the labor and material cost columns. In Excel, such graphics can be created using tools from the toolbars visible above the spreadsheet.

However, if you are not ready to move to Excel, you can create graphic elements for word-processed sheets and, using colored ink, even for handwritten spreadsheets.

it is embraced even by remodeling companies, and even for projects as small as a bathroom renovation. The key feature of the developer format, as the sidebar on the previous page suggests, is this: Each trade is given a division of its own. The divisions are arranged in the order in which the trades are most likely to first appear on the job site. For example, while the plumbers will make several visits to the job site, and may even install a last bit of trim after the painters have departed, their first visit to the site will be to install waste and drain pipes. In the spreadsheet, they will come before all the trades that typically make their first appearance after waste and drain is installed.

The traditional format is best suited to builders who continue to favor the business model that was widely used before the developer model gained ground. As shown in the sidebar, all carpentry from foundations through finish, as well as closely related work such as deconstruction and flashing, is done by the builder's crew, i.e., carpenters and laborers. Specialty trades such as plumbing and tile are handled by subcontractors. The work of the crew is separated in the list from the work done by subs. But the work of both crew and subs is listed in the order in which it is first initiated.

Whichever business model and estimating format you choose, you will likely tailor it to fit your preferences, the particularities of your operation, and the area in which you work. You might choose to reorder the divisions of work handled by your crew. For subcontracted work, you might choose to combine specialties—for example including media and communication with electrical if the same subcontractor performs all that work. If you operate in Massachusetts, you might choose to combine roofing and siding, which are often done by a single sub (or so it appears from the signs I see on pickup trucks when I visit there), whereas in California, where I operate, the two trades are practiced by different subcontractors. You might choose to add other divisions of subcontracted work. If you are building in an area where rooftop solar has come into wide use, you may decide to add a division for photovoltaic and thermal solar installation.

In other words, there is no one-size-fits-all format. In fact, some builders use a format originally developed by insurance companies for the purpose of estimating the cost of repairs to buildings owned by their clients. Rather than rows being organized by trade or division of work, in the insurance spreadsheet the rows are grouped room by room. The insurance format is now in use, not only by insurance firms and their estimators, but also by builders, especially residential remodelers. I am not a big fan of the format. But I know builders who swear by it.

Assemblies, items, & the recap sheet

If are building a spreadsheet for a company organized in the traditional way, you will want to develop extensive rows of assemblies and items for the work to be done by your own forces. In the next two chapters, after we complete our discussion of the overall design of spreadsheets in this chapter, I will offer starter lists of assemblies and items you can make use of as you like.

Just how many rows you will include in your spreadsheet is your call. However, to maximize its function as a checklist that will help you to avoid overlooking any of the work in a project, you will likely want your spreadsheet to, at the very least, provide rows for all the assemblies and items that you typically build. You would do well to go further and include even those you run across only occasionally. My own spreadsheet has grown to approximately 20 pages—some 640 rows. I thought that pretty comprehensive until I ran across a homebuilder who told me that his is 40 pages long—and that even with that comprehensive a spreadsheet/checklist, he still misses a small percentage of the work in a project.

A spreadsheet can, however, become too long. If you included just half the items listed in one of the comprehensive catalogues of items available from construction information publishers, your spreadsheet would run to hundreds of pages. When estimating a project, you would find yourself scanning page after page of irrelevant tasks.

A spreadsheet can also become unnecessarily long if it covers too wide a variety of projects—for example, residential remodels, new homes, tenant improvements, restaurants, and other commercial projects. If you do such a variety of work, your will estimate more efficiently if you:

- Create a master spreadsheet that includes every item of work you do regularly or occasionally.

- Make copies of it for each of your types of work.

- Winnow down the rows in each copy so that it serves for one or another of your specialties.

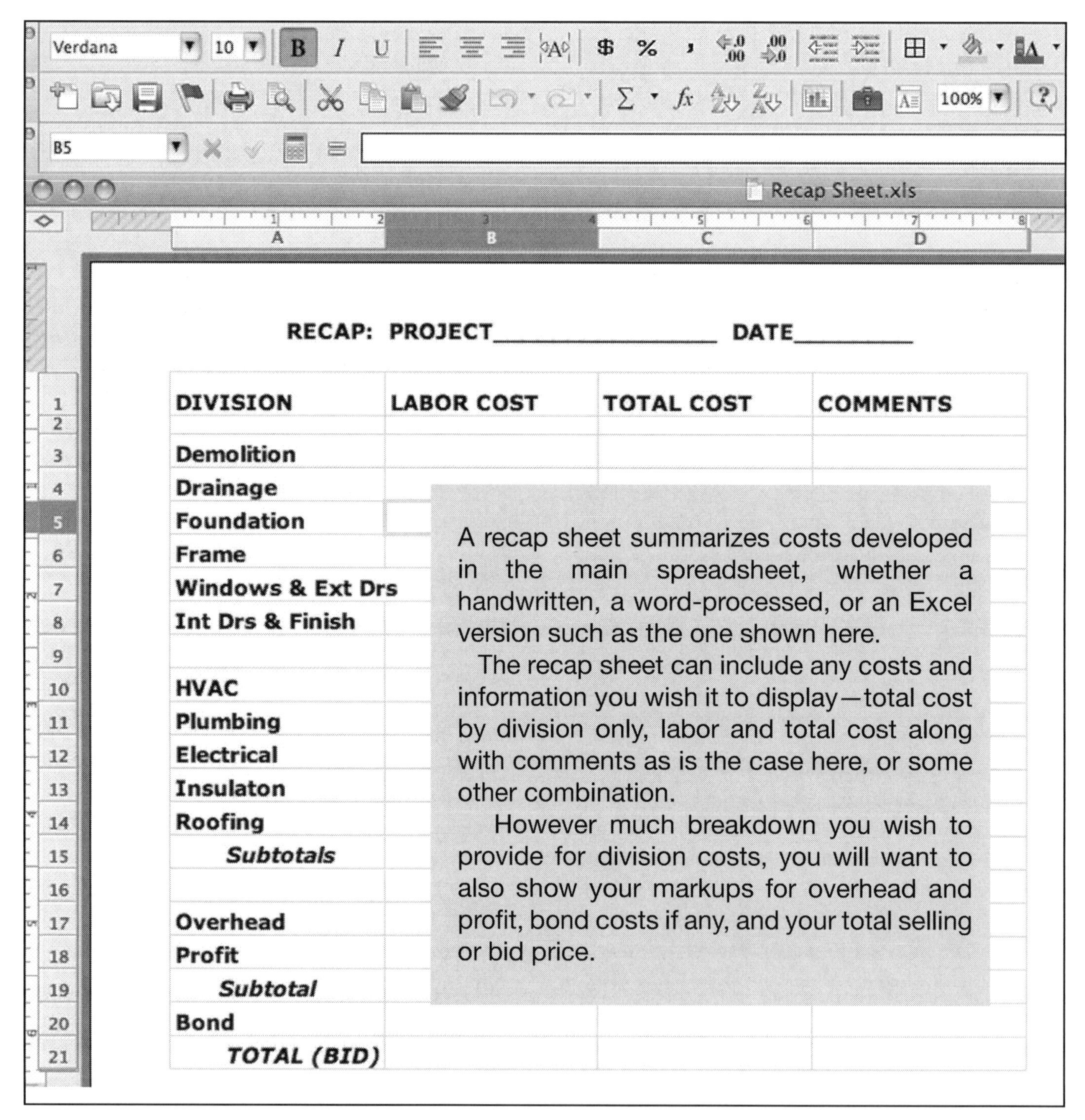

You will then have a distinct spreadsheet for residential remodels, for new homes, for tenant improvements, and so on. (Such a range of spreadsheets can easily be produced once you have graduated to the electronic spreadsheets.)

The final step in the creation of your spreadsheet is the setup of a summary sheet, also known as a "recap sheet" in the world of construction estimators. An electronic spreadsheet also facilitates the setup of the recap sheet, as we will discuss in the final chapter. But you can certainly produce one for a word-processed spreadsheet or even a handwritten one. For a recap sheet is simply a condensed one-page summary of the costs itemized in your spreadsheet. To create a recap sheet such as the one shown in the above illustration:

- List all the divisions in your spreadsheet on the recap sheet.

- Total the cost for each division of work in the project. For example, for Framing you would total labor and material costs for all floor, wall, and roof framing together with costs for subs and services, if any, involved in framing.

- Forward each division total to the recap sheet.

- On the recap sheet, total the division costs to arrive at a grand total for all the direct costs of construction for the job.

- Figure your overhead and profit (additional important tasks we will be discussing in detail in subsequent chapters) and record those figures as well on the recap sheet.

- Add the total of the division costs to the overhead and profit figures.

- Figure and add the cost of a bond if any is required, to arrive at a selling or bid price for the project.

A recap sheet has several uses. It shows you where the costs are concentrated and where you might most successfully look to cut costs if need be. For certain competitive bids it may be required. When you make the move to charging for your estimates, you will find that a recap sheet is appreciated by clients. It enables them get an overview of your bid before you guide them through the sometimes mind-boggling detail contained in your long spreadsheet of item and assembly, costs.

As you create your estimating spreadsheet with its multiple columns, its many rows for assemblies and items, and finally its recap sheet, your overriding goal—as with every other facet of estimating—is to create a net that will help you catch nearly every cost in the project being bid. In a word, one often used by professional estimators, in order to nail your numbers you want to prevent "slippage." Or, as I like to say, you want to snare all the "slipperies," those costs that can easily get away from you whether they are associated with work done by your crew, work done by your subs, or with quotes from suppliers or service providers.

Slipperies are a hazard in estimating costs for every division of work. But there is one division that is especially liable to slippage. In fact, start-up builders (including myself when I was beginning my career) are often entirely unaware of it, even though it can include costs that total up to an amount as great as the profit they hope to earn on the job. That division is General Requirements, and it is the subject of our next chapter.

CHAPTER 6

Spot the General Requirements

I don't know what you are talking about

As I was gearing up to write this book, I talked with a builder who has worked for 30 years in the construction industry. He had started out in the trades, become a skilled carpenter, and then served as editor of a journal for architects and builders before returning to carpentry and becoming a general contractor. When I asked him how he estimated for General Requirements, he answered, with his characteristic honesty, "General Requirements? I don't know what you mean by that."

Right then and there, perhaps, I should have given him the definition that I will suggest to you at the beginning of the next section. But I was startled by his admission and just gave him the first example of a General Requirement (GR) that came to mind. "Well, for example, the daily cleanup at your job site," I said, without reeling off the long list of other GR items, from permits to site-built ramps, scaffolding, and sani-cans, that you must capture in your estimates so that they are complete and you nail your numbers.

COMING UP

- The importance of General Requirements
- General Requirements defined
- Preconstruction General Requirements
- Job site General Requirements
- The most costly General Requirement of them all
- Judgment calls and controversies

"I think of that as marketing," he answered, meaning that he did the cleanup for free, as a way of pleasing his customers and earning their references—not that he charged for it as part of his overhead markup. Using that logic, he could have decided to do everything for free—his tight framing, thoughtfully laid out electrical runs, and precise finish work. Why not? All of it pleases his customers. All of it wins references. Why choose to give away only the cleanup? Heck, why not do the whole project without charge? That would surely bring in a whole lot of new work!

Another builder bumped into a friend of mine down at the lumberyard and asked, "Say Bob, do you charge for these trips down here to the yard to pick up material? They sure can eat up a lot of time." My friend answered, "Of course.

These trips are part of what's called 'mobilization.' You have to include the cost in your estimate unless you want to work for free."

It seems that many experienced builders, never mind guys just starting out, fail to cover General Requirement costs in their estimates. Wayne DelPico, author of several textbooks on estimating (see Resources section, page 382), has found that the failure to spot and charge for General Requirements extends even into larger companies. "Many of the items [in] General Requirements," writes DelPico, "such as scaffolding and the hookup and dismantling of temporary power, are not production-related and are sometimes mistaken as having no real cost." However, he emphasizes, "there is not only a definite cost," but for many General Requirements, the longer the project runs the greater the cost will be.

Just how costly are the General Requirements in total? Here's a sample of experience-based opinions:

- A professional estimator who produces cost projections for architects of extreme houses, 10,000 s.f. and up and costing $1,300/s.f. or more, allocates 8% of construction costs to General Requirements.
- For my own larger remodel and new construction projects, I have found that General Requirements consistently amounted to close to 10% of direct construction costs.
- A builder licensed as both an architect and a general contractor, who largely builds residential additions and kitchens, reports General Requirement costs running 8% to 12% of his total costs.
- The cost catalogues from a variety of publishers who collect their data from builders across the country suggest a range of figures. One says 4.5% for production housing. Another, 5% to 15% for residential construction. Still another, 10% to 12% for bath and kitchen remodels. And another, 20% for repairs and remodels under $200K.

As you can see, the figures vary widely. In fact, an estimating manual published by DeWalt states that the General Requirements can run from 20% to 40% of costs! The manual provides no evidence

No matter what the name, they matter

General Requirements are sometimes referred to by other names.

"General Conditions" is a widely used alternative term. "General Conditions," however, is also the name of a major section of American Institute of Architects (AIA) contracts, and confusion can result from the same term being used for a division of an estimate.

Some estimating manuals use the term "Indirect Overhead." That, too, can cause confusion, in this case with "Direct Overhead," or just plain "Overhead," namely the costs of running a company, as opposed to the on-site costs of building a project.

One provider of estimating software uses the term "Supporting Services" in place of General Requirements. The problem with that term is that some supporting services, like a special inspection, are one-time events and not General Requirements extending across a whole job or a major portion of it. Such supporting services are most reliably estimated with the phase of construction with which they are associated.

Whatever you might choose to call them, General Requirements are a major cost. You must identify them. They will amount to a sizeable chunk of your total project costs.

Selected General Requirements

PRECONSTRUCTION
- Plan check
- Building permit
- Sewer permit
- Electrical and gas permits
- Water company permit
- Street use permit
- Fire marshal permit
- School district fee
- Labor for obtaining permits
- Liability insurance
- Builder's risk insurance
- Local business license
- Sales commission

DURING CONSTRUCTION
- Initial setup (mobilization)
- Removal and storage of owner's possessions
- Setup of temporary facilities for owner
- Project sign
- Site office and supplies
- Equipment storage
- Fuel for equipment
- Material storage
- Material protection
- Travel allowance
- Parking fees
- Daily setup
- Daily rollup
- Daily cleanup
- Safety meetings
- Safety rails
- Hole covers
- Trench covers
- Path maintenance
- Plant protection
- Erosion control
- Dust control
- Noise control
- Rodent and other pest control
- Building interior protect
- Building exterior protection
- Site-built staging
- Temporary stairs
- Temporary ramps
- Material handling
- Material movement by elevator
- Material movement on a steep site
- Weather protection
- Dewatering
- Snow and ice removal
- Heating
- Cooling
- Lighting
- Recyclables storage
- Site fencing and gates
- Barricades
- Security personnel
- Traffic control
- Sani-can
- Phone (landline)
- Computer modem
- Power pole
- Temporary power hookup
- Temporary water hookup
- Power
- Water
- Rented scaffolding
- Cranes
- Hoists
- Conveyors
- Lifts
- Crew labor to support service providers
- Recyclables removal
- Waste removal

POSTCONSTRUCTION
- Final rollup
- Final cleanup
- Building performance audits
- Providing of operation and maintenance manuals to owner
- Occupancy permit
- Service calls

PROJECT MANAGEMENT
- Material takeoffs and orders
- Subcontractor scheduling
- Responding to sub RFIs
- Review of shop drawings
- Coordinating work of subs at job site
- Resolving disputes between subs
- Service provider scheduling
- Scheduling of inspections
- Temp labor orders
- Meetings with client
- Meetings with designers
- RFIs to designers
- Communicating with office and with builder
- Change orders
- Safety management
- Crew morale
- Time cards
- Creating photographic record of job
- Home office support for project

for that startlingly high figure. I can imagine General Requirements running that high only for an extraordinarily inefficient and top-heavy company. However, what is undeniable is that General Requirements, as one experienced estimator puts it, "make up such a large percentage of the project [costs], it is imperative that they be given a section of their own in each estimate." If you want to earn a decent living as a builder, you must understand exactly what General Requirements are, learn to spot them in each project, and build coverage for their costs into your estimates.

Preconstruction and Job Site General Requirements

I have seen some fancy definitions of General Requirements. My own definition is simple: General Requirements are all the items that don't become a permanent part of the project, but that are necessary to get it built.

If you look over the potential General Requirements in the list on the previous page, you will see what I mean. Aside from a few instructional manuals or copies of permits left behind for the owner, there's not a single item that will remain at the job after you complete final cleanup. Yet, every single one of the items comes with a cost for labor or material or services or all three.

Certain of the items may be especially difficult to spot. They tend to get away from us during estimating. If we let them, then we will miss significant costs in our project. They deserve not just listing but discussion. Among these "slipperies," as I like to call them, are some that are incurred even before the job starts, during preconstruction.

Design/build companies often employ salesmen who work on *commission.* Some builders pay a commission to any employee who brings in a customer. Such commissions could be considered overhead, but they are costs specifically associated with a particular job, and are typically based on the charges to the customer for that job. Therefore, they are appropriately categorized as a General Requirement cost of the job.

Permits are especially slippery. Over the years, permits have steadily proliferated, in part because local governments, deprived of other tax revenue, have loaded fees onto building projects to fund the services that their constituents demand. It's easy to overlook one or another of the required permits, especially if you are working in an unfamiliar jurisdiction. The building department personnel probably won't volunteer to inform you of the complete panoply of permits. You will have to dredge them up with persistent questioning.

Likewise, it's easy to overlook or underestimate the *labor of obtaining permits.* You may have to procure them from half a dozen agencies. A recent project of mine required separate permits from a zoning board, the building department, the sewage treatment people, the fire marshal, the school system, the gas and electric company, and the water district.

The various agencies that require permits may be in widely spread-out locations and available at different times. Likely most will require that you or an employee show up in person and stand in line to obtain a permit. As a result, you or someone in your company can easily spend the better part of several days chasing permits. Each permit is a GR item. You need multiple rows for permits in your spreadsheet.

During construction, the GR items mount up rapidly. Those items fall into two groups. In the list of GR items on the previous page, the groups are divided by an imaginary line at the mid-point of the list. Before the line are items that you, your crew, or project manager will take care of directly during construction. After the line come those items that are more likely to be handled by outside vendors, such as equipment rental companies. The GR work requiring labor by your crew begins with mobilization—and not only the *initial mobilization*, but also the *daily setup and rollup*. Remobilization may be necessary several times a day as you constantly move

The access factor

A contractor who specializes in wall covering was pleased when he won the job of removing and replacing all the wallpaper, top to bottom, in a hotel in downtown San Francisco. But his excitement turned to anxiety when he began to set up on the job. He realized he had not factored in the cost of moving equipment and material up and down the building on a slow freight elevator. That poor, time-consuming access turned a potentially profitable job into one on which he lost money.

Similarly, I have learned that material handling grows from a minor to a major cost when my crew and I are rebuilding a multistory home on a tight, steep site, as opposed to building on a wide, level lot.

Access can affect the cost of multiple General Requirements, including interior and exterior protection, cleanup, recycling, and waste removal. In your own estimating spreadsheet, you might reasonably add rows for difficult access at all those items and more. They will remind you to add sufficient cost into your estimate when a job with bad access comes your way.

Access is a big, big deal!

equipment from one space to another and to and from your vehicles or a staging area.

While likely to be more intense on remodel jobs, such remobilization is a cost on new work as well. In fact, when I worked as a framer on large multiunit projects, a laborer on our crew spent his entire workday making sure each carpenter had the tools and materials he needed, where and when he needed them. With the laborer's ability to anticipate the carpenters' needs so well, we rarely had to call out for a power cord or a fresh box of nails; he was like a top-notch point guard, feeding you the ball at just the right time and place for an open shot. He kept us set up, on the move, mobilized. The full carpenter's wage he was paid for his work amounted to a substantial General Requirement cost.

Like mobilization, cleanup work mounts up steadily, and along with it a variety of other site maintenance tasks including: *Protection of the building interior. Protection of the building exterior.* And, especially, the *maintenance of safe and efficient access* to work areas via construction of temporary paths, ramps, stairs, and staging—along with, very important, policing for tripping hazards.

Policing a job site takes time and costs money, and is often neglected. The cost of failing to attend to site policing can, however, end up being greater than the cost of doing it. If a site is littered with debris, or if maintenance of a safe site is otherwise neglected, productivity will drop off as workers pick their way past hazards. Accidents will happen, worker's compensation insurance rates will go up, and potentially ruinous insurance claims or lawsuits will be filed.

I know this from almost experiencing that fate. Years ago, I noticed a hole in the subfloor of a home we were remodeling. But I neglected to ask my crew to cover it. The owners' nine-year-old boy stepped into the hole. Fortunately, he was not seriously hurt, and his parents were forgiving. But I won't ever forget the incident. The boy could have been badly injured and might never have gone on, as he did, to become a professional tennis player.

Over the last couple of decades another set of General Requirement items have proliferated, namely environmental protection items—*plant protection, erosion control, noise and dust control*, and more—as clients, their neighbors, and new regulations have required it.

Other GR work done by your crew may not come so regularly. But when items such as installation and maintenance of artificial lighting or weather protection are necessary, they can eat up labor rapidly.

Certain of the requirements typically fulfilled through outside vendors, especially *rental of equipment* such as a crane, conveyor, or forklift, may be optional, appropriate to one job and not another. During estimating, you often face a choice between relying on manual labor and or renting equipment for various items of work. (Here's where estimators who understand construction can really earn their keep. They can visualize the options and crunch the numbers to determine which will give the better overall results.)

General Requirements that are fulfilled by outside service vendors will entail additional work, beyond what is provided directly by the vendor, that is easily overlooked during estimating—namely the increments of incidental *labor provided by your crew to service the servicers.* For example, will you be ordering a sani-can for your job site? Likely the provider will need a bit of your crew's time when he arrives, if only to ask where to locate the sani-can and to requisition a carpenter to drive a few stakes to hold it in place. Will the job need steel scaffolding all around the perimeter? When they arrive, the scaffolders, too, will likely make requests of the crew. Every GR service provider will, at a minimum, distract the crew and sometimes will directly interfere with the crew's workflow, requiring crewmembers to step aside, break off one task, and mobilize for another. Your estimating spreadsheet, if well designed, will provide you with a place to capture that work and its cost. Each item, in and of itself, may burden you with only a minor cost. But the costs do add up, as shown in the illustration of the GR division of a spreadsheet on the next page.

Are you are having trouble capturing all General Requirements in your estimates? Next time you pull over to check out someone else's job site, try looking past the construction and, instead, attempt to spot the GR costs. Soon you will be seeing them in your mind's eye as you crank out estimates.

In this photo, there are at least five GRs that are clearly visible. A few more are hinted at. See them?

(Hey, no peeking at these answers till you have spotted the GR's for yourself. The five I see are: *Sani-Can. Scaffolding. Recycling. Temporary Power. Gate.*)

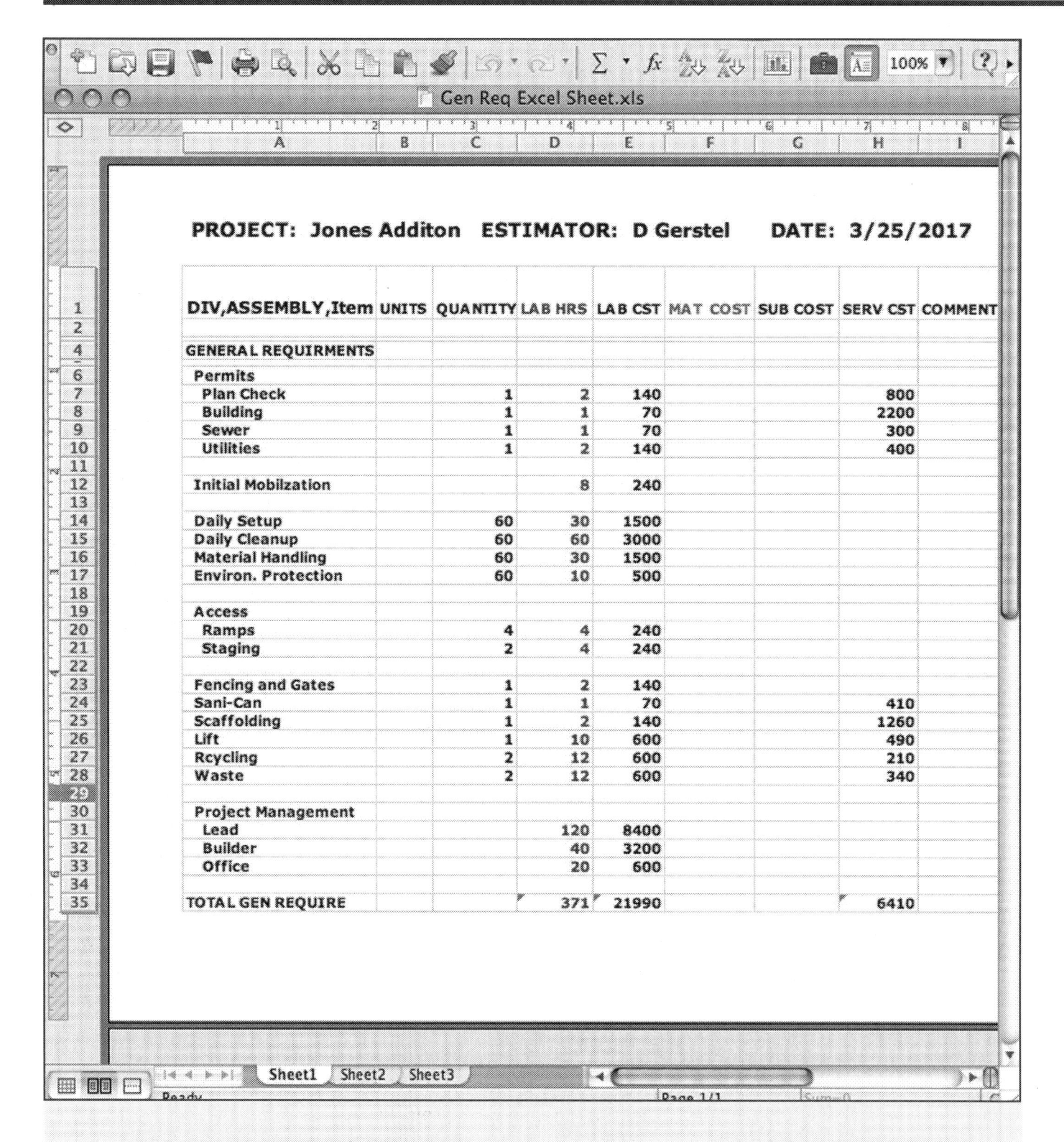

PROJECT: Jones Additon ESTIMATOR: D Gerstel DATE: 3/25/2017

DIV,ASSEMBLY,Item	UNITS	QUANTITY	LAB HRS	LAB CST	MAT COST	SUB COST	SERV CST	COMMENT
GENERAL REQUIRMENTS								
Permits								
Plan Check		1	2	140			800	
Building		1	1	70			2200	
Sewer		1	1	70			300	
Utilities		1	2	140			400	
Initial Mobilzation			8	240				
Daily Setup		60	30	1500				
Daily Cleanup		60	60	3000				
Material Handling		60	30	1500				
Environ. Protection		60	10	500				
Access								
Ramps		4	4	240				
Staging		2	4	240				
Fencing and Gates		1	2	140				
Sani-Can		1	1	70			410	
Scaffolding		1	2	140			1260	
Lift		1	10	600			490	
Rcycling		2	12	600			210	
Waste		2	12	600			340	
Project Management								
Lead			120	8400				
Builder			40	3200				
Office			20	600				
TOTAL GEN REQUIRE			371	21990			6410	

This abbreviated spreadsheet illustrates two key estimating issues:

1. Outside services, such as permits, entail not only the vendor's cost, but also a cost for crew labor
2. Project management is a huge GR cost!!!

The most costly General Requirement

Among all the General Requirement items of work by your crew, one stands out. Its cost is never minor. Typically, it is the largest of the GR costs. That is the cost for *project management*, whether handled by a person with the title of "lead person," "project manager," "supervisor," or "superintendent." A good lead is critical to the success of any project. In fact, a qualified lead may even be explicitly called for in the specs. Savvy designers know that the single best thing they can do to ensure that their projects are properly built is to require that a qualified lead be at the project full time, running the show.

As you can see from the list of General Requirements displayed at the beginning of this chapter, a lead's project management duties are numerous. A few deserve emphasis: The lead, more than anyone else, will be aware when extra work crops up that is not called for in the plans and specs. While he may or may not produce the paperwork (i.e., the *change orders*, a crucial extension of estimating and bidding we will discuss in detail in Chapter Fifteen) for that additional work, he must at least make note of it. And he must make sure the information gets to the person who does produce the change orders.

Equally critical is maintenance of *time cards*. Accurate time cards showing just how many hours were spent by each member of the crew on each assembly or item of work are the foundation of labor productivity records—without which accurate estimating is not possible. We will go into time cards and labor productivity records thoroughly in the Part III. Here it is enough to emphasize that the lead is responsible for making sure time cards are complete and accurate. Getting an entire crew to fill them out reliably is a nearly impossible task. In fact, motivating your lead to do so is challenge enough; but it's a challenge that can be met.

All told, project management tasks easily consume a couple of hours a day even on a small remodel. They can take up the lead's entire day on large, complex projects. Each of the subcontractors must be scheduled not once but repeatedly and more exactly as the time they are needed on the job grows closer. Detailed material takeoff and ordering likewise must be done repeatedly. A good lead even gives some time each day to keeping up crew morale, handing out pats on the back, giving thoughtful tips and encouragement to apprentices, maybe even brewing up a pot of coffee for the crew on a cold morning. Yeah, on top of everything else, he has to play the role of den mother from time to time.

You may or may not want to provide a row for each of the lead management tasks on your spreadsheet. For my own estimates, I have generally found it sufficient to allocate a row for "Lead Time." That's for the project management hours put in by the lead himself or herself. If I anticipate contributing significantly to project management myself, as I will to support a relatively inexperienced lead, I will add another line, labeled "*Builder Project Management*," for the hours I will put in managing the job. (I've designated that item as slippery, with italics, because for years after I delegated project management largely to my leads, I neglected to estimate for the hours I was still contributing. I bet I am not the only builder to make that error.) A third line, "Home Office Project Management," might be appropriate for larger companies, where some of the tasks directly related to running a project are performed by office-based managers rather than the on-site project lead.

On the other hand, you may find it useful to go beyond those few lines and list out a substantial number of the specific project management tasks. Such a list may help a client understand why your costs for lead project management are so large. Several builders have told me they encounter a good deal of resistance when they show the full cost of project management time in their General Requirements division rather than "including" (or in reality, hiding) it in other categories. One builder, whose larger projects run upward of $10 million and absolutely require a full-time, on-site project manager, reports that his clients for those projects nevertheless balk at the cost. They object when they

I will never get a job!

If you are a builder still in the start-up phase of your career, congratulations! You are reading this book. That means you have recognized that you must put in the hard yards and master the business skills needed to run a construction company.

However, even though you may feel pretty good about coming to grips with estimating and bidding, you may also be thinking, "If I try to charge for all these General Requirements, I will never get a job."

Let's be frank. You are probably right. At least you are right for now. In fact, there's a lot you may not be able to charge adequately for at the beginning of your career—not just General Requirements but also overhead, profit, and the full value of your labor. That was certainly true for me when I started out on my own.

Your solution, however, should not be to simply pretend those costs do not exist and leave them completely out of your estimate. Instead, identify them. Provide for them as you build up your spreadsheet rows. Then, at the outset of your career, undercharge for GRs, overhead, and profit as necessary to get work.

Put in a low plug figure for profit just to begin the habit of charging for it. Reduce your overhead charge by including only a nominal salary for all the work you do running your company. And to reduce your GR costs for project management, do it yourself at apprentice's wages, or less.

In time, as you establish a reputation as a builder, you will be able to push your charges for GRs, overhead, and profit toward the levels you need to sustain your business and prosper. And when you get to those levels, you will find yourself operating with a comprehensive spreadsheet/checklist that itemizes all those costs so that you don't forget to capture them.

see charges in the estimate of several hundred thousand dollars for project management of their two-year-long projects. But in actuality the charges are reasonable for a major metropolitan area ($70/hr × 40 hrs a week × 50 weeks/year × 2 years = $340,000, or 3.4% of a $10 million project).

Another builder tells me that the architects he works with resist the General Requirement costs altogether, insisting—who knows on what basis?—that they should not exceed 5% of constructions costs, and object most strenuously to his project management costs. Perhaps if such clients and designers saw a fuller listing of project management duties—and the list can be detailed out even beyond what I have included in the bulleted list of selected GRs—they would understand why the costs are so substantial.

And so may you if you are one of the guys who still tend to blow off GR costs!

Ambiguous and controversial items

There are a number of items I have not included in the list of General Requirements that can reasonably be covered under GRs or in another division. Among them:

- Layout
- Special inspections, testing, and submittals
- Vehicle costs
- Communication costs
- Post-completion inspections and service

Layout of the foundation, framing assemblies, cabinets, finish work, and other building components can reasonably be viewed as a project management and GR cost. However, I prefer not to do so for two reasons. First, the project management charge is already so large and off-putting to clients that I would rather not make it larger yet. More importantly, while layout does go on all through a project, it also breaks down into separate tasks—each of which is closely associated with the installation of one or another specific item or assembly. In that way, layout is different from a GR like daily cleanup. Therefore, in my spreadsheet, I group each layout task with the work with which it is associated—batterboard installation with foundation work, wall layout with wall framing, and so on—thereby sticking with the natural order of construction.

Similarly, I group task-specific requirements, such as concrete testing, welding inspections, and submittals of shop drawings, with the larger task to which they are connected. For sure, they could be characterized as General Requirements. Such tasks do not remain as a permanent part of the job any more than does a permit. But they are highly task specific. So again, to stay with the natural order of construction, I feel it makes sense to cover them together with the work to which they are related.

As for vehicle costs: For a small-volume operation, operating a single pickup truck and using it for multiple tasks from sales calls to hauling material to projects, it is simplest to include costs for the truck in company overhead. For larger companies doing bigger projects and providing project leads with a pickup truck, the costs of owning and operating the vehicle during a project can be reasonably included as a GR cost for that project.

Similarly, if you need a landline or computer modem installed for a project, that is a General Requirement item. On the other hand, if you are supplying your crew with smartphones or tablets that they use for multiple projects, you can logically include the cost in overhead.

Postconstruction service calls can be properly included as a GR cost. I am not talking here about callbacks to correct defective work. That's what I call a "profit cost," a reduction of profit that can occur days to years after a job is completed (more about profit costs in Chapter Fourteen). What I'm talking about here is the courtesy calls—visits to the site by the builder or lead after a project is complete, to make sure everything is shipshape. The time spent on those calls can be considered General Requirement costs. On the other hand, because such calls are great marketing, they can also be reasonably categorized as marketing, i.e., as an overhead cost, rather than as a GR. Take your pick.

There are other items on your estimating spreadsheet, however, whose placement is not merely a matter of circumstance or personal preference. In particular, including bonds under General Requirements is a dubious decision, even though it is recommended by some textbooks on estimating. Bonds are more like a markup than a direct job cost, because their cost is typically figured as a percentage of the total of direct costs, overhead, and profit. Therefore, it is most logical to include them not as a General Requirement but as the very last item on the Recap Sheet (i.e., the estimate summary sheet described and illustrated at the end of the last chapter).

To my dismay, I have seen industry consultants who hold themselves out as experts on estimating include the following costs in their list of General Requirements: Workers' compensation insurance. Unemployment insurance. Social Security contributions. Union contributions.

That is a serious mistake. Such labor burden items are not General Requirement costs. They are part of the cost of the labor for each item and assembly within a project. If you include them in your estimate as a General Requirement, you can cause real havoc with both your estimate and your tracking of job costs as the project proceeds. For example, when you figure the cost of labor for wall framing, you will have to first figure the wages you pay out for the framing in your Framing division. Then you will have to move over to General Requirements and include figures for each of the labor burden items. While that might be possible to do, it would be cumbersome and inefficient, and would increase the chance of error.

The longer I work at building, the longer grows my list of General Requirements. In part, that is a result of my increasing ability to spot GR costs not visible in the plans and specs. In part, it is because of the increasing complexity of construction, with each added complexity bringing in new GRs. Something similar may be said for our next several divisions of costs: Foundation, Frame, Finish, and any other work done by my own crew—or yours. All that work grows steadily more complex, with more items and assemblies. And all of those must be called out in on your estimating spreadsheet.

Listed first, estimated last

General Requirements are traditionally listed as the first division in a spreadsheet. Here I have stayed with that convention.

The fact is, however, it's a bit odd to list GRs first. It would make more sense to list them last.

Why? Because many of the items within the General Requirements division are time sensitive. That is, their cost is determined by the length of time the job takes. Daily mobilization will grow more costly as the project grows longer. So will cleanup, recycling, waste removal, site maintenance, rental of sani-cans or scaffolding, and, most significantly lead time for project management.

You can't really nail down the cost of the time-sensitive GRs until you know how long the project will take. And you will not know that with certainty before you have estimated the labor hours and duration of all the other divisions of work from foundation through finish.

Therefore, during estimating, you will have to leave a good many rows of your General Requirements division blank until you have completed the other divisions and even made out a schedule for the project. With that done, you will be able to return to the time-sensitive General Requirements, cost them out accurately, and complete your estimate.

CHAPTER 7

Corral Your Crew's Work

An opportunity

To begin this chapter, I'd like to look back at what we have already covered in Part II and then forward to what we will be covering next. We are discussing a pretty involved task, building an estimating system. I want to make sure we keep our bearings as we work through all the steps in that task.

So far, as we have gone about building an estimating system, we reviewed (in Chapter Four) the process of figuring the quantities of work in a job. Next (Chapter Five) we discussed designing an estimating spreadsheet, primarily the setup of columns and rows for divisions of work such as Concrete and Framing.

In Chapter Six we reviewed an often neglected division, General Requirements, and looked at many of the items of work that GR includes. Thereby we began the detailed work of building an estimating spreadsheet into a comprehensive checklist that will help you capture the costs of all work in a project.

In this chapter and the next we will go further with that detailed work. Here in Chapter Seven, we'll build up a list of the items and assemblies that are typically produced by the crew of a construction company that handles site, demo and deconstruction, foundation, framing, and finish work in-house—that is, with its own crew. Then in Chapter Eight, we'll get into the gathering of quotes for subcontracted work in such a way that no work (or to be realistic, almost no work) falls between trades and is left out of your estimate.

One more thing: As we proceed with building an estimating system here in Part II, we won't be getting into the dollar costs of items or assemblies, much less of divisions. You may be impatient to get on to that. After all, the purpose of estimating and bidding is to generate costs for a job. We will get to that juicy topic in the next part (Capture Your Costs: Part III).

Of all the estimating challenges, corralling the work to be done by your crew is the most daunting. It is also your greatest opportunity. If you can cover all that work and reliably project costs for it in your spreadsheet,

COMING UP

- An opportunity
- Deconstruction and demolition
- Site work
- Concrete and drainage
- Framing
- Finish work
- Windows and exterior doors
- Water and moisture management
- Finish

then the chances are you will nail your numbers for your entire estimate. Capturing the other work in a project—material, subs, and services—is relatively straightforward. Get your crew work right, you'll likely get your estimate right.

Additionally, at least in my experience, if you have the skills necessary to accurately estimate the work done by your own crew, you can keep the work in-house (if you can't estimate work, you have to sub it out to avoid unacceptable financial risk). And by keeping the work in-house, you strengthen the possibility of producing quality work. A carpentry crew that takes a project from foundation to finish has a greater stake in the quality of the work they perform at each stage of construction because they will be building on it at the next stage. As we say at my job sites, "finish work begins at the foundation." A crew that is on the job from installing batterboards to the last stick of trim work owns the job. They can take pride in it in a way that a crew that comes in only briefly to install a foundation or frame cannot. They can say, "That's my project; I built it."

Successful estimating of crew work requires a comprehensive checklist/spreadsheet that lists all the items and assemblies of work regularly or even occasionally done by the crew—whether that is just you or ten laborers and carpenters you employ. As I discuss items and assemblies here, I will, just as in the chapter on General Requirement items, focus especially on slippery items that tend to get away from us during estimating. To begin with, there are a number of potential slippages that extend across several or even all the divisions. A good checklist/spreadsheet will include multiple rows that remind you to allow for them in your estimate (if you need to refresh your visualization of spreadsheets before proceeding, please take a look at both the word-processed and electronic spreadsheets shown in Chapter Five):

Seemingly simple, i.e., "minimalist," designs, such as the interior of this house in Vienna, Austria, require exacting horizontal and vertical control, meticulous installation of flashings, and ultraprecise fitting of finish work.

One pro estimator, highly experienced with "simple" contemporaries, told me that she provides generously for labor even at the foundation, because "I have seen projects where the foundations must be within a tiny fraction of an inch" so something on the third floor fits."

As a lead carpenter in my company once said to me, "We have to remember, simple is hard." Yes, and simple is expensive; and the cost must be allowed for in the estimate.

Photographer: Hertha Hurnaus
Project: cj_5 by Caramel Architects

- *Tie-ins* are often a major cost factor in remodeling and come into play during deconstruction, concrete work, framing, waterproofing, and finish. Examples: Setting dowels into old concrete work to tie it to new work. Blending the plumb and straight wall for a new addition into an existing out-of-plumb wall of an existing structure, a task that can require labor-consuming installation of control lines and shims. Tying the wall wrap for a new exterior wall into the fragile old felt underlying existing finishes.
- *Unusual details.* Some designers are driven to create "innovative" details. When you encounter one of their innovations during an estimate, you must first break it down into components and cost out each component on the basis of your records for similar items you have built in the past. Then, as best you are able, you will need to construct an additional charge for the time your crew will need to pause, think, and figure out how to build that undulating concrete column or that roof for a bus stop that looks like an upside-down boat.
- *Contemporary "minimalist" designs,* such as that shown in the photo on the previous page, require ultraexacting installation from the foundation up. A common feature of contemporary design is structural work doubling as finish work. This demands control far more precise than that needed for traditional work, for without it, visible surfaces will fail to join in the clean, crisp manner called for in the plans and specs, and eaveless roofs will leak at their flush junctions with the walls.

A final kind of slippage that can readily occur at every division of the estimate is what my mentor, the great builder Deva Rajan, taught me to call, in his characteristically optimistic fashion, "completion." By that he meant all the irksome unforeseen items that inevitably come up toward the end of any division of work in a project. No matter how thorough your estimate, you will, here and there, miss something—like the screws one builder assumed would be packaged with the custom hardware required for the many interior doors in one of his projects. As it turned out, the screws were not included. Worse, the supplier was not able to provide them. As a result, the builder spent hours scouring the shelves of hardware stores in search of the screws he needed. Those "completion" hours, since they were not included in his estimate, went unpaid.

This sort of stuff comes up not only during finish, but also regularly during all divisions of work. Examples: Digging out a collapsed footing trench. Cleaning up an unusual amount of spillage from a concrete pour. Culling out twisted sticks from a unit of studs. Correcting excessively crowned joists. Re-hanging a bunch of badly assembled doors.

Completion can readily amount to 2% of the cost of a project. If you do not provide for it, your potential profit can be seriously impaired. You lose at multiple levels. First you lose the 2%. Then you lose the overhead markup on the 2% completion charge. Then you lose the profit markup on the 2% completion charge and on the overhead markup. Finally, you lose the chance to invest all that lost income in your retirement account, where it can grow.

I will leave the math to you (it is worth doing). But I will digress from items and assemblies for a moment and tell you that if you habitually leave completion costs out of your estimate and your overhead markup is 20% and your profit markup is 10%, you can end up half a million bucks short at retirement. It is hard enough to earn a profit and a secure retirement as a builder without throwing away money honestly earned for completing your jobs. Enough said for now (more about this sort of thing in the chapter on overhead and profit). Let's get back to our items and assemblies.

In the sections and accompanying bullet lists that follow, I have again emphasized slippery items of crew work by italicizing them. Experienced builders may find it worthwhile to scan the lists for items that might escape their attention. Start-up builders, on the other hand, may wish to give the sections and lists a closer read and use them as a source for expanding the rows of items and assemblies in their developing spreadsheet/checklists.

Deconstruction and demolition

Should you prefer, like me, to estimate in the natural order of construction, your first division for crew work will be that for deconstruction and demolition. Here, one by one, I will go over deconstruction and

Deconstruction and demolition

- Temporary protection of existing surfaces to remain
- Removal & storage of appliances & equipment
- Hazardous waste removal
- Exterior plaster & lathe
- Removal of exterior doors & windows
- Removal of exterior moldings
- Deconstruction & salvage of decks and fences
- Deconstruction of frame & salvage of framing lumber
- Salvage of metals
- Salvage of plumbing fixtures
- Deconstruction & salvage of interior doors, trim, & hardware
- Prepare salvaged material for reuse
- Store/catalogue salvaged material for reuse
- Dust control for demolition
- Staging
- Chutes
- Conveyors
- Concrete cuts horizontal
- Concrete cuts vertical
- Concrete core drilling
- Slab removal
- Curb removal
- Driveway removal
- Walkway removal
- Footing removal
- Concrete wall removal
- Removal of concrete of greater than usual thickness and/or with greater than usual reinforcing
- Interior plaster & lathe
- Drywall
- Multiple layers of plaster & drywall
- Insulation
- Concealed ducts
- Concealed pipes
- Concealed wiring
- Roofing
- Items penetrating fragile roofing
- Suspended ceilings
- Wall frame
- Floor frame
- Plywood subfloor glued in place
- Roof frame
- Nail removal
- Equipment rental: jackhammer, concrete saw, other equipment
- Rental equipment pickup & return
- Rental equipment fuel, maintenance, blades & bits
- Completion

demo items that I have found to be most likely to slip by during estimating.

Temporary protection—for example, of ceiling paneling or tile to remain in place even as the walls below are stripped to the frame—can readily escape attention during estimating. *Removal and storage of appliances and equipment* likewise is work often required as a prelude to demo that can consume significant crew labor, but that is easily overlooked because it is not normally called out in the plans and often not in the specs.

With protection in place, deconstruction, including *salvage*, begins. Whereas outright demo goes rapidly, deconstruction—taking apart trim and other existing work, then storing and even labeling it for future use—is labor-intensive. For example, pulling apart old wooden window casings carefully enough to preserve them for reuse can take more time than would installing new casings.

Once items that must be salvaged have been removed and stored, demolition can begin. Though it can often be done at a much more rapid pace than deconstruction, it too can include work that can slip by an estimator. Especially in built-up urban areas, a builder may be required to provide for *dust control. Staging and chutes* may be needed. Concrete of greater than usual thickness as well as *hidden or partially hidden multiple layers* of material, such as a layer of plaster beneath newer drywall, can easily escape attention during estimating. Likewise, hidden ducts, pipe, and wires may be overlooked.

True, you may be able to charge your client something extra beyond the original contract price for "hidden conditions." But you may also run into resistance from an owner who argues that you overlooked evidence of those conditions—such as vent pipes or duct caps visible at the roof or chases suggestive of hidden duct runs. And charging for demolition extras as a project is just getting underway may create tension and distrust that can undermine the project further down the road. It's better to capture the slippery items during estimating. At the very least, you want to give the owners advance notice of possible extra costs that may come up during demo.

From time to time you will rent equipment for demo that you do not use often enough to make owning it cost-effective. Rental, too, entails slippery costs. They include the crew hours consumed by *pickup and return of the equipment* and such *operating costs* as fuel, extra blades and bits, and maintenance.

When deconstruction and demo look to be done, wrap-up items, such as *cleaning nails off framing* that has been stripped of plaster or drywall, may remain. Those items, too, need to be provided for in the spreadsheet.

Always during deconstruction and demo there will be some niggling details that even the most comprehensive spreadsheet does not cover. Often they are not noticed till demo is done and production of new work is underway, with the result that your crew will lose momentum as it pauses to take care of them. The cost of such items can be captured in a line for "*completion*," which, as already emphasized, is slippage that should be provided for at the end of every division of work.

Site work

For builders involved in large-scale industrial or commercial projects, preparation of the site is a major division of their estimates. In fact, entire books are devoted to estimating only the costs of excavation for such projects. For the typical light-frame project, if there is a large amount of site work, such as clearing and grading the lot for a new home, it is probably best to sub it out to a company that operates the necessary heavy equipment. For limited site work that your crew can handle efficiently, slippery items you need to keep an eye out for include *stump removal, root removal, other plant removal,* and *grubbing*. Removal of a small tree will be obvious. It is sitting right in the middle of the planned construction. Cutting it down, chopping it up, and hauling it away are obvious costs. Taking out the stump may not, however, automatically come to mind during estimating. Removing the stump, along with any roots that would interfere with trenching for the foundation footing, may require more labor than removal of the tree itself.

Even if there are no trees that have to be removed, if large trees are near the construction area you may hit their roots during excavation for the foundation. Pausing, getting out the Sawzall, putting in a fresh blade, and cutting through the roots take time—time that a comprehensive checklist/spreadsheet will remind you to allow for during estimating. The same holds for *grubbing*. Scraping away grasses where you plan to build may not appear to be much of a task. But their roots can form a dense mat, sometimes extending several feet below the surface of the soil, that will make excavation difficult. Like a lot of the slippery items, removal of tree stumps and matted grass roots may not impose huge costs. But as we have noted before, small costs can add up, to the point where they undermine the chance for decent earnings.

Transplanting, by contrast, is a slippery site item that can be costly in and of itself. In the case of a fig tree I was once required to save, transplanting involved excavating a three-foot-deep trench about 30' in circumference around the fig, digging beneath its root ball, lashing 4x4s to its trunk, mobilizing myself and three laborers to carry it across the yard to its new hole, backfilling both holes, and then hand watering for months to keep the fig alive. The many hours of labor involved were not in my estimate for the project. But by the end of the project "Transplanting" had become a new line item in my spreadsheet; I won't overlook it again. (The fig, by the way, is happy, producing abundant fruit 10 years after I gave it a free ride to its sunny new location.)

Another potentially slippery and costly site item: *shoring*, both of cuts and trenches. In some soils and in some seasons, it may not be necessary. The soils will hold. Or the rains that can make shoring necessary may hold off. But your spreadsheet should nudge you to anticipate those conditions when shoring may be required. I once failed to anticipate shoring for an eight-foot cut along the back side of an existing garage with a $200K sports car parked inside.

Just prior to that project and at another site, I had made an even deeper cut in similar soil, and the soil had held. But that prior job was done in the summer, whereas the second one was delayed into winter. The rains came. On New Years Eve during a torrential downpour, the cut began to collapse. My crew and I, assisted by four heavy equipment

Site work

- **Protection of existing structures & plants**
- **Silt barriers**
- **Tree removal**
- **Stump removal**
- **Root removal**
- **Transplanting**
- **Surface grubbing**
- **Below surface grubbing**
- **Other plant removal**
- **Grade**
- **Cut**
- **Shoring of cuts**
- **Fill**
- **Place gravel**
- **Place sand**
- **Place soil**
- **Compaction of placed material**
- **Trenching**
- **Trench shoring**
- **Storm drainage**
- **Sanitary sewer tie-ins**
- **Equipment rental**
- **Equipment pickup & return**
- **Equipment operation & maintenance**
- **Completion**

The brown cardboard boxes visible in shallow trenches below the rebar cages are "void forms." Heavily ribbed to support concrete during the pour and curing of the grade beam, the boxes will later decay, leaving a four-inch void below the beam. When the earth swells during the winter rainy season, it will expand into the void rather than swelling against and possibly cracking the grade beam. Concrete work involves many such unusual items that can escape detection during estimating.

operators, spent our holiday working in the mud till sundown to install a shoring system of steel H-beams and 8x8 wood lagging. Now my spreadsheet includes separate rows for shoring in dry conditions and additional shoring that may be necessitated by rain. I might not include the latter in my bid, but if not I will make a note to spell that out in my contract so that the owner knows it could come up as an extra.

Like deconstruction and demolition, site work often will require *rented equipment*, and you do not want to let the related costs—fuel, maintenance, blades—slip away from you. Finally, as with all divisions of work, unanticipated items will turn up during site preparation. Provide for them at a row for *completion.*

Concrete and drainage

Frequently these days, light-frame builders sub out concrete work rather than doing it with their own crews. That may be a mistake. Other than the pouring and finishing of slabs, doing the work in-house is, in my experience, rewarding and profitable. It provides an opportunity for close control that cascades through a project, for a well-built foundation facilitates good framing, which in turn fosters clean finish work. Slippery items abound, however. Especially for contemporary designs, *layout* must be exacting, both horizontal and vertical control dead-on. Foundation trenches must not only be accurately cut; they must often be *trimmed* and *cleaned* by hand prior to installation of concrete. Similarly for slabs, *fine grading*—especially to achieve a consistent slope to a drain line below the slab—will require handwork after the rough grading has been done by heavy equipment.

Below-slab insulation is also easily overlooked because it may be handled apart from the bulk of the insulation, which occurs after framing. Recently, I saw a light-frame builder leave it entirely out of his estimate for construction of a new cottage. Likewise, I have seen *equipment pads*, such as those for a forced-air furnace, overlooked during an estimate, only to be noticed during construction. A well-developed estimating spreadsheet will remind you to include both items.

Occasionally at badly run construction sites forms give way for lack of adequate bracing. An estimate should allow for the generous *bracing*, especially on taller walls, that will prevent such mishaps. The taller a concrete form, the more bracing and staging it requires, and the more labor-intensive construction becomes. Your spreadsheet should

Concrete and Drainage

- Set batterboards
- Set vertical control
- Layout for formwork
- Excavate for footings
- Trim footing trench
- Shore footing trench
- Clean trenches
- Grading for slab
- Fine grade for slab
- Trench for drainage below slab
- Trench for plumbing drain lines below slab
- Form slabs
- Below-slab insulation
- Install expansion joints
- Set screeds
- Form equipment pads
- Place gravel
- Vapor barrier
- Place sand
- Form footings
- Drill piers
- Install rebar cages for piers
- Pour piers
- Install void forms
- Form grade beams
- Form and brace walls <4'
- Form and brace walls >4'
- Form and brace walls >8'
- Form columns
- Shoring for suspended floor
- Form and set reinforcing for suspended floors
- Install sleeves
- Detail forms: chamfer strips, relief, or other decorative details
- Form steps
- Form walkways
- Form driveways
- Install dowels from old to new concrete
- Anchor bolts
- Hold downs
- Welded wire
- Rebar
- Install ufer bar
- Build chute
- Install concrete
- Short load charges
- Stand by time
- Vibrate
- Strike off
- Float and trowel
- Cleanup spillage
- Patch concrete
- Finish cured concrete (grind, sandblast, other)
- Cure concrete
- Strip forms
- Clean form lumber for reuse
- Seal foundations
- Install precast concrete
- Ratproofing
- Trench for drainage
- Install drainage fabric
- Filter cloth
- Drain rock
- Backfill
- Remove excess dirt
- Completion

provide for walls of different heights at different rows, rather than providing only a single line for walls, to ensure that the higher costs for the taller walls do not slip by you.

Beyond forming and bracing and other obvious work such as installing of rebar, setting of anchoring hardware, and pouring and finishing, concrete work often includes obscure specialty items that can easily be overlooked during estimating. For example, in areas with highly expansive soils, void forms such as those shown in the photo a couple of pages back may be required. Formwork may include placement of *sleeves* for pipes, *chamfer strips, detailing for decorative purposes,* or of a *Ufer bar* to serve as an electrical ground. If you are bidding a project that requires less than a full truckload of concrete, you may have to allow for a *short-load charge.* And for any pour that is likely to consume more time than your supplier allows for in his base charge, you want your estimating sheet to include a row that provides for the costs of *stand-by time.*

Once concrete has been placed, there are still many small tasks to be done. A number of them can easily slip away from you as you build your estimate. Proper *curing* of the concrete in very cold climates may require heating it. In warm climates, you may have to keep it covered for several days or longer with damp cloth—definitely so if you have installed concrete with a high percentage of fly ash in the mix. If you give the concrete time to cure properly, *stripping and cleaning* the form lumber for reuse should go smoothly, but it will still require labor, and your spreadsheet must call for that labor. Finally, after including rows for a few more slippery items such as *removal of excess dirt,* you will, once again, want to put in a line for *completion.*

Framing

When I worked as a union carpenter, I spent a year speed-framing an apartment complex at the edge of the San Francisco Bay. The guys I worked with were super fast and taught me about efficient movement and production. Quality, not so much. As they threw down joists and subflooring or stood, plumbed, and lined their walls, they would call out, "Good enough for government work" or "Can't see it from my house, nail the sucker."

Later, the project supervisor moved me over to installing the exterior trim, and I had to install it against the framers' work. I saw the problems they caused for tradesmen who worked right behind them. And eventually I even saw the problems they had created for guys who worked at the complex four decades later. I was playing tennis there, and I bumped into a repair contractor renovating one of the apartments. He told me with pride how he had managed, after working on his knees for hours, to compensate for the severely out-of-plane subfloor in the kitchen he was working on.

These days, I see many builders subcontracting their framing. They complain about the shoddy work they get. They say the framers leave out stuff that their crew must later pick up. One builder says it amounts to 5% of the work. I do framing in-house for the same reasons that I do concrete with my own forces: control, quality, pride, the pleasure of being a "complete builder"—a guy who can take a project from start to finish—and a better shot at not missing any of the work in my estimate.

When estimating framing, I rely on my spreadsheet to remind me of all the work that goes into a good framing job, including the slippery items—the very sorts of things builders who subcontract the work say the subs tend to overlook.

For remodel jobs, ceiling framing must often be *furred* to create a level plane that will parallel the line of the wall cabinets. Walls, too, can require furring because old studs were badly installed in the first place or have gone out of line over time. Sometimes in remodels or repair projects, *temporary shoring* must be installed while old framing is removed and replaced.

Often overlooked in framing estimates—in fact, not even mentioned in any of the widely used estimating manuals that I have reviewed—is the

Framing

- Furr ceilings to level
- Furr walls
- Interior partition walls in existing building
- Temporary openings in existing frame
- Shoring
- Tie-ins of old to new work
- Precise control for contemporary design
- Mudsill
- Pony walls
- Girders
- Girder posts
- First floor joist system with ledger, blocking and doublers
- First floor subfloor
- First floor exterior walls
- First floor interior walls
- First floor exterior door & window openings
- First floor interior door openings
- Second floor joist system
- Second floor subfloor
- Second floor walls
- Second floor door & window openings
- Floor framing for bays
- Wall framing for bays
- Posts
- Beams
- Post caps & bases
- Wall sheathing
- Drywall nailers, horizontal & vertical
- Soffits
- Chases
- Rafter system including ridge, rafters, collar ties, purlins, & blocking
- Roof trusses including blocking & bracing
- Crane to place trusses
- Staging for roof framing
- Dormers
- Eyebrows
- Rafter tail cuts
- Intersecting gable roofs
- Parapets
- Roof sheathing & blocking
- Ceiling frame including ledgers, soffits, chases, skylight wells
- Stair carriages & landings
- Blocking for cabinets
- Blocking for pipes and ducts
- Blocking for fixtures
- Blocking for accessories
- Blocking for shear panels
- Fire blocking
- Holdowns
- Hangers
- Anchors, clips, straps
- Bolts & screws
- Adhesive
- Nails
- Decks
- Porches
- Entry stairs
- Trellis
- Fence
- Restocking charges & labor
- Pick-up framing (Completion)

difference in cost for framing at the first floor and the second floor (or higher). The latter takes more labor per unit of work for several reasons including: The need for workers to move materials and tools up and down stairways or scaffolding. The fact that workers can't use the ground below as staging for joist installation and other tasks. The greater danger of working high up, which can (should) result in carpenters moving more cautiously and slowly. To capture the difference in costs of framing at different floor levels, you must provide for them on separate lines of your spreadsheet, just as you provide different lines for different-sized concrete walls.

Similarly, in order that you don't miss significant work and cost, separate on your spreadsheet the framing of *bays* and of *door and window openings that occur at unusually high frequency* from the balance of wall framing. It takes significantly longer to frame a wall with bays or with numerous openings than to frame a straight wall with few openings. The difference needs to be captured; and to capture it you need to first call it out in your spreadsheet. It should also provide separately for *interior and exterior walls*. The exterior walls may require larger-dimension lumber, 2x6s instead of 2x4s, for example. They will impose somewhat greater cost for labor and significantly greater cost for material.

When estimating the roof framing, you encounter additional slippery items, notably *staging* and *cutting of rafter tails*.

A huge slippery occurs all across the framing, namely *blocking*. Allow me to repeat that for emphasis: *blocking*. One professional estimator allows a cost for blocking that amounts to 4%(!) of the total cost of framing. Because costs for blocking are so substantial and so easily overlooked, I personally like to be reminded of it by many rows, including one for each of several types of bathroom and kitchen accessories, for the junction where kitchen cabinet backs meet backsplash surfaces, and for fire blocking in walls and floors and along stair stringers.

Likewise for hardware. It used to be a negligible item, so minor your estimate would hardly be affected if you entirely overlooked it. When I was a young framer, the only hardware I saw on job sites were boxes of 16d and 8d nails and joist hangers—though a few other items from a new company called Simpson Manufacturing were on display at the lumberyard.

Now frames are festooned with an ever-proliferating array of *clips*, *straps*, *bases*, *caps*, *hangers*, *hold-downs*, *bolts*, and *special screws*. The Simpson catalogue is nearly as thick as this book. Just a partial selection of Simpson's offerings now occupies 100 feet of shelving at my lumberyard. Nails that come loose in a box are obsolete. Instead, we use gun nails shipped to us in coils and straps. All that hardware is costly, and the cost must be corralled in the Framing division of your spreadsheet.

And finally, you must provide for what we used to call "*pickup framing.*" Now I label it "*completion.*" The completion row here provides for all the miscellaneous framing items, invisible on the plans but necessary—from culling or straightening excessively crowned sticks, to cutting out the plates at doorways, to pulling bent nails and shiners, to a last bit of blocking or backing—that inevitably crop up during construction but tend to elude you during estimating.

Windows and exterior doors

Installation of windows and exterior doors and associated work deserves a division of its own in your spreadsheet. It is costly, treacherous, and infested with slipperies that are barely implied in the plans and specs and that can hide out of sight during estimating. The *water management systems* that must be integrated with the doors and windows demand particularly close attention. Even on simple installations, water management, with its interwoven flashings and membranes, has become so involved that estimating for it can be far more time-consuming than estimating the installation of the windows and doors themselves.

The task is made more difficult by the sad fact that designers often do not understand water management, cannot design it, get it wrong if they try, and sometimes even leave if off their plans altogether. Nevertheless, it must be visualized and specified during estimating. Among items that may be needed but not mentioned in the plans are these: A *sill bevel* (created by fastening a piece of bevel siding atop the rough sill plate or ripping a bevel into the sill plate itself), that will direct any water that does get past other barriers away from the building. *Testing of windows,* necessary because a significant percentage—20% according to one experienced builder who has paid especially careful attention to window installation—are already leaky by the time they are delivered to job sites.

Window *flashing* can require weaving together a veritable origami of papers, plastic materials, and adhesive tapes together with galvanized metal, copper, stainless steel, or aluminum devices. On a recent project of mine, the lovely interior window trim, which elicited much oohing and aahing from visitors, actually consisted of only seven sticks of clear pine, all easily installed with modern power finish tools by an able apprentice. In contrast, the water managements systems—which no visitor even noticed—required the perfectly sequenced and controlled installation of 17 floppy and sometimes sticky pieces of material with only the aid of a utility knife, stapler, and roller. The work required close supervision from the project lead, as does the installing of all water management systems; for it is a terrible mistake, though a frequently made one (which is one reason why so many buildings leak badly), to delegate the work to lower-skilled crew. An estimating spreadsheet must be sufficiently detailed to allow for that high-cost labor.

Use of liquid flashing materials has simplified installation of water management at doors and windows. But it still involves slippery items that an estimator must provide for. They include: *Precoating* of fasteners. *Double-coating* of all surfaces even though the manufacturer of the liquid flashing might claim one coating will do. And *integration of the liquid flashing with building wraps.*

Windows and Doors

- **True up framing**
- **Bevel sill**
- **Test windows**
- **Prime backside of window jambs**
- **Window pans**
- **Head flashing**
- **Liquid or other flashing systems**
- **Install windows**
- **Perimeter backer rod & sealant**
- **Jamb extensions**
- **Install window hardware**
- **Integrate window flashing with building wrap**
- **Skylights**
- **Skylight flashing**
- **Prehang swinging doors**
- **Door sills, pans, thresholds**
- **Prime door jambs & heads**
- **Hang doors**
- **Jamb extensions**
- **Stops**
- **Locksets & deadbolts**
- **Weatherstrip**
- **Specialty hardware including closers, mail slots, peep holes,**
- **Assemble and install sliding doors**
- **Door flashing**
- **Integrate door water management system with building wrap**
- **Install screen doors**
- **Storm doors**
- **Garage doors**
- **Garage door opener**
- **Pet doors**
- **Tune up doors & windows**
- **Completion**

Sadly, at some point in time, the originals of the windows shown here had been torn out, discarded, and replaced with poorly installed French doors that were leaking and rotting away by the time I got to the site.

Restoring the openings to their original look with newly crafted windows required, along with much else, installation of a complex water management system including copper head and side flashings and careful integration of new with the existing antique building wrap.

Fortunately, by the time I took on the project, I had developed a comprehensive checklist/spreadsheet. Every step of the work was covered in the spreadsheet. As a result, the restoration was not only enormously satisfying but a financial success—and it has never leaked.

Beyond water management, there are other elusive details associated with exterior door and window installation that a comprehensive spreadsheet will snare during estimating. They include *jamb extensions*, *mail slots*, *peep holes*, and *closers*. All are small, almost incidental, items. But their cost adds up and can easily be overlooked if your spreadsheet does not prompt you to check for them.

Along with the smaller details, your spreadsheet should encourage a check for all the various types of exterior doors—not only swinging doors, but roll-up garage doors, barn-style doors, sliders, screen doors, folding doors, and even *pet doors*—that may turn up in your projects. Likewise, you should provide for an end-of-project tune-up of doors and windows. They do tend to get out of whack during construction. And finally of course, you want to provide for *completion* in your Door and Window division just as in all others.

Water and moisture management

As you have noticed, I use the term "water management," not "waterproofing." I've come around to the way of thinking espoused by a builder who is truly expert at what we used to call "waterproofing." It is impossible, he says. We cannot waterproof buildings. Raindrops, especially wind-driven raindrops, are too damn smart. They will find a way into our structures. What we can do is manage the movement of liquid water—direct it off and away from buildings, and give it a way out if (when) it does get past the eaves, flashings, diverters, pans, sealants, building wraps, and other barriers we put in its way.

Much the same can be said for moisture in the form of vapor. Coming from both interior and exterior sources, it is going to get into our buildings—including into wall and roof cavities. Vapor is even cannier at slipping through barriers than are raindrops. Because it is going to get in, we have to give it a way out as well.

My own preference during estimating is to corral the items and assemblies that make up water and moisture management systems in the order in which they naturally come up during construction. For that reason, here water management at exterior doors and windows is itemized in the Door and Window division, You might, however, choose to provide for water management differently, for unlike other work covered in a spreadsheet, it occurs all through a project, during concrete, framing, and even finish phases. Because that is the case, even if you generally stick with the natural order of construction in creating your spreadsheet, you might reasonably elect to provide for water management items and assemblies in a separate division all their own. During estimating you can then comb through the plans and specifications, providing for water management in a single pass, rather than estimating for the items one at a time as you would if you were estimating in the order of construction.

Whichever way you refer to provide for water management in your estimates, whether in order of construction or in a separate division, it is critical that you do not skimp on the work during estimating, so that you will not skimp on it during construction. The work is far more critical than it was in those bygone years when we mistakenly thought we were actually waterproofing buildings with our rolls of felt and paper flashing.

Now, in the interest of energy efficiency, we stuff our buildings with water-retaining insulation materials. We sharply reduce their ability to breathe and to dry out by closing them up with sealants. We frame them with plantation-grown lumber not nearly so dense and water-resistant as the lumber used just a few decades back. As a result, if we trap moisture in our structures, they decay far more rapidly than in the past. To build durable structures, we must build into them skillfully crafted systems that can stop, divert, and get rid of moisture—and we have to allow for that complex work, including all the items in the bulleted list shown here, in our estimates. Two particularly slippery items

Water and Moisture Management

- **Deck drainage**
- **Moisture seal foundation walls exterior**
- **Moisture seal foundation walls interior**
- **Bathroom moisture barriers including shower pans, tub wall membranes, & sealants**
- **Window water management systems**
- **Exterior door water management systems**
- **Building wrap**
- **Integration of building wrap with door & window water management systems**
- **Integration of new with old building wrap (remodel projects)**
- **Backer rod & sealant at doors, windows, & corner trim**
- **Skylight flashings**
- **Roof flashings**
- **Gutters**
- **Rain screen systems**
- **Integrate rain screen with other water management at exterior doors & windows**
- **Leak detection & inspections**

Exterior finish

- Shop drawings
- Building wrap
- Window sills
- Window casings
- Window aprons
- Door casings
- Shutters
- Water tables
- Vents
- Back prime siding
- Siding
- Prime siding cuts
- Prime nail & other penetrations at siding
- Starter strip for siding
- First piece of exposed siding
- Notch siding around rafter tails
- Scribing of siding to soffit or exposed eave decking
- Frieze board
- Cornice
- Rake board & molding
- Fascia
- High work adjustment
- Corner trim
- Corner trim sealant
- Awnings & canopies
- Security bars
- Ornamental iron work
- Mailbox
- Address numbers
- Match existing trim
- Contemporary details
- Joints between dissimilar materials
- Structural work doubling as finish
- Completion

you want to provide amply for in your estimating spreadsheet: *Integration of building wrap with flashing at doors and windows.* And *integration of new into existing building wrap at remodeling projects.* That's not glamorous work, but it demands skill, and it is labor-intensive when done properly—i.e. meticulously.

Exterior finish

Often building wrap installation is done together with the first stage of exterior finish, the installation of siding. Much of the siding now in use is highly processed stuff—consisting, for example, of short lengths of finger-jointed boards thickly coated with primer. When that primer is interrupted by a cut, it must be supplemented with site-applied primer. The siding *manufacturers' requirements for preparation* of their products at the site can, however, go way beyond priming of cuts. While builders are often aware of the requirement to reprime at cuts, they are sometimes oblivious to the other requirements.

An appropriate line or two in your spreadsheet will remind you to check for manufacturer's installation requirements when estimating for an unfamiliar siding product and to provide for any necessary costs in your estimate. Failure to estimate for the work may well be a prelude to failure to perform it in the field. And failure to do the work and abide by the requirements during installation can result in—you guessed it—voiding of the manufacturer's warranty while increasing the risk of failure of their product.

Outfitted with the right tools, a skilled and attentive apprentice can install the bulk of siding rapidly. However, the work slows and requires greater skill for installation, and therefore greater attention during estimating, at two points: First, at the beginning of each run a *starter strip* must be installed and, along with it, *the first visible piece set to a horizontal line* (or one that closely parallels window sills, as in a remodel of an old building that has settled and sagged). Second, at the top of a run, the siding must be fitted in various ways: *notched around rafters, scribed to an exposed eave deck, or scribed to a soffit.* A comprehensive spreadsheet provides rows not only for the bulk area of siding to be installed but for the points where the more exacting work must be done.

How often do these labor-consuming items slip by estimators? There is no way to know for sure, but I can tell you that they are not even mentioned in widely sold catalogues of items, assemblies, and costs. Instead, siding work is provided for simply on a square-foot basis as though all the square footage took the same amount of labor when, in reality, it does not. In fact, for a siding job that involves careful detailing at the bottom or top of an area, labor for that small amount of square footage can amount to a substantial portion of the labor for the whole task. In estimating, as in just about everything else, the devil is in the details, and a well-built estimating spreadsheet will ensure that you catch them.

Siding, of course, goes faster at the lower part of a wall than in higher areas that carpenters must access from staging and scaffolding. An estimating spreadsheet can allow for the difference with separate lines for *progressively higher work*—say, above 7' when it can no longer be installed by a carpenter standing on the ground, above 10' when it can no longer be reached from planks set atop sawhorses, and so on.

Beyond the closing-in of a building with wrap and cladding, we come to a host of other exterior finish details. They include *address numbers* and another that bedeviled me for a time—*mailboxes.* I overlooked them repeatedly before inserting "Mailboxes" into my checklist/spreadsheet.

Estimating for the *matching of existing trim* can be fussy work, requiring you to figure carefully just where and how existing work will need to be matched. But you do not want to overlook the work during estimating. It can be costly. It entails not only linear-foot costs for installation, but costs for *milling*—including for getting a sample over to your millwright and for getting his product to the job site. And if old and new work are to be tied together, that may involve some finicky tooling at the joints.

Yet another slippery item: the exterior finish *details for contemporary designs*—especially those that involve the joining of dissimilar materials such as cementatious panels and cedar plank siding or structural work that does double duty as finish work. Then obsessive control, as stressed earlier, will be required during construction. Consequently, it must be generously allowed for during estimating.

And finally, at the end of the exterior division we come, as in previous divisions, to *completion,* a line in your spreadsheet to allow for all the little cost events that inevitably turn up during construction and that are all but impossible to fully anticipate. Examples: The new type of fasteners that turn out to require predrilling so that they do not bend when driven in. The error in an architect's plans that resulted in a bolt poking through a level-five drywall finish and that calls for some clever finish work to successfully capitalize on the error. A customer who thinks the few small knots in the decking you installed are unsightly and insists that several boards be removed and replaced.

Interior finish

There was a time when finish work was the ultimate expression of carpentry, the most demanding work and the work requiring the greatest skill. However, modern tools such as compound sliding saws, jigsaws with coping heads, impact drivers, and nail guns have made finish work far easier. Meanwhile, because of the control demanded by contemporary designs and by increasingly complex structural work, concrete and framing have become vastly more demanding. On many projects, it is far more challenging than the finish work.

The arrival of all the wonderful new finish tools has not, however, made the estimating of finish work less challenging. There are still slippery items and assemblies within the division that can get away from an estimator. In very hot, cold, dry, or damp climates, materials used for finish work must be *acclimatized*—stored at normal indoor conditions—before installation, lest they shrink or swell afterward. In any climate, because we instinctively focus on floor plans when sizing up a project, it is easy to forget to look up and remember the work that is high on the walls, or above them, such as picture moldings, crown moldings, ceiling paneling, and *attic access doors and trim.*

Baseboard can get away from you during an estimate because it can appear to be so easy to estimate. Just turn to your takeoff sheets. There you've

Interior finish

- **Shop drawings**
- **Acclimatize material**
- **Picture moldings**
- **Crown moldings**
- **Ceiling paneling**
- **Prehang and hang interior swinging doors**
- **Fire doors**
- **Pocket doors**
- **Bi-fold doors**
- **Shower doors**
- **Attic access doors**
- **Interior door passage hardware**
- **Interior door kick plates**
- **Other interior door hardware**
- **Fireplace mantle**
- **Paneling**
- **Chair rail**
- **Window sills paint grade**
- **Window trim paint grade**
- **Window trim stain grade**
- **Door casing paint grade**
- **Door casing stain grade**
- **Attic access door trim**
- **Baseboard paint grade**
- **Baseboard stain grade**
- **Baseboard pieces**
- **Baseboard joints**
- **Match existing trim**
- **Wall Cabinets**
- **Base cabinets**
- **Tall cabinets**
- **Appliance panels**
- **Islands**
- **Vanities**
- **Cabinet hardware**
- **Cabinet corrections**
- **Countertops**
- **Stair skirts**
- **Stair treads & risers**
- **Stairway balustrade**
- **Handrails**
- **Bathroom accessories (towel bars, paper holder, medicine cabinets, safety bars, clothing hooks)**
- **Tub surrounds**
- **Mirrors**
- **Shower doors**
- **Registers**
- **Grilles**
- **Fire extinguishers**
- **Laundry storage & accessories**
- **Chalk Boards**
- **Tack boards**
- **Other specialty items for commercial work: flagpoles, bike racks, toilet partitions, lockers. benches, corner guards, signage, etc.**
- **Specialized equipment & furnishings for medical clinics, labs, libraries, churches, etc.**
- **Tune interior doors**
- **Completion**

doubled the linear footage of the interior walls, added the footage of exterior walls, and arrived at the total quantity of baseboard. That's all you need to estimate the cost, right? That's what the cost books available from publishers of construction cost data would have you think. They give you a cost per linear foot and nothing more for baseboard installation.

The trouble is that focusing only on linear footage will give you only a crude and sometimes wildly inaccurate figure for your baseboard labor cost (though it may be okay for determining material cost if you add a percentage for waste). Here's why: Two projects might have roughly the same linear footage of baseboard. But if one project is much more cut up than the other, it will involve substantially more *baseboard pieces* and *baseboard joints*. The quantity of pieces and joints, just as much or even more than total linear footage, will determine labor cost.

Individual *bathroom accessories* can likewise require little labor or a great deal and should, therefore, be estimated by the item, not in one gross group. In general, interior hardware options for residential work are increasingly complex and require attention to make sure some new twist does not slip by during estimating. If you move from residential to commercial work, you will find that the finish hardware items you encounter will explode in number and variety, with different kinds of hardware coming into play for differing types of projects, from retail outlets to churches. I have included just a small sample in the list on the previous page. Should you move into such work, you will want to develop your own more detailed checklist (see the Resources section at the back of this book for helpful references).

And finally, for interior finish, as in all previous divisions, you must provide for *completion*! Do that, develop a comprehensive spreadsheet/checklist for all the divisions of work you do in-house, and you will be well on your way to corralling your crew's work.

Now, to complete construction of a strong basic estimating system, take one last step. Create a systematic process for assigning your subs (and service providers) their responsibilities. We will turn to that task in our next chapter.

CHAPTER 8

Gather Subcontractor Bids

The expansion of subcontracting

A few years ago, I visited a large nineteenth-century plantation house near Savannah, Georgia that had been maintained in its original condition. Other than a stone foundation and fireplace and glass windowpanes, it was constructed entirely of wood from its floor systems through the roof shingles and exterior and interior finish. Though I saw no record of just who had built the house, I imagined it to have been constructed by a talented and highly skilled master builder and his crew—with no subcontractors needed. I thought that if I had been able to ask the builder how he went about getting bids from his subs, likely I would have gotten a blank stare in return.

COMING UP

- The expansion of subcontracting
- Basic practices and principles
- Preventing omissions and slippage
- Slippery items in multiple phases of work
- Slippery items in demo and site work
- Slippery items in structural work
- Slippery items in mechanical, plumbing, and electrical work
- Slippery items in siding through exterior finish
- Slippery items in interior finish work
- Catching the specialties

Flash forward a century and a half (just two carpenter's lifetimes): I mention subcontractors to a twenty-first century master builder who uses the developer model—i.e., subcontracting all on-site work other than project management and cleanup. He tells me that he works with 300 different subcontractors. Yes, 300 (though not all of them on every project, of course). Had I asked him whether he had ever considered building one of his large houses without subs, with an in-house crew that did all the trades, he would have asked me if I was kidding. During estimating, he focuses primarily on making sure that every item in a project is covered by one sub or another; and makes use of an elegantly organized system of tabbed folders and checklists (one folder per trade) to assure that they are.

Even if you do not embrace the developer model and contract with subs for all trade work, but instead stick with the traditional model and employ a crew of carpenters, you will depend in good part on subcontractors. For even with the traditional model you will probably call on subs for many of the specialty trades from plumbing through drywall and wall coverings. As the bulleted lists in this chapter suggest, the number of trades handled by subs has exploded since that Savannah farmhouse was built. They continue to proliferate. You cannot become expert and efficient at all of them. Even so, when you are estimating a project that does include subs—and on a complex project there can be upward of two dozen—you must make sure, in order to nail your numbers, that all work is included in someone's price, i.e., that one or another sub has the work covered if your crew is not handling it. You want as little as possible to slip between the trades.

Basic practices and principles

Subcontracting begins with finding reliable subs. The best source, in fact the only source, in my experience, is other builders who do good work. When it comes time to send out invitations to subs to bid a new project, a cordial e-mail will do for your regular guys. For new subs, a phone call to introduce yourself will likely get better results, especially during strong markets when subs are busy. Once subs have committed to bidding the project, you can prevent slippage by making certain that each sub has received all documentation relevant to his work.

If the plans and specs are produced on paper, ideally you will provide complete copies of both to every sub from whom you are soliciting a bid. Yes, distributing complete plans and specs costs money. And, yes, there are some subs who do not need every page. But the cost is a trivial percentage of all the costs of producing an estimate. And the cost is dwarfed by the costs of even a few minor items slipping between the trades and being left out of a bid because a sub has not been given a relevant page of documentation.

As the Associated General Contractors (AGC) put it, "If vendors or subs miss something because

When to subcontract

During estimating for a project, you may occasionally need to make a judgment call about whether to subcontract a particular phase of work or keep it in-house and have it done by your crew. Your decision will depend largely on the capabilities of your crew and the circumstances of the job.

For my own projects, I consider keeping the work in-house if:

1. My crew is capable of producing high-quality work. Where we cannot match the quality achieved by our subcontractors, for example, on work such as finishing a concrete slab, I won't consider doing it with my crew. If it is work that we can do well, such as running electrical circuits, I will.
2. There is such a small amount of work that doing it in-house will be less costly than paying a sub's minimum charge. For example, my drywaller's minimum charge is $2,000. If there are 30 sheets to hang and finish to level five, paying the minimum makes sense. If there are only 10 sheets, estimating for the work to be done by the crew makes sense.
3. I need the crew to do the work to keep them employed.

Occasionally I have been unbusinesslike when deciding whether to sub work out. Sometimes I keep it in-house just because my crew or I badly want to do it, because we want the thrill of accomplishment that comes with, say, rewiring a house, then flipping the switches and experiencing the miracle of lamps lighting up and illuminating a room.

they have not received complete documents, the consequences can be that the [general] contractor may be deemed liable for the missing items." For example, you are bidding for the construction of a new home. You send a plumbing sub a floor plan and elevations but, because you are focused on his work within the house, you send no site plan. He then has no way of knowing about the hose bib to be installed at the rear of the property. He will not be liable for the cost of trenching, running pipe, and connecting the hose bib. You will be.

Plans and specs are more likely these days to be sent to you as PDFs than on paper. That has made distributing complete sets to subs far easier. Now you can distribute complete documents as e-mail attachments at no more cost than partial plans—or even less, since you don't have to do the work of taking them apart. Whether you initially distribute plans and specs in paper or PDF form, you must make sure to also distribute any addenda that arrive during the bidding process. With plans distributed, require of your subs that they let you know exactly what they have received. And as their bids come in, make certain they affirm the dates and the pages of the documents on which the bids are based and that all relevant documents are mentioned.

Save all that communication with the subs, whether in tabbed three-ring binders or computer folders. Then, during a project, if there is uncertainty as to who is responsible for an item of work, you will have the record that gives you the answer. Either the work is shown on documentation you provided the sub or it is not. You don't end up paying for work the sub has missed. That happens a lot to builders with relaxed estimating procedures. As one says of his company, they end up "splitting the difference with subs a lot," or "going to clients hat in hand" to ask for extra payment to cover the overlooked work. (Geez. How pathetic. How embarrassing.)

Your thoroughness will protect your subs as well as you. It will strengthen your relationship with them. One builder whose estimating practices are rigorous rather than relaxed says of his subs, "I am nothing without these guys. We are all one company." He counts on his subs taking care of him, and he looks out for their interests. He wants their loyalty, and he understands that the thorough estimating that saves him from having to ask them to "split the difference" fosters that loyalty. Squeeze

My crew and I can handle basic sheet metal installations. But when it comes to complex work, we cannot approach the quality or efficiency of our star sheet metal subcontractor, Les Williams, shown here with the two halves of the 90° corner of a copper gutter that he has prepared for soldering.

Gathering Sub Bids: Procedure and Principles

- Host a walk-through for subs who have committed to bidding a project.
- Provide each sub with complete plans and specs for the entire project.
- Send subs addenda to the plans and specs that arrive during the bid period.
- Obtain a note or an e-mail confirming receipt of each set of documents distributed to a sub.
- Ask subs to state in their bid the date and page numbers of all documents on which the bid is based.
- Determine whether each sub bid includes all work—permits; necessary staging, scaffolding, and ramps; necessary trenching; etc.—that the sub might assume would be taken care of by others.
- Make certain to include in your estimating spreadsheet, along with each sub bid, labor and material charges for incidental work your crew will need to perform for the sub.
- Require all subs to provide an Included/Not Included (NIC) checklist with their bid.

your subs too often and word gets around. The good subs will not want to work for you.

As sub bids come in you will focus primarily on making sure that no work is falling between the subtrades or between a subtrade and yourself. Is the sub providing his own *permits* or is he assuming his work will be covered under an all-in-one permit that you will pull? Will he be taking care of all the equipment—*scaffolding, staging, ramps*—he needs to gain access to his work, or is he assuming you will have it in place when he arrives at the job? Ditto for *trenching,* including *backfilling, tamping, and off-hauling* of excess dirt. Is the sub going to do the work, or is he assuming you or another sub will handle it? The variety of such potential gray areas and overlaps is endless. And the only way to prevent slippage occurring within them is to go after it systematically, determining with each sub just what he is covering and not covering in his bid.

Included or NIC? That's the question

With every bid you take from a sub, you should collect a written statement of just what work is included and what is not included (NIC) in the bid. More than anything else, these statements will enable you to avoid disagreements with subs over division of responsibilities during a job—and to avoid being stuck with costs you have not been anticipating.

Unusually well-organized subcontractors may provide an Included/NIC checklist of their

A Generic Included/NIC Checklist

Subcontractor:________________________ Signature: ________________

Project:____

Our bid is based on the following plans and specifications for the above-named project:
Architectural Plans, pages: ________________________ Date:__________
Engineering Plans, pages ___________________________ Date:__________
Specifications, pages ______________________________ Date:__________
Addenda, pages___________________________________ Date:__________

Our bid includes all work and costs regularly, though not necessarily always, covered by our trade excepting items circled below to indicate they are not included (NIC).

Building permit	Provide roof jacks
Shop drawings	Provide straps
Liability insurance	Provide roof caps
Workers Comp	Install jacks, caps,
Insurance	& straps
Trench & backfill	Material handling
Off-haul excess soil	Remove waste
Demolition	Recycle
Concrete cut/core	Protect (E) work
Cut and patch	Blocking
Scaffolding/Staging	Firestops
Ramps	Insulation
Ladders	Sealants & caulk
Damage repair	Paint Touchup
Utility hookups	Touchup drywall
Flashing	Daily cleaning
	Final cleanup

own. For subs who do not, you can provide a generic checklist like that illustrated on this page. A generic checklist will not fit all subs perfectly, as it will cover items irrelevant to some trades. For example, trenching does not come up with drywall subs. However, by using a generic checklist, you

Sub Work: Project________________ Client________________

DIVISION	Sub Bid	Crew	Material	Services	Comments
Demolition	4200	120	20		
Foundation	9100	300	100	120	Provid waste remov
HVAC	7232	120	20		
Plumbing	6850	200	20		
Electrical	4617	200	20		
Insula & Air Seal	2740	60	10		
Roofing	4100	120	40		
Totals	*38839*	*1120*	*230*	*120*	

Every subcontractor who comes to a project site will use your crew's time for odds and ends of work: Helping to move a heavy fixture. Clearing a work area. Lending a ladder. Getting out of the way. Installing a block. Adjusting for a mistake. The costs add up. Catch them by providing columns for crew labor and material, as in this spreadsheet. The total may not amount to much in percentage terms. Here the amount for crew labor is just over 3% of the sub bids. But the dollar amount is more than pocket change.

may be able to encourage your subs to develop trade-specific checklists of their own. If they won't, the generic checklist will go a long way to avoiding slippage and conflicts over responsibility for work out at the job site.

In addition to the specific items that can fall between you and/or your subs and that can be caught with NIC/Included checklists, you will want to be especially on the lookout for another type of slippage, one that is harder to quantify or pin down. It is the all but invisible crew labor and time eaten up by subs whenever they show up at your job site. That cost, too, must be allowed for in your spreadsheet and captured in your estimate if you want to nail your numbers.

Matrix for Comparing Sub Bids

Project: Jacobs Addition **Date: __/__/____**

TRADE	SUB 1	SUB 2	SUB 3
Plumbing	McVey	Kerson	Pages
Base Price	16,500	15,200	17,300
Permits	No +900	No +900	Yes
All Plans & Specs	Yes	Yes	Yes
All Addenda	Yes	No +600	Yes
All Mat & Labor	Yes	Yes	Yes
NIC Items	Protect. +300	Recycle +200	Recycle +200
NIC Items	———	Trench +600	Trench +600
Redundant Items	None	None	Vent Jacks -200
ADJUSTED TOTAL	17,700	17,500	17,900

Every subcontractor who comes to a project eats up crew hours. The subcontractors distract the crew with shoptalk and chitchat. They occupy space the crew also needs. To make way for a sub, the crew sometimes must stop working on one task and mobilize for another, then remobilize and return to their original task. A sub's trash is often mingled with the waste produced by the crew, who must clean it up along with their own. Most importantly, labor gets eaten up by tasks the crew performs for the subs, such as adjusting the framing for an electrical panel, giving the plumber a hand moving or setting a heavy fixture, or cutting holes in the flooring for heat registers. Therefore, for each subcontracted trade called out on your spreadsheet, you want to provide for labor by your crew, and along with it, for any incidental material and services you provide.

As a further step toward preventing slippage, if you are taking bids for a project from several subs in the same trade, you would ideally like to see their bids submitted on the same basis. In other words, if three plumbers bid on a project, you will have an easier time evaluating their bids if their Included/NIC checklists are identical. You won't have one sub excluding trenching and final cleanup while the other two exclude only insulation of hot water supply lines. If the NIC checklists are significantly different, a sheet (aka "matrix") for comparing sub bids, such as the one shown in the accompanying illustration, will enable you to compare dissimilar bids and determine which one to accept.

With your Included/NIC sheets in hand and comparison matrix completed, you still face one problem. How do you make sure that items of work are not being covered twice or more as a result of the work being picked up by subcontractors in different trades (rather than its not being covered at all because no one is including it in their bid)? For example, how do you know that plumbing vent jacks are not being included by both the roofer and the plumber?

I have no good answer. It seems I'm not alone. I have never seen the issue of such potential redundancy even raised in other sources of information about estimating. Apparently, the best we can do here is to keep an eye out for redundancy while we focus on making sure that no items escape us as we estimate the subcontracted work from demo through finish.

Demolition and site work

At the outset of our careers, when we are dong only very small jobs, site work and demolition—items such as trenching for a water line or cutting and patching drywall for a few electrical boxes—are likely to be just a very small part of our projects. We take care of the work in-house and estimate for it as best we can. As we strengthen our companies, our jobs often grow in size. Demo and site work grow to such volume that subcontracting them may be to our advantage. By taking bids from subs who specialize in the work, we can create more accurate estimates, for a good sub's

Demo and site work by subs

- **Hazardous material removal**
- **Mold abatement**
- **Fire damage remediation**
- **Demo existing buildings**
- **Demo accessory structures**
- **Demo exterior skin of structures**
- **Remove windows and ext. doors**
- **Demo interior structural work**
- **Demo interior finish work**
- **Remove exposed nails**
- **Cleanup**
- **Traffic management**
- **Dewater site**
- **Clear and grub site**
- **Stump removal**
- **Chipping**
- **Grade**
- **Bulk excavation**
- **Trenching and trimming**
- **Stockpile topsoil**
- **Erosion control**
- **Backfill**
- **Compaction**
- **Fine grading and rolling**
- **Drilling**
- **Shoring**
- **Drainage**
- **Sewer**
- **Catch basins and tanks**
- **Manholes**
- **Swimming pools and hot tubs**
- **Retaining walls**
- **Concrete paving**
- **Bituminous concrete paving**
- **Walkways and curbs**
- **Fencing and gates**
- **Outdoor installations: benches, awnings, flagpoles, ramps, railings**
- **Landscaping**
- **Irrigation**
- **Excess soil off-haul**
- **Recycling and waste removal**

Structural work by subs

- Concrete forming
- Concrete reinforcing
- Concrete installation and finish
- Form removal and cleanup
- Precast concrete
- Concrete masonry units (CMU)
- Masonry embedments
- Brick
- Stone
- Chimneys
- Masonry reinforcement
- Masonry restoration
- Masonry cleaning
- Masonry insulation
- Lally columns
- Steel columns
- Steel moment frames
- Metal joists
- Metal decking
- Delivery and storage of structural steel
- Steel to wood frame connections
- Wood framing including hardware, blocking, and furring
- Framing pickup/completion
- Outdoor structures (decks, gazebos, greenhouses, etc.)
- Staging and scaffolding
- Cleanup, recycling, waste removal

number will be reliable, while our own may not be. Subbing the work may also be more cost-effective, for subs may be able to do it for less cost and better than our crews.

Certain aspects of subcontracted demolition and site work are particularly prone to slippage. If a project site contains only a very small amount of a *hazardous material,* you may be able to have it taken out by a subcontractor who is handling general demolition. If not, your estimate needs to provide for a specialist to handle it. As you move to general demolition, whether of entire buildings or of parts of a building (gutting an interior to the frame, for example), slippage is most likely to occur around completion. You may be assuming in your estimate that the sub will leave the *framing clean of nails, the floor swept, and all waste and recyclables removed.* He may assume in his bid that such "incidental" work will be taken care of by your crew.

Comparable disconnects can occur at various phases of earthwork for a project. You may assume a subcontractor is bidding to do not only the bulk excavation of trenches, but also to do necessary *handwork, such as trimming of the trenches* and *cleaning loose soil off their bottoms.* He might be thinking your crew will handle such touchup. Likewise, you may assume he will be providing *erosion control* as well as *backfilling and compaction* on a second trip out. Meanwhile, the sub has figured only on excavating earth, and believes that once he is done with that task he has completed his work on your job.

All demolition and site work entails similar potential for slippage, particularly at the *completion phases* of tasks. Is the sub who is going to clear and grub the site also taking care of *stump removal*? Is he going to *chip* logs and branches of trees that he has taken down? Is the paving guy on board for the *fine grading and rolling,* or is he assuming that it will be done before he arrives? Is the guy who puts in the walkway and curbs going to monitor his installation during *curing,* or does he assume you will babysit it? Is the landscaper going to *care for the plants* she puts in until they can survive on their own? Or does she assume the owner will hire a maintenance gardener for that work? Do any of the demo and site subs plan on doing a *thorough cleanup* and on *handling waste and recyclables* generated by their work? Or are they leaving that to you?

Structural

Over the past few decades, I have noticed that many builders who began with the traditional model have since moved partway to the developer model. They continue to employ a crew of carpenters, apprentices, and laborers; but on larger jobs especially, they now subcontract the structural work—concrete forming, reinforcing, and installation, as well as much of the framing. Several such builders whom I interviewed for this book complained that as they have moved to subcontracting structural work, their crews have repeatedly had to handle leftovers of the work and that the crew labor is "unbillable" because they had not included it in their bids. In other words, their slippage around structural work has gone up sharply. One builder says his framers get only 95% of the work done, leaving the rest to be picked up by his carpenters. Another refers to such work as "slush," items that have somehow melted away during estimating but then reappear during construction.

Their problem is clearly the result of inadequate estimating procedures. If they were to insist their subs provide thorough Included/NIC lists, they would have no more than an occasional tiny puddle of slush showing up during a job. Virtually all work would be mopped up during estimating and then folded into the contract price with the client, and would, therefore, be billable.

Concrete subs would either include in their bid or call out as NIC and explicitly leave to you—so that you know to corral the costs in your estimate of crew work—all such slippery items as: *Removal of concrete spillage. Stripping of forms. Cleanup of form lumber. Filling of voids in a pour. Removal of form ties.* And *curing of the concrete.* Framers would know that their bid must include—or explicitly exclude—not only the obvious stuff from joists to roof sheathing, but also the incidentals: *Staging. Cutting out plates* at doorways. *Blocking* for cabinets and fixtures. *Fire blocking.* All *structural hardware. Furring.* And anything else required by codes or reasonably implied in the plans and specifications.

Your regular structural subs should understand that if they leave *any work regularly associated with their trades* out of their estimates, fail to inform you that it is NIC, or fail to take care of it at the job site, they—not you—will bear the burden. Either they will be called back to take care of the work immediately. Or, they will be back-charged for the time your crew spends picking up the left-out items.

Why not? You take good care of your regular subs. You steadily update them during projects as to when they will be needed at the job site. You don't surprise them with a 5:00 p.m. text message that you need them on the job at 9:00 a.m. the next morning (as do badly organized general contractors). You make sure the work site is ready for them to do their work when they arrive, so that they can proceed efficiently. You pay them promptly. You communicate with them respectfully. You even provide them with an NIC/Included form! They, in turn, must do their part. That includes making sure that all incidental items are covered in (or explicitly excluded from) their bids so that it will be covered in yours. No damn slush allowed!

On the other hand, when you are working with new subs, especially when soliciting bids for work with which you are not familiar, then extra effort on your part will be necessary to catch the slippery items. For example, in my area, we do little masonry work. (It's not a good idea to construct walls of blocks stuck together with cement in a metropolis located on the "Ring of Fire," the landmasses bordering the Pacific Ocean that are subject to massive earthquakes. The blocks tend to come unstuck during shakes and fall on people.)

On the rare occasion when my projects have called for structural masonry, I have had to go over in detail with my subcontractor just what items associated with his work he would be leaving to me. Is he thinking *electrical boxes and conduit* will be surface-mounted? Or does he realize that we will need to coordinate with the electrician so that he can embed boxes and conduit in the masonry as it is installed? Is he taking care of all other *embedments*? How about *grouting, patching*, and *cleaning* of the masonry surfaces? Is he providing his own *scaffolding and staging*? Or does he expect me to have it in place when he arrives? And, most important of all: Is there any other item that might slip between the trades that I am not aware of because of my lack of experience with his trade?

Similarly, if you begin to expand from residential to commercial construction, you will encounter unfamiliar requirements for structural steel. Your estimate may have to provide for such items as Lally columns (steel columns filled with concrete), moment frames, or metal decking. You will need to ask questions about the interface between the structural steel sub and the work of your crew and other subs. A few brief examples: Will the sub be at the site to accept *delivery* of his material? Or does he expect your crew to deal with the delivery driver and put his material aside for him? Will he *remove scrap*? Does he provide hardware for *connections to the wood framing*? If the answer to that last question is yes, does he make the connections or is he leaving

Mechanical, plumbing, and electrical work by subs

- Gas supply
- Heating systems
- Air cooling systems
- Duct sealing
- Prefabricated fireplaces
- Wood burning stoves
- Air exchangers
- Exhaust fans
- Central vacuums
- Plumbing supply and DWV (drain/waste/vents)
- Connections between heating and plumbing systems
- Sewer line connection
- Hot water heaters
- Plumbing fixtures
- Fire suppression (sprinklers)
- Sump pump
- Floor and roof drains
- Fountains
- Solar thermal (hot water)
- Solar photo-voltaic
- Temporary power
- Site conduits for cable, electrical, phone
- Site lighting
- Electrical panels
- Electrical wiring
- Electrical ground
- Electrical finish and fixtures
- Telephone connections
- Audio system connections
- Visual and data system connections
- Security systems
- Electrical connections to fans, equipment, and other systems
- Smoke and carbon monoxide detectors
- Door chimes
- Utility hookups
- Appliance connections
- Hot tubs and saunas
- Equipment and fixture delivery
- Cleanup, recycle, and waste removal

that work for your crew? Thus, as always with sub bids, the key to nailing your numbers is to determine what is included and what is not, especially at areas of overlap between trade responsibilities.

Mechanical, plumbing, and electrical

The greatest risk of slippage with subcontractor bids comes with those from mechanical, plumbing, and electrical (MPE) subs. Their work is costly. And it is complex, woven into the structural work and interconnected in such a way that items are easily obscured and overlooked. Example: a *Ufer bar*—a stick of rebar placed in the foundation to serve as an electrical ground.

That's an item that got away from me on one occasion. My electrician reasonably assumed I would install it, since I claimed to know what I was doing as a builder. When I did not install it, he had to drive a costly copper rod into the soil to act as a ground. And I had to pay for the work, which, of course, I had left out of my estimate. When I discovered my omission, I did what I always do to steadily strengthen my spreadsheet/checklist. I made a note to myself to put "Ufer bar" in my estimating spreadsheet. Back at the office I put the note in my in-box so that I would be sure to act on it. And I did.

Slippage with MPE bids occurs especially because of these factors:

- Associated carpentry such as blocking, cutouts, and chases.
- Connections between work handled by different subs, such as a combi-system that provides both hot water and space heating and therefore can involve work by both plumbing and HVAC subcontractors.
- The tendency of designers to designate MPE as "design/build," barely hinting at the work in their plans and specs and leaving it you and your subs to design as well as install the systems.

To gather complete numbers for MPE work, you must be asking questions such as these about associated carpentry: Who is responsible for the *earthquake strapping* for the hot water heater? You or the plumber? Who is making the *cutouts* for heat registers? You or the HVAC sub? Who's got the staging? You or the sub?

And these, about *connections*: Has the plumber included hookup of his waste lines to the sewer in his bid, or is he expecting the sewer contractor to make the connection? Do you need to clarify expectations with both the plumber and the sewer contractor? If separate subcontractors are installing a boiler and an under-floor hydronic heating system, who handles the connection between them? Is the electrician providing for connection of control wires for mechanical and plumbing devices such as fire suppression systems and recirculation pumps? Or, again, is he assuming other MPE subs will take care of the connections to their work?

Who is responsible for appliance connections? Is your appliance company supplying and installing the flex line that connects the stove to the gas valve? Or is the stove being delivered only as far as the driveway, so that you or your sub will have to make the connection as well as *get the stove into the building, uncrate it, and dispose of the packaging*?

As for MPE work designated as design/build, you want to nail down answers to questions like these: If the project requires a forced-air heating system, does the sub's bid include thorough *duct sealing?* Who pays for a *third-party expert to test* the ducts and certify that they are adequately sealed, you or the HVAC sub? (HVAC installers historically have been so negligent about sealing that a third of the heated air escapes the ducts before reaching the registers.)

How about *ventilation devices* such as the *bathroom fan*? Has either the electrical or the mechanical sub included it in their bid, and if so does the fan specified meet the requirements of the construction documents? Or have both the electrician and the mechanical sub included the fan in their bid, so that you have redundancy? If the electrician is installing the fan, who is responsible for providing the *vent roof cap*? Is it the electrician or the roofing sub?

Close-in work by subs

- **Roofing**
- **Roof flashing**
- **Roof jacks and caps**
- **Gutters**
- **Fenestration (doors and windows)**
- **Water management systems at doors and windows**
- **Insulation panels**
- **Wood siding**
- **Vinyl siding**
- **Shingles**
- **Cementatious siding**
- **Exterior insulation and finish system (EIF)**
- **Brick veneer**
- **Stucco**
- **Ornamental ironwork (security bars)**
- **Painting and sealing**

With MPE design/build work, especially, there is another level of potential slippage, one having to do not with the interface between trades but with whole categories of work: Temporary power. Security systems. And telephone, audio, visual, and data systems that may be installed by a single sub but that may also be broken up, with each system being handled by a separate specialist. Your task during estimating is to make sure that no systems have been overlooked, and that all have been gathered into one or another subcontractor's estimate—and not more than one. And while you are at it, have you made sure someone has picked up the *smoke and CO_2 detectors*? And the *doorbell*? And even *HVAC*, which, as mentioned in an earlier chapter, does, incredibly enough, get entirely left out of bids.

Close-in

Mercifully, subcontracting the work associated with closing in a building, from roofing through exterior finish, entails relatively little potential for slippage, because the overlap and interface between subs is not extensive. There is, however, some risk that an assembly might slip away altogether. You might assume that your roofer is including *gutters* in his bid. He might assume, meanwhile, that the sheet metal sub has that item. You might overlook the mailbox or some other item attached to the exterior.

However, the items associated with close-in that are most liable to slippage have to do with *water management*. Poor water management causes especially severe problems now that we build with rapid growth lumber and fill our framing bays with sponges—aka insulation. The new lumber, with its high proportion of weak and wide-grained summer growth, has little resistance to water penetration or fungus (aka dry rot and mold). And the insulation traps water that makes its way into the frame, assuring that the lumber will remain wet for extended periods and that the fungus will thrive.

If we make dumb mistakes in installation of our water management systems so that water does make its way into the frame and is held there by insulation, dry rot accelerates. Catastrophic decay can happen much faster than in older frame buildings built with denser lumber and insulation-free frames that readily air dried. As the building scientist Joe Lstiburek says, the distance "from stupid to sorry" as a result of inattention to water management is much shorter these days.

Attention to water management should start during estimating. Inattention can lead to failure to cover the costs of water management in a bid. That can lead to failure to install it properly—especially as a result of delegating the work to low-cost unskilled subcontractors (or employees) because there is not enough money in the bid to have it done well. That in turn leads to rapid building failure, angry clients, lawsuits, and sorrow. For example, a few years ago I wrote an expert witness report on a failure at a large home that was skillfully crafted except in one respect: installation of building wrap and flashings was inept. The home was only two years old. But so much moisture was trapped in the walls that fungus had not only attacked the frame but broken through to the surface of the drywall. Two years after the house was built, mold so thickly coated its interior surfaces that it was uninhabitable. Repair costs ran to over a million dollars, (and legal fees resulting from the homeowner's tenacious lawsuit against the builder and his insurance company nearly that much).

During estimating, if the work is to be subcontracted, take care to ensure that one sub bid or another covers all water management detail including, from the top to the bottom of your project: All roof-to-wall flashings. Flashings at skylights and any other roof penetrations. Fascia flashing. Water table flashing. Sill pans, vertical flashing and head flashing at windows and exterior doors. Frame-to-wall flashings, membranes, and scuppers at balconies and decks. And don't overlook the need for and cost of shop drawings for atypical water management details including those at joints between dissimilar materials.

I have experienced leaks a few times and have had to tear into my completed projects to correct flashing errors and omissions. It's been embarrassing. And costly. Fortunately, I stumbled into a seminar on water management and learned how the details should be built. During estimating, I bore down on them, making sure that all the necessary work was anticipated and assigned to either my crew or capable subs.

Build-out and specialty work by subs

- **Light gauge metal framing (LGMF)**
- **Insulation**
- **Air sealing**
- **Soundproofing**
- **Vapor barriers**
- **Drywall**
- **Lathe and plaster**
- **Stairs**
- **Ladders**
- **Passage doors**
- **Shower doors**
- **Cabinets**
- **Countertops**
- **Precast lintels**
- **Wood trim**
- **Other finish carpentry**
- **Acoustical ceiling systems**
- **Glass block**
- **Ceramic tile**
- **Wood flooring**
- **Carpeting**
- **Resilient flooring**
- **Paint and stain**
- **Wall covering**
- **Window treatments**
- **Appliances**
- **Furnishings**
- **Equipment**
- **Storefronts**
- **Specialty items: chalkboards, toilet partitions, wall and corner guards, medicine cabinets, coat racks and wardrobes, bathroom accessories, and much more**
- **Touchup**
- **Cleaning**

More recent projects have not leaked at all. That feels good. And, as icing on the cake, because the work was covered in my estimates, I got paid for it.

Build-out and specialties

As you move from estimating for close-in to estimating for build-out of a project interior, slippage is most likely to occur around two issues:

First, are subcontractors providing for the *protection of one another's work*? Subs will be installing their work on top of a predecessor's work. Drywallers are fitting wallboard and taping around electrical boxes. The trim carpenters are applying their sticks over drywall. The flooring subs are placing their material against (or under) baseboards and casings. The countertop sub installs atop the cabinets. Water closets are put in over finish flooring.

Subs must include protection of previously installed work in their bid. Otherwise, the protection, or damage repair necessary because of lack of protection, may end up as an unbillable item. You'll have to take care of it; but you won't get paid for it.

Second is the matter of *cleanup*. Are the subs taking care of it, both *daily and at the conclusion of their work*, or is it going to be the responsibility of your crew? A few years back, I contracted—too casually—with a new drywall sub. He and his crew did a beautiful job, applying an unusually delicate orange peel finish. But having laid no protective paper over the subfloor, they left behind a film of joint compound over its surface. When I protested, the drywaller apologetically explained, "I thought you were going with carpet," meaning that he assumed the carpet pad would be laid right over the mud film with no harm resulting. In fact, I *was* planning on carpet—not, wall-to-wall however, but rather carpet tile held in place by mastic. The film of mud would have prevented it from adhering.

"Whoops. My bad, I will take care of the cleanup," I told the drywaller. I was at fault. I had skipped over the usual Included/NIC form that I should have provided to him when he bid the job. The cleanup required seven hours of grueling labor.

Along with cleanup, there are areas of overlap in interior build-out where items can fall between the trades. Two notable items: *Preparation for finish flooring*. *Touchup of drywall* after application of the paint primer highlights drywall defects.

Some finish flooring products must be *acclimatized*, stored within the building after it has been heated to normal operating temperature. Other flooring products require preinstallation of particular kinds of *substrate*. Still others tolerate no dip or rise in the subfloor, so that it must be *leveled*, or at least *brought into an even plane* by one means or the other—sometimes by grinding, sometimes by labor-intensive elimination of joist crowns. Such tasks can go to any one of several trades.

As for drywall, it will almost inevitably get nicked repeatedly before the painter arrives. Who is responsible for the cost of touching it up? Ideally, the trade that caused the damage, but in reality, it is generally impossible to determine that. So it comes down to the drywaller or the painter. If you leave the choice till construction rather than taking care of it during estimating, division of responsibility can become problematic. I once had to hold back a drywaller from slugging my favorite painter, who had festooned the crisply finished wallboard with hundreds of little strips of blue tape, each marking a tiny defect, and was demanding that the drywaller fix each one.

You need to clarify responsibility for drywall touchup during estimating. One possibility: Ask the drywaller and/or painter to allow for a specific and reasonable number of hours of touch up in their bids. If they go over that amount, the extra hours get paid for out of your pocket. Why? Because the excess damage is likely the result of your having failed to see to it that other subs protected the drywall as they did their work.

Beyond the usual build-out work there are innumerable items that can be described as "*furnishings*," "*accessories*," and "*equipment*" or, alternately, lumped together as "*specialties.*" The categories include such items as: Closet storage systems. Window treatments. Fold-down ironing boards. Floor-to-ceiling

mirrors. Conveyors. Elevators. Lifts. And, of course, ever more numerous and elaborate kitchen devices. If you move from residential work to commercial or industrial work, you must be alert for unfamiliar specialty items that go with each step into a new niche. Examples include: Pews in churches. Heavy-duty ventilation systems and grease traps in restaurants. Projection equipment in theaters. And goodness knows what in dental or medical facilities, clean rooms, gyms, ice rinks, tennis clubs, and pet hotels.

All in all, there are a staggering variety of items that can slip between the sub trades as you gather up their bids. Moreover, they vary from one geographical area to another in parallel with the variations in technologies in use and codes in force in the different areas. If you build in an area that is earthquake-free but subject to extreme temperatures, you will need to catch slipperies that are not of concern to builders in seismically threatened but climatically moderate zones, and vice versa. There is no checklist I have seen (and because I don't want this section to go on forever, I will not attempt to create one here) that captures even a fraction of all the possible slippage in subcontracted work.

So far as I know, there is only one way to prevent items from getting away from you during estimating of subcontracted work. Capitalize on experience. Together with your subs, steadily build awareness of potential slippage. Create thorough Included/NIC checklists, encourage your subs to do the same, and use them. Otherwise, you, too, may find yourself mopping up "slush" forever, and that can't be fun. On the other hand, with a comprehensive spreadsheet and NIC checklists you will be well positioned as you move to actually costing out your projects, the subject we will take up in the next chapter.

A Note to Readers:

You have now come to the end of Part II. You have worked through the fundamentals of building an estimating and bidding system from qualification of projects for a bid, to systematic takeoff procedures, and on through to design and development of a comprehensive estimating spreadsheet.

Now you will find it worthwhile to pause and make a Task list of the ideas and procedures you think that you should incorporate into your own system. That might be development of a lead sheet. It might be creating a checklist of takeoff speed tricks. Or it might be making sure that your spreadsheet covers the slippery items and allows, at multiple lines, for completion. It might be something altogether different.

Whatever you choose to focus on, do tighten your grip on knowledge you've gained to this point and create an action plan -- and act on it. You will then be in a strong position to make use of the information in the following chapters.

PART THREE

CAPTURE YOUR COSTS

CHAPTER 9

Into the Money

A rare skill

In previous chapters, we have discussed the tools and the processes a well organized builder uses in the two initial phases of creating a reliable estimate and bid: First, taking off the quantities of work. Second, extending those quantities into a detailed spreadsheet/checklist. Here in Part III we will take the next logical step: Figuring the dollar costs for constructing the project. Since figuring costs for material, subs, and services is relatively straightforward, we will focus first and especially on nailing the labor costs for items and assemblies built by your own crew.

COMING UP

- Getting down to dollar costs
- An industry weakness
- Uneasy adjustments for uncertainty
- Intuition-based estimating and its drawbacks
- A preview of labor productivity records

If you do develop the practices that will enable you to reliably project your dollar costs for crew work, you will, from what I have seen, be unusual among light-frame builders. During the interviews for this book, I was often impressed by the overall bidding strategies of the folks I talked with. Equally, I was surprised at how deficient many of them were at estimating the costs of crew labor for their projects.

Some were very capable at gathering subcontractor bids. Some even had a NIC/Included protocol in place so that they were able to prevent items from slipping between the trades. Many seemed competent at the straightforward work of quantifying materials needed for a project and getting quotes for those materials from their suppliers. But with only one exception, they had not developed methods and resources for closely estimating the productivity rates of their own crews. And of course, without being able to figure how much work their crew could produce per hour or day, they could not figure the dollar cost of work.

Uneasy adjustments for uncertainty

I am not talking about flakes here, but about guys who have been in business for years, have strong marketing programs, do good work, sustain loyal crews, and take estimating and bidding seriously. Often they were aware of their deficiency, and had made uneasy adjustments to compensate for it. One, a builder with a small company that specializes in repair work (with almost all of it done by his own crew and very little by subs), carefully lines out all the items of work in a project, does his best to imagine the number of hours each will take, totals up the hours, and multiplies them by an hourly labor charge to arrive at his costs. And then, he adds 30%! That's to make up—more or less—for the shortfall, coming from where he knows not, that he has learned will be there in his numbers.

Another builder, one employing two dozen people and specializing in high-end residential work, goes to his lead carpenters for help with figuring labor for upcoming projects. "Do the hours we have in the estimate look about right?" he asks them. Often, he says, they think his initial projections are "crazy low." But when he asks them how many hours they think an assembly might take, their answers are not much help. "Everybody has a different answer," he groans. He finds himself stuck between his "crazy low" numbers and his leads' contradictory guesses.

He finds himself stuck between his "crazy low" numbers and his leads' contradictory guesses.

To protect himself against shortfalls, he has programmed his computer to automatically add 20% to the figures he inputs for labor. The 20% is for "uncertainty," he says. He's not happy with relying on such a rough adjustment; he knows his process needs improvement. He knows he needs "historical records," of how long it has taken his crews to produce foundations, framing, and finish work. Such records would, he thinks, give him a sound basis for estimating labor costs for new projects. Prodded by a controller he recently hired, he plans to get around—after 40 years in business—to finally creating those records. Perhaps he will even use the format for historical records I have used for several decades and look forward to sharing with you in Chapters Eleven and Twelve.

Intuition-based estimating and the better way

Not all builders who proceed without historical labor records are dissatisfied with their results. Some feel confident that they can nail their numbers without reference to records of past experience. Their three-step procedure involves looking over a job, breaking it down into assemblies or phases of work, and then "intuitively," that is, on the basis of memory, projecting their costs for each one. "The costs are pretty much in Enrique's head," says one such builder, speaking of the employee who produces his estimates. Several others I spoke with similarly report that that they rely on "experience" to estimate costs. They feel certain that they know, from years of building, just how long the various phases of a job are going to take; and they claim that when they check their estimates against their results in the field, the variance is generally only a few percent.

I am a little dubious about their claims, for when I "intuitively" project costs for a job and then crank out a record-based estimate, my intuitive take often turns out to be way off. However, I do a wide variety of complex projects while those builders who tell me they successfully practice "it's all in my head" estimating tend to specialize in one or another narrow range of projects—and relatively small projects at that. One replaces house foundations and installs perimeter drainage systems. Another removes and replaces mid-cost bathrooms. A third builds cookie-cutter additions.

However, even for builders who specialize in one or another kind of work and who are satisfied with and confident in their results, relying on memory for labor costs poses hazards. Supposing the memory a builder is relying on is his estimator's, and the estimator decides to move on? Then where will the costs come from? Supposing the builder has been doing his own estimating, but wants to step away and delegate it to an employee or hire an estimator? Where should the new estimator get the costs? From guesswork? From the cost experience of the builder he previously worked for? From his boss, the builder who just delegated estimating to him in order to concentrate on other tasks? And what if an opportunity came along to take on a larger, more complex, more challenging, and potentially more profitable job? Would the builder be able to estimate it confidently, right out of his head? Or, would his lack of records require him to take a pass on the job?

Lack of reliable historical records and reliance on all-in-my-head (or someone else's head) estimating can lead to other problems:

- Guessing too low at the costs for a project and losing money on it, or guessing unnecessarily high and losing the project.
- Being forced to work on a time-and-material basis to protect yourself against low estimates and, thereby, taking on the risks that T&M work brings, including the greater likelihood of lawsuits.
- Having no choice, because you cannot reliably estimate how long it will take your crew to do it, but to subcontract work that you otherwise would have been able to do enjoyably and profitably in-house.
- Compromising your case that you should get paid for your preconstruction estimating rather than bidding for free. That is a subject I will go into in detail in Part IV. For now, it is enough to note that your preconstruction services are more valuable if your estimates are based on records rather than intuition and guesses.

Hierarchy of cost sources

- Worst: Hearsay
- Not much better: Tables and mechanical relationships
- Surprisingly poor: Cost reference books (i.e., cost catalogues)
- Limited: Constructive imagination
- Best: Written records of labor productivity at your past projects.

In his classic books on estimating and bidding (see Resources), Paul Cook suggests a hierarchy of sources for costs. The best bases for estimating costs of future work, he emphasizes, are item-by-item records of costs for similar work on past jobs, and that's on your jobs, not someone else's. As we will see, all it takes to build up such records are procedures, patience, and persistence.

As the accompanying sidebar, "Hierarchy of cost sources" emphasizes, there are a variety of ways to determine costs for the work in a project. But written records of your own company's past costs—*in particular, records that detail out labor hours item by item and assembly by assembly, and that are organized in such a way as to be easily accessible*—are the professional estimator's gold standard. In the long run, you will be much better off if you create and use such records rather than relying on rough percentage adjustments or numbers plucked intuitively from memory to determine the labor costs for your projects.

I wish that I could show you a variety of ways of keeping and organizing such records. Unfortunately, I cannot. There must be light-frame builders out there who have devised excellent methods—perhaps better than my own—for maintaining labor productivity records. But so far, I have not found them. Despite

Labor Productivity Record

ITEM OR ASSEMBLY:

QUANTITY & DEGREE OF DIFFICULTY of ITEM:

PROJECT:

TIME OF CONSTRUCTION:

CLIENT:
DESIGNER:

ACCESS:
WEATHER:
OTHER FACTORS:

CREW CAPABILITY:
Lead:
Journeymen:
Apprentice:

HOURS PER CREW PERSON:
Journeymen:
Apprentice:
Laborer:
TOTAL HOURS:

HOURS PER UNIT:

In upcoming chapters you will find a series of completed labor productivity records from my projects. While I advise you to create your own records rather than relying on my numbers, my records should help you get started down the path to compiling your own.

Do so and you will greatly strengthen your ability to nail your estimating numbers. Fail to do so, and you may end up as yet another builder who has to rummage around in job cost records or guesstimate to figure the hours his crew will need to form a foundation, frame walls, install trim or do the other work required for a project.

How pathetic would that be?

interviewing many builders, I have only come across two methods, the one I use and another employed by a builder of small housing tracts. In the last chapter of this section, I will explain both his method and mine. I will also provide a sample of my own labor productivity records, which you can use as a starting point for developing your own.

Here, with the illustration of my labor productivity records form, I have given you a sneak preview of my almost laughably straightforward method. But before going over the form and my method in detail—as well as going over the inclusion of material and subcontractor quotes in an estimate and bid—I want to review a number of shortcuts to estimating costs that are widely used but that are, as I see things, bush league.

CHAPTER 10

Dubious Paths and Building Blocks

Choices

Why, I've wondered, is estimating in the light-frame world saddled with so much uncertainty and reliance on off-the-cuff calculation of labor costs? You don't encounter that in manuals describing the estimating practices of companies operating in the world of larger-scale commercial and industrial construction. Instead, you find confidence that methodical procedures can and do produce reliable estimates.

COMING UP

- Bad choices
- Softheaded consultants and software salesmen
- Lessons from the world of big-time construction
- Ridiculous sources of cost information
- A skeptical view of cost catalogues
- Time cards

Why the difference? Maybe this: Builders in the light- and heavy- construction worlds choose to get their guidance from very different folks. Among the limited number of people operating as educators and consultants in the light-frame world there are few who appear to have actually worked in, much less owned and operated, a construction company that has endured and been financially successful over a long period. The others tend to give short shrift to the down-in-the-dirt aspects of estimating, preferring to focus on juicier topics like markup and profit.

One troubling example: A popular consultant, who has provided guidance to many hundreds of builders, assured the attendees at a seminar on business management that if they just made certain to include high markups in their bids, they would be on the road to prosperity. A builder in the audience objected. Supposing you badly under-estimate your costs for constructing the project, he asked. In that case, isn't it likely that even with a high markup you could end up losing money on the job? Isn't it possible that that you could run so low in your estimate for labor, material, and sub costs that your shortfall could wipe out your profit markup and then some? The consultant airily dismissed his concern and moved on to another topic.

Other consultants in the light-frame world attempt to persuade their clients that they can off-load responsibility for estimating their costs to outsiders. These consultants peddle estimating programs that come with databases purporting to provide costs for thousands of items and assemblies. The programs are beguiling. They require only that you plunk quantities from your takeoff sheet into the program's spreadsheet. The program will fill out the rest of the spreadsheet, telling you how much to charge for each item and assembly, and it can even dictate figures for overhead and profit to you. Presto, you've got a bid. But there's a problem: The charges aren't based on your costs. They are based on someone else's—or, supposedly, on the average of a lot of someone elses' costs. These "averages" of other builders' costs, however, may vary widely, perhaps seriously and dangerously, from your costs.

In the world of heavy construction and large-scale companies you encounter much better advice. It is the opposite of what you will hear from too many of the light-frame consultants. For example, Paul Jenkinson, the chief estimator for a large general and mechanical contracting firm, emphasizes that you must figure costs strictly on the basis of in-house information: "You must know your crews" and produce estimates "on the basis of historically correct data derived from *your* company's experience," not someone else's experience.

Similarly, the Associated General Contractors (AGC) handbook on estimating and bidding insists, "The estimator must beware of being influenced by labor cost data from outside the firm…[and] should *never* be concerned with labor costs someone else says they have achieved." In short, if you want to produce reliable estimates and bids, you can't let some software package take over the costing of your jobs. You have to rely on your own numbers. They must be carefully recorded, then organized and maintained so that you can access them efficiently.

Unfortunately, builders in light-frame construction too often rely on other sources, some even worse than the average costs pumped out by software databases, such as:

- Newspaper articles. I have read reports in my local newspapers of construction costs, derived from God knows where, that cite prices about one-third of those typically charged in my area by established companies.
- Hearsay, namely vague reports or claims about what other builders are charging. Even if true, the information is useless. You have no idea whether the builder will be in business or bankrupt a year down the road.
- Designers. How many builders have gone broke attempting to build for a price suggested or insisted on by a designer? I can tell you of one who nearly did. She "won" a job by submitting a bid that matched the figure proposed by the designer. During construction, her costs soon exceeded her estimate. Demonstrating impressive integrity, rather than walk away from the job, she funded completion of the project by taking out a mortgage on her home.

The cost catalogue temptation

Other sources of outside cost information appear, on the surface, to be less dubious, but they, too, should be viewed with skepticism. Their seeming respectability can lead you into spending money and relying on them when you should instead concentrate on recording your own costs. Those sources are cost catalogues, often called "cost reference books" or just "cost books," provided by a variety of publishers.

The cost catalogues come in two forms, large-format paperbacks and digital databases. Like the databases that come with estimating programs, they provide costs that are, or at least claim to be, the average of costs achieved by a large sample of builders. From either version, paper or digital, you can easily retrieve cost numbers for a vast array of items and assemblies.

The publishers of the cost books ardently promote their products. One of the largest publishers tells readers (though not in the fine print, which we will get to shortly, and which tells a very different story) that their books are among "the most powerful construction [business] tools available" and provide the very "foundation" of a "reliable estimate" based on "current construction costs." At the other end of the spectrum, one of the smaller publishers boasts of employing a "research staff" that provides builders with "pinpoint accuracy" on costs "in over 250 areas of the U.S. and Canada."

You may find a cost book containing data closely corresponding to your own costs and that will help you to produce "dependable and accurate" estimates. I have not reviewed every single cost book on the market. However, the only builders I have ever encountered who express enthusiasm about cost books (or databases) are those who appear in ads for them, make money off them, or are beginners.

Otherwise, the endorsements I have run into are lukewarm at best. One consultant widely relied upon by remodelers and homebuilders reports that while he has found most of the cost books useless, he has seen acceptable remodeling estimates produced from data provided by one of the smaller publishers. Another industry educator expresses some satisfaction with a different publisher, writing that their figures tend to run high, but "better high than low." Maybe so, up to a point, though it's also true that any inaccurate estimate is a bad estimate, whether it is high or low. Excessively high estimates can cost you by losing you good jobs, and low estimates can lose you a lot of money.

> ***"The estimator must beware of being influenced by labor cost data from outside the firm . . . (and) should never be concerned with labor costs someone else says they have achieved."***

Apart from a few such qualified approvals, builders' reactions to cost books range from dismissive to downright disgust. When I asked one veteran estimator whether she ever made use of cost books she replied, "Only when I am desperate."

Probably she should not use them even then. One of the best selling cost books for a recent year suggested a square-foot price for "one of a kind" homes that was one-seventh of the average construction cost of the homes designed by the architects that the estimator works for. Yes, one-seventh, *14%*. That's not a typo.

A builder who operates at the other end of the cost spectrum, specializing in renovation and remodeling of middle-class homes, is even more dubious about cost books. When I asked him about them, he spun his desk chair around and yanked a cost catalogue off his bookshelf. "Look at this piece of crap," he said, holding up a large paperback for which he had paid a hundred bucks. "I wish I could shred it, but my shredder won't swallow it."

The skepticism about cost books extends into the world of large-scale builders. The AGC warns builders against being lulled into a false sense of security by the impressive volume of data included in the computerized versions of cost books, for they can obscure the "conditions of a specific situation."

In my own experience, the cost books' figures for labor are sometimes roughly accurate for simple items such as the time needed to install 4'× 8' sheets of plywood shear panel in new construction. But otherwise they diverge so widely from my reality, I make no use of them for estimating. Why is it that, for all the marketing claims vendors of cost catalogues make for their products, knowledgeable builders and estimators view them with such skepticism? I feel that question needs to be answered in detail, with the deficiencies of cost books called out one by one.

Problems with cost catalogues

The problems begin with the sources of the cost data in catalogues. One publisher relies in part on questionnaires it sends out to builders, enticing them to respond with promises of a discount on a future purchase. Is cost data collected in this way at all reliable? The only builder I have found who actually filled out a questionnaire told me that he put little effort into it and just jotted down information off the top of his head. Maybe he is not typical, but how much time would you spend filling out a form about your costs in exchange for a discount on a book you might or might not buy in the future?

A publisher who caters primarily to remodeling and custom home contractors reportedly goes beyond questionnaires, and interviews builders about their costs. But again, is the data obtained dependable? A builder who is also the author of a prominent construction business blog points out that builders—at least those in the light-frame world—are "hugely unreliable" on the matter of costs. Too often they don't have a handle on their own costs. So how can they be counted on to contribute to a pool of reliable cost data for other builders?

A third method, employed by one of the larger publishers, may seem more promising: time-and-motion studies conducted by a staff of "cost engineers" dispatched to actual job sites. However, the publisher is primarily focused on large-scale construction. The costs their engineers gather are for the large volumes of repetitive work typically done at big commercial and multiunit projects. Are those costs of much use for smaller-volume, light-frame builders remodeling homes or apartments or constructing single-family houses or small multiunit buildings?

The publishers would have you think so. They put out a catalogue of costs for repair and remodeling contractors. But when I spoke with one of their sales reps, to my amazement he warned me away from his company's publication for repair and remodeling work! The costs it contained, he said, were derived from the publisher's data from large projects; so for smaller-scale repair and remodeling projects, it wasn't good for much beyond figuring the price of patching a hole in drywall.

Actually, it is not even good for that. I checked on the costs the remodeling cost book suggested for patching a hole in drywall. It allowed about six minutes to patch a one-square-foot hole, apparently including setup and cleanup (though that's not entirely clear). Know anybody who can set up, trim a hole, cut and set a drywall plug, tape, top, and texture, then clean up in six minutes? Maybe guys patching dozens of holes all at once on a large commercial renovation could average six minutes a hole. But does the figure have any relevance at all for the contractors to whom the repair and remodeling cost book is peddled?

Problems with cost catalogues

- The sources of the costs
- Small-scale residential costs derived from large-scale projects
- Out-of-date costs
- Averaged costs
- Use of prevailing wage labor rates
- Many adjustments necessary to translate a cost book number into a number for an actual estimate
- Shifting of responsibility back to user of cost book and away from publisher

Aside from the questionable reliability and relevance of their sources, there are a number of other concerns about cost books. They include:

- *Date information collected.* The cost books promise up-to-date cost information. That is just what you need for figuring the costs of a job. Construction costs do fluctuate wildly, rising or falling by some 20% in a single year. But it is questionable just how closely the cost books track fluctuating costs. One widely sold cost book, claiming to provide reliable data for the current year, was actually copyrighted the previous year.

- *Averaging.* The costs listed in the books are average costs for large areas or even the entire country, and you must adjust the costs to fit your exact locale. For example, with one major cost catalogue, if you build in my area you are required to *increase* the cost book averages for framing by 35% and for doors and windows by 27%. If you are operating in Kansas City, you are instead required to *decrease* costs by similarly large percentages. Do you think you can nail your numbers using this sort of crude adjustment?
- *Reliance on prevailing wages.* Some cost books base their numbers on compensation levels for unionized tradesmen. If your company is not unionized, you must adjust the costs to correspond to compensation levels in your own operation.

The cost book tease

To hear the typical publisher tell it, their cost catalogue will be your reliable partner in producing dependable estimates. But when you actually start using the cost book, you constantly find yourself being told, "Wait, hold on, there's just one more step you have to take to get to the Promised Land."

First the book tells you to look up the publisher's cost for your item. So you do that. Then the book starts with the addenda: Wait, no, don't use the number as is. Instead, go to the index for location, find your zip code and the associated modification factor. Then multiply the modification factor by the cost you looked up.

But, hold on, don't just plunk that modified number into your estimate. Make sure your labor rates, the amount it costs you per hour to put guys on a job, match up to those assumed here in the book. Oh, they don't? Your labor rates are different than the prevailing wage rates assumed here? Okay, you better figure out how to adjust the cost again so it aligns with your rates. How? Oh, we don't really get into that. You'll figure it out.

Well done! Now you are really getting close. But there are a couple of other items you better check on. Did you make sure that our cost includes all the items of work that you might assume it would cover? No? Layout is not included in wall framing? And window installation does not include head flashing. Wow! Better add something for that sort of stuff. There could be a lot of it. Better look closely. And don't forget to add sales taxes. Our costs don't include them either.

And by the way, you might want to adjust for the season of the year and weather conditions. And for quality because we assume normal, sound quality (no, sorry, we don't tell you just what that is). Then there's productivity. That is a tough one. It's really up to you to make sure the cost you have arrived at by making all those adjustments to our initial figure matches up with your company's actual productivity. Good luck with that. We hope you are successful, because that will just prove once again what a great estimating partner we are!

- *Lack of clarity about what is covered by a cost.* For example, one book intended for remodeling work includes only plates, studs, and fire blocking in its per-foot cost for framing new walls. Headers, rough-in of openings, bracing, and other wall-framing items are listed separately. But it's not clear whether the costs for those items are intended only for changes to existing walls or are also intended as add-ons to the per-foot cost for new walls.
- *Omission of critical items.* In a cost book from one large publisher, the costs for window flashing are not included with either window installation or in the moisture management section. In other words, the cost book encourages the cardinal sin of estimating, i.e., leaving something out—and in this case, a particularly important something.
- *Abstract labor cost numbers divorced from the conditions at an individual project.* As consultant Martin King points out, "There is no procedure so simple that production costs cannot increase from two to 10 times because of difficult conditions." Most cost books ignore that truth. One book attempts to account for the variance by providing for a range of labor hours for installation of different items. It suggests, for example, between 8 and 33 hours to install 1,000 b.f. of girder material. Useful? How do you determine where, within such a wide range, costs for your project with its particular conditions are likely to fall?

So, if not from newspapers, if not from designers, if not from hearsay about what some builder across town is supposedly charging, and if most definitely not via the torturous process (described in the sidebar) necessary to extract even dubious information from cost books, then where should a builder get the costs, especially the labor productivity figures, he needs to create his estimates? My answer will come as no surprise: Rely on complete quotes from subs and suppliers. And most especially: Build up records of labor productivity from your own projects. You can create a basic set rapidly—even during construction of a single addition—if you conscientiously record the labor hours for all work from foundation through trim. Over the course of a year you can create a quite comprehensive set of records. Keep at it, and—with steady effort but surprisingly little use of your time—you'll create a truly awesome set of records. Your first step is to put in place the essential building block of reliable labor productivity records—time cards.

Time cards

So long as your projects are limited to modest-sized jobs such as build-outs of small commercial spaces or residential additions, an all-in-one time card such as the one shown on the next page may be all that you need. For larger projects, such as a year-long reconstruction of a larger building, you may want to use a series of time cards, one for each of the primary divisions of work—whether concrete, framing, or finish—done by your crew. For both the all-in-one card and the division time cards, you can create templates and tailor them for individual projects using your word processor.

Why, you might ask, should you take the trouble to revise time cards for use on different projects? That's an important question. The answer: A static time card may be all you need for payroll and for insurance reports. But with job-specific versions, you will be able to focus on collecting hours for just those items requiring new or updated productivity records. For example, do you lack a good record for building stairways and have a job coming up that includes one? Then create time cards with lines for all the steps in the stairway's construction from framing of the carriages through setting of treads and risers to installation of the balustrade. The next time you have to estimate the labor for building a set of stairs, you won't have to rely on "constructive imagination." You will have reliable numbers to use as the basis for your estimate.

All-in-one time card

Employee Name: ____________________

Project: ____________________ Date: ____________

DAY AND HOURS	Mon.	Tues.	Wed.	Thurs.	Fri.	Totals
TASK						
Project Management						
Setup						
Cleanup						
Material Pickups						
Other. Gen. Requirm'ts						
Deconstruction						
Demolition						
Site Excavation						
Tie-ins to (E) Fnd.						
Footing						
Foundation stem						
New Floor Frame						
Floor Tie-ins						
New Wall Frame						
Wall Tie-ins						
New Roof Frame						
Sheathing						
Window Install						
Window Flashing						
Ext. Finish Carp.						
Baseboard						
Int. Window Trim						
Other Int. Fin. Carp.						
Sheet Metal						
Mechanical						
Electrical						
Drywall						
Cabinets & Tops						
Flooring						
Roofing						
Paint						
Other						

TOTAL for WEEK ________

Regardless of what time card format you choose, the likelihood is that your greater challenge will be not the creation of your time cards, but getting them filled out accurately and reliably. Many builders give up on the task. One told me that he has his crew fill out time cards broken down into phases of work, but does not really know if the results are fact or fiction. As a result, he has no reliable labor productivity records and must re-imagine the labor needed for each job. The outcome, he candidly admits, is that he regularly misses the mark by some 25%.

Another builder insists, speaking not just for himself but for builders in general, that it is not possible to create accurate labor productivity records. The guys on the crew, he complains, can't "get their time cards right. They fill them out on Friday when they have to turn them in to get a paycheck." By then, of course, they have no idea how much time they have spent on different tasks.

Sal Alfano, a builder who became editor-in-chief of the *Journal of Light Construction*, has observed, "An employee who is skilled at working with wood isn't necessarily good at working with its more refined form…paper." Alfano's observation leads us to the first rule of time card maintenance: *Do not have the individual members of your crew fill out their time cards themselves*. If you do that, you will have to police the time cards of each employee separately. Success will be unlikely.

Therefore, instead of asking each member of your crew to separately fill out his or her own time card, have the crew lead fill out all time cards. And insist that he or she fill out a card for each member who has worked on the project by the close of each workday. If you are the lead, then you fill out the time cards, including one for yourself.

If it is your leads who will fill out the time cards, do not assume you can casually hand off the task to them. That won't cut it. That does not work even in big-time construction. Even there, left to their own devices, supervisors tend to focus on workflow and coordination of subs and crew. Unless the estimator demands it, they won't bear down on time cards. You, likewise, must train your leads and make sure they understand this algorithm:

- Reliable time cards = reliable labor productivity records—the records of how long it has taken guys on the crew to build each of the items and assemblies needed for foundations, frames, and finish work.

- Reliable labor productivity records = reliable estimating and realistic (good) bids.

- Good bids = the company stays in business.

- Conversely, bad time cards lead to bad labor records, which lead to bad bids, which lead to the company going under, which results in you, your leads, and the rest of the crew not having a job.

That approach has worked for me. I entrust time cards to my leads and to my leads only. Leads are the guys who can handle that "refined" form of wood—the paperwork. They read plans, they make out material orders, and they may even write change orders. They are certainly capable of accurately keeping time cards. If not, they really do not have what it takes to be leads. If they do have what it takes, then you must make it clear

Framing time card

Employee Name: ____________________

Project: ____________________ Date: ____________

	DAY AND HOURS					
	Mon.	Tues.	Wed.	Thurs.	Fri.	Totals
TASK						
Project Management	____	____	____	____	____	____
Setup	____	____	____	____	____	____
Cleanup	____	____	____	____	____	____
Material Pickups	____	____	____	____	____	____
Other. Gen. Requirm'ts	____	____	____	____	____	____
Mudsill	____	____	____	____	____	____
Pony Wall	____	____	____	____	____	____
Posts	____	____	____	____	____	____
Girders	____	____	____	____	____	____
First Floor Joists	____	____	____	____	____	____
First Floor Subfloor	____	____	____	____	____	____
First Floor Wall Frame	____	____	____	____	____	____
First Fl'r Wall Sheath	____	____	____	____	____	____
Stairway Landing	____	____	____	____	____	____
Stair Carriages	____	____	____	____	____	____
Stairway Blocking	____	____	____	____	____	____
Second Floor Joists	____	____	____	____	____	____
Second Floor Subfloor	____	____	____	____	____	____
Second Fl'r Wall Fr.	____	____	____	____	____	____
Secon Fl'r Wall Sheath	____	____	____	____	____	____
Roof Ridge	____	____	____	____	____	____
Roof Rafters	____	____	____	____	____	____
Dormers	____	____	____	____	____	____
Ceiling Joists	____	____	____	____	____	____
Roof Sheathing	____	____	____	____	____	____
Rear Deck Ledger	____	____	____	____	____	____
Rear Deck Posts	____	____	____	____	____	____
Rear Deck Beam	____	____	____	____	____	____
Front Porch Frame	____	____	____	____	____	____
Other ( )	____	____	____	____	____	____
TOTAL for WEEK						______

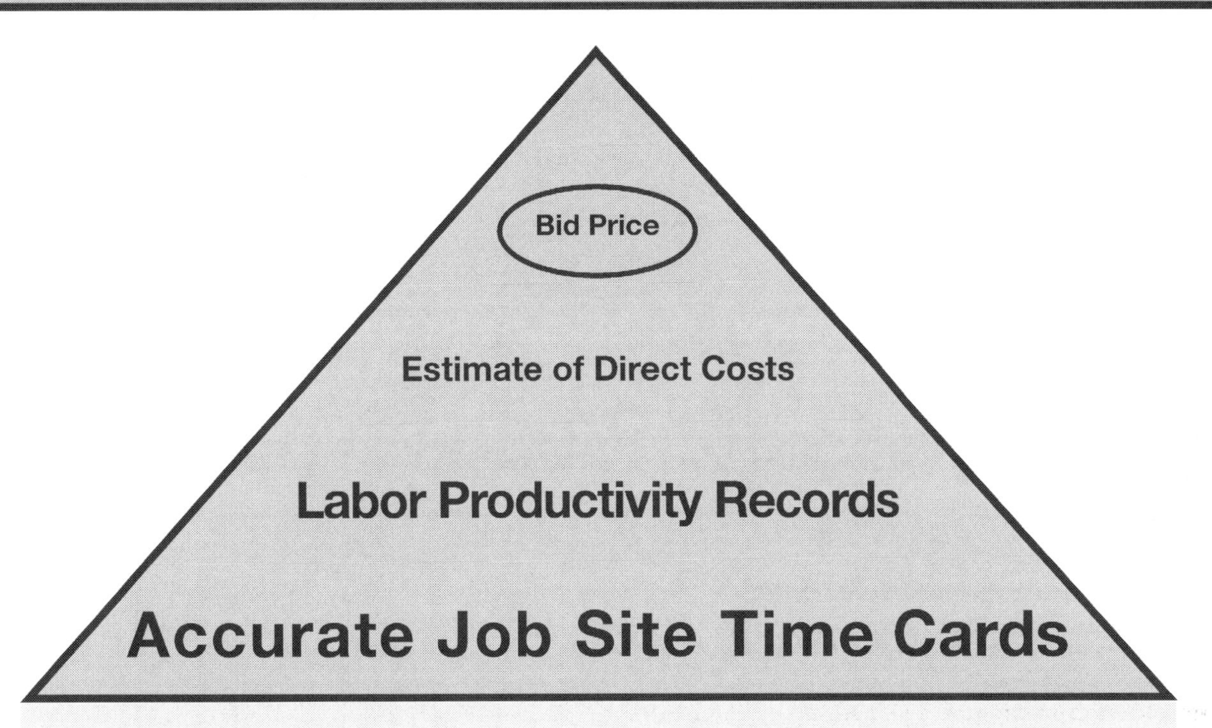

A construction company cannot survive if it cannot produce reliable bids.

It cannot produce reliable bids if it cannot produce accurate estimates of direct costs.

It cannot produce those accurate estimates without reliable records of its crew's productivity. And it cannot create such records if it cannot produce reliable time cards.

In short, accurate job-site time cards are a foundation of a construction company's long-term financial viability.

to them that paying attention to timecards and getting them right is a requirement for continuing employment.

I have been fortunate to work with outstanding leads. I drive home to my leads the importance of time cards to the survival and prosperity of our company and, therefore, to the security of their employment and the viability of our profit sharing program. Even so, I have not gotten perfect performance with time cards. Now and then I have collected a set of time cards that looked fishy, that showed hours for a task that were way out of line with previous records for that task. Here's one horrible example: A set of time cards that showed six hours per foot for kitchen cabinet installation, which was roughly three times our usual rate.

When a discrepancy like that turned up, I would point it out to the lead responsible for it. Always he would assure me that he had recorded the hours for the work meticulously, filling out time cards at the end of every workday, never putting off the irksome task till the end of the week. I would try to find ways to indicate my disbelief without being insulting. Though I never did get an admission from a lead that he had slacked off on the time cards, my insistence on their importance ultimately produced good results. The great majority came in with believable numbers. Over time, these have served as solid building blocks for records of the hours my company has needed to build a wide range of items and assemblies. Those records have, in turn, enabled the production of the accurate estimates that have been the foundation of my company's prosperity. My experience has led me to believe that reliable labor productivity records—the topic we will focus on in the next chapter—are the single most important ingredient of good estimates and, therefore, are critical to a construction company's long-term viability.

CHAPTER 11

Move Up to Labor Productivity Records

From time cards to job costing

Rather than compiling item-by-item records of labor productivity such as the one I have been encouraging, many builders opt for so called "job cost records" as a source of estimating data. I think that is a mistake. But it is certainly better than relying on intuition or wild-assed guesses (WAGs) to figure labor for a job. And it is also true that the use of job cost records as sources of estimating data is widespread. So it's necessary to take a look at both the value and limitation of job costing.

COMING UP

- Job costing
- The use of job cost records for estimating
- The limitations of job cost records for estimating
- Moving up to historical labor productivity records
- The resistance to compiling labor productivity records
- Breaking through
- Rewards

Job costing is simply tracking actual costs of construction for a job and comparing them to estimated costs, typically by division of work. Builders initially turn to job costing as a way of monitoring financial performance on projects during the course of construction. That is the case for Gordon Reed, a young builder who successfully started a one-person design/build company specializing in residential remodels. Each day Gordon records the hours he spends at various divisions of work. Periodically he totals the hours, compares them to his estimated hours, and evaluates his performance. For example, if he sees that he has spent 56 of his estimated 74 hours for framing an addition, he thinks to himself, "Good," because he is confident that he can complete the framing in the remaining hours.

As their companies grow, builders often step up to more sophisticated job costing, typically making use of accounting software to create regularly updated records along the lines of, though more elaborate than, the simplified report illustrated on the following page. From such records builders can see whether their estimates are proving accurate or whether they are experiencing cost overruns on a project. Once alerted to overruns, they

Work	Estimat Hrs	Estimat $$$	Hrs/$$$ to date	Remain Hrs/$$$
Site Clear	24	1400	22/1200	2/200
Drainage	34	2600	36/2900	-2/-300
Foundation	78	5800	62/3100	16/2700
Frame 1st Flr	62			
Frame 2nd Flr	58			
Frame Roof	42			
Ext. Drs&Winc	18			
Ext Finish	56			
Moisture Prot.	28			
Roofing	Sub			
Plumbing	Sub			

The purpose of job costing is to track actual hours and costs on a project against estimated hours and costs. This simplified report performs those basic functions using only five columns. Heavy-duty software job costing programs produce more elaborate reports with more columns.

may be able to figure out why they are occurring and fix the problem.

Unfortunately, however, builders often instinctively turn to their job cost records for more than project tracking purposes. They draw on them for estimating data. Why? Because it is the only data they have other than cost books and hearsay, which they have realized are really unreliable sources. Thus, Gordon Reed now goes to his job cost records to figure labor for new projects: How long, he asks, did the foundation take at the Dorado job? Hmmm, looks like I spent 57 hours on it. This new job has 12 feet less of foundation, so I will figure on 46 hours. That feels about right.

The problem here, fairly apparent even at Gordon Reed's level, is that though they may be useful for project tracking, job cost records are crude and inadequate sources of data for estimating costs on future projects. The lack of precision, as well as the inefficiency of this method, is even more apparent in well-established larger companies, such as a respected forty-year-old company I will call David Bass Construction. A Bass estimator begins an estimate by listing all the major phases of work in the job. Then he creates an initial job schedule, or timeline—a first rough guess, based on intuition and experience, of how long the project will take. Next, he browses through job cost records looking for similar past jobs and noting how close (or how far off) Bass's estimate was on those jobs. Using that information, he attempts to adjust his intuitive timeline and the resulting cost projections for the new job.

David Bass describes his company's process as "iterative." What we do, he says, "is go back and forth from schedule to estimating history, to schedule, to history, adjusting as we go." Often, Bass adds, an estimator discovers that estimates for previous jobs were woefully low, so in an estimate for a new project, he bumps the hours upward.

Bass defends his system. "It's primitive," he says, "but pretty effective." His system is similar to the methods used by other builders I know. Some, like Bass, are happy with the approximations to actual costs their process produces. One says, "If you have experience, you can rely on intuition." But, she adds, she goes to her job cost records to check on her intuitive estimates. "If intuition and numbers match up," she says, "you're okay. If not, you keep working till you feel comfortable with the estimate."

Limitations of job cost records

The iterative, intuitive, approximating, back-and-forth process that David Bass and other builders embrace may be to your liking. There are, however, serious downsides to relying on job cost records for estimating data. In the first place, job cost records are a poor source of labor costs for new jobs that are substantially different in scope or type than those you have done before. You may, however, find that you want to go after different types of jobs as you move further into your career. You will want to go after them when you are ready to expand beyond a familiar niche. You may have no choice but to pursue them to survive when work becomes scarce during a recession.

One huge problem with using job cost records of past jobs for very different projects is this: Job cost records typically break down costs and labor hours into large divisions, such as foundations, framing, or finish carpentry. Such gross records are just too broad to be of much help when you are trying to dial in your costs for new jobs of a different type or level of complexity. For example, one builder still in the early years of his career and working on his job sites with one helper, had successfully completed a series of economy bathrooms. Then, he had a chance to bid on his first luxury bathroom. From his earlier economy bathrooms, he had cost records showing total time spent for each project. He attempted to draw on those records to estimate labor for the new project. The economy bathrooms had taken five weeks. Noting that the luxury bathroom was about twice the size of

Job cost record limitations for estimating purposes

- Unreliable for estimating jobs of markedly different type and scope
- Useless for estimating material costs
- Useless for subcontractor charges
- Lack descriptions of the conditions at the job
- Lack descriptions of the crew who did the job
- Do not provide costs for discrete items and assemblies but only for large divisions of work
- Do not provide an estimator with efficient access to usefully close costs but instead require the estimator to rummage around in search of vaguely similar costs

the economy jobs, he decided to allow for 10 weeks of labor by himself and his helper.

The builder felt very unsure about his estimate. He should have been, for the estimate for the luxury bath was based on a crude comparison with a very different kind of work. The luxury bathroom would involve different scope, specifications, detailing, and, most likely, client expectations, at virtually every step from general conditions through finishes.

To nail down labor for the luxury bathroom, the builder needed to go beyond records of total hours for past jobs and break down the new luxury bath project into items and assemblies—removal of fixtures, stripping walls to the studs, removal of floor covering, adjustments to existing wall frame, new wall frame, leveling of ceiling frame, and so on, all the way through to drywall and installation of granite and plumbing fixtures.

Next, to estimate the cost of those items correctly, he needed records of the labor it took him to build similar, even if simpler and less complex, components in the past. For example, a record of the labor hours per foot that it took to frame the partitions and closet in an economy bath would have given him a good basis for estimating wall framing labor in the luxury bath. If the framing in an economy bath required an hour per foot, he could have allowed an hour and a half per foot in the luxury bath, due to two factors: The luxury bath imposed a requirement for greater precision because of its modernist detailing. And it was on the third floor of a large home, not the first floor of a small one, as had been the case for his economy baths, meaning that mobilization and access would chew up a lot more time. (Of course, during the luxury bath job he would have wanted to keep accurate time cards for framing and all other items and create labor productivity records from them. Then, when he bid his next luxury bath he would be able to quickly and closely dial in his likely labor costs).

The same principle—namely the inadequacy of job cost records and the need for item-by-item labor productivity records—applies to larger jobs. That became clear when builders whose experience had been with remodeling bid to replace houses destroyed by the firestorm that struck Oakland, California in 1989. They learned that their overall costs for installing foundations for additions did not translate to the costs of constructing pier and grade beams foundations with high retaining walls on steep hillsides. To achieve accuracy, they needed to break the new work down into items and to then estimate from records for similar past work—so many hours per foot to excavate footings of a certain size, so many hours to form them, so many hours to install rebar, and so on. But their job cost records did not provide them with such discrete item-by-item data. As a result, their cost estimates were often seriously off.

Perhaps worst of all, job cost records do not report conditions at the job site. And they do not report the experience and efficiency of the crew that did the work. Both factors are crucial in determining productivity. As we shall see, both factors are moved front and center in labor productivity records.

If job cost records are generally too broad in scope and too sketchy to be useful for accurately estimating labor costs, they are utterly useless for estimating material costs. Material costs fluctuate wildly. The costs of lumber and hardware for a project built a year ago, or even just a few months back, may be much higher or lower than those that your suppliers will offer for a new job. The same goes for work to be subcontracted. Subcontractors regularly change their labor rates and their profit markups as market conditions permit or require. The price you get from your plumber during a recession will likely be very different from the one he will give you as work picks up.

Even successful builders who have long made do with job cost records for estimating purposes are sometimes unhappy with them. One, who was looking to improve his estimating, told me that he had initially started keeping job cost records for billing purposes and then assumed they could do double duty as sources of estimating data. That, he discovered, had been unduly optimistic. He and his estimators were forced to "rummage around" in the records of past jobs trying to come up with comparable work that might give them some idea of how much labor would be required for a new project. "The data is cumbersome," he says. "It's not useful."

In my discussions of estimating from job cost records, the word "rummage" has come up repeatedly. David Bass insisted to me that his process was not mere rummaging around, though at a glance it might look like it. Another builder commented that if you rely on job cost records for estimating, you "have to rummage through a heap" of records trying to find comparables—say a two-story addition that you did quite a few years back and that is similar to the new one you are now bidding. You get caught up, he lamented, in digging through the past jobs, analyzing, then extrapolating, then analyzing some more. You can, he says, really "go down the rabbit hole."

Moving to historical labor productivity records

If a "primitive iterative" process and "rummaging" aren't up to snuff as an estimating methodology, what is? If not job cost records, what is it that you do need for the labor side of your estimate? (Material and subs we will get to later.) My answer is, you need records that:

- Provide labor hours in an item-by-item and assembly-by-assembly format.
- Are applicable to differently sized units of work from small items like anchor bolts to whole assemblies, such as all work from excavation through stripping and backfill for a two-story foundation.
- Present the hours of work it has taken you and/or your crew to install a range of items and assemblies.
- Clearly and concisely record the quality of the crew and the conditions they encountered at their job site.
- Can be assembled with reasonable, if persistent, effort.
- Are organized, like your checklist/spreadsheet, in the natural order of construction, so that you will be able to proceed through your spreadsheet and records in tandem.

Characteristics of good labor productivity records

- Logically organized in parallel with natural order of construction
- Easily accessible
- Cover differently sized units of work
- Derived from your experience on your projects
- Created with modest investment of time

Such records are formally called "historical labor productivity records," or just "labor productivity records." One builder defines them simply as records that "tell you how much time it has taken someone to do a specific unit of work." What you are doing when you compile such records, he adds, "is creating a set of usable numbers for the future" so that "estimates are a lot easier."

Say, for example, you are estimating the cost of framing walls for a new project. You open, whether in a computer folder or a three-ring binder, to your wall framing records, perhaps displayed on a form very much like the blank form shown in Chapter Nine, or perhaps on an improved version that you have created for yourself. Looking at your records for wall framing, you see one—like that in illustration on the following page (though for framing, not lead time)—for a similar project built under similar conditions. There, you note, your crew needed 1.25 hours per foot to frame the walls. For the new project, since job site

Labor Productivity Record

ITEM OR ASSEMBLY: Lead Time
Included: Reviewing plans. Communicating with client, designer, & inspectors, and myself; ordering material. Managing subcontractors.
NIC: Layout of work for crew (that time is included in hours for items or assemblies installed).

QUANTITY & DEGREE OF DIFFICULTY of ITEM: Fifteen week project. Straightforward work.

PROJECT: Basement Conversion to Family Room. Moraga Av., Lafayette.
TIME OF CONSTRUCTION: Spring
CLIENT: Barnfield. Fairminded. Appreciative of crew.
DESIGNER: Homer. Sketchy plans but prompt responses to questions.

ACCESS: Easy.
WEATHER: Good mostly. A few showers.
OTHER FACTORS: Pleasant place to work. Crew morale high.

CREW CAPABILITY:
Lead: Daniel M. Third year as lead. Capable.
Journeymen: NA
Apprentice: NA
Laborer: NA

HOURS PER CREW PERSON:
Lead: 120.5 hours
Journeymen: NA
Apprentice: NA
Laborer: NA
TOTAL HOURS: 120.5

HOURS PER UNIT: 2 hrs./day

conditions are going to be about the same, and the crew's experience and skills similar, you decide to stick with 1.25 hours per foot for the wall framing.

Alternately, you might consider two records, one for 1.25 hours per foot, another for 1.5. You note that access is a little more difficult for the new project than it was at the project where you needed 1.25 hours per foot, But the complexity of the new project is a notch below that of the 1.5 hours per foot project. You decide on a figure between those for the two jobs, settling on 1.375 hours per foot.

Note that with this process, there is no rummaging around. You are able to go right to your records for wall framing because they are organized—along with records for other items—in the natural order of construction. You spot the optimal record or records for the item of work at hand. Next, you multiply the number in that record by the quantity of work you are estimating, which gives you the total labor hours—for example, 1.25 hours per foot for wall framing × 200' of wall = 250 hours for wall framing. To get to an actual dollar cost for the wall framing, you simply multiply the total hours by your labor rate to determine your labor cost for framing walls in the new project: 250 hours × $50/hr.= $12,500. Done! You are ready to move on to the next item or assembly in your spreadsheet.

Every record of labor hours has a story behind it – a place, circumstances, and group of characters including the craftsmen, the designers, and the clients involved with the project.

If you read the books by the estimating experts who serve builders from the world of large-scale construction, you will see "historical labor productivity records" or similar terms used regularly. For example, Wayne DelPico writes: "The best measure of a crew's productivity is its own itemized historical cost data... Records for labor hours to complete a *specific task* offer the best guidelines for predicting future performance." The Associated General Contractors (AGC) insists, "Over a period of time an estimator should accumulate optimum production rates for *items* the firm repeatedly executes."

Rummaging around in and estimating from job cost records after the fashion of light-frame builders is never mentioned. Of course not! Can you imagine figuring the cost of framing or any other division of work for a 370-unit apartment complex by extrapolating from gross costs for the last complex you built? Would you say to yourself, "Framing on that last one ran 1.4 million. This one's a little smaller, so it might be about 1.3 million. Hmmm, that feels a little high. Let's go with 1.2 million."

Heck no! If you missed on the low side by just 8%, you'd be out around $100,000. You would want to take off the quantity of work in the new project by specific items and/or assemblies—and then determine your labor hours from productivity records for identical (or at least closely similar) items. Example: You are bidding a new job, a straightforward moderate income apartment complex with 29,600' of 8' wall frame At the last such box you built, labor ran .85 hours per foot for walls in a very similar configuration—i.e., same height, same frequency of corners and door and window openings. Therefore you estimate the new job will require 25,075 hours for wall framing and cost you $1,258,000. Here's the math: 29,600 × .85 = 25,160 × $50 = $1.26 million (rounded up $2K for completion items).

In following pages I will move beyond the blank form you saw in Chapter Nine and the descriptions I have provided here and go into the detail of my own labor productivity record-keeping system. Frankly, I think that my system, though simple, is powerful. It meets all the requirements for job cost records described above, and then some. It has served me well for many years. Perhaps it will serve your needs "as is," or better yet, in a modified form tailored to your own way of operating.

Narrative labor productivity records

Because every record has a story behind it—a place, circumstances, and a cast of characters including the craftsmen, the designers, and the clients involved with the project—I call my records "narrative labor productivity records." I want my records to narrate the full story, because the cost of work at different projects, which might appear at first glance to be nearly identical, can vary several hundred percent depending upon

At this job site, a client who enjoyed telling tales of his adventures running huge construction projects in the Mideast, a playful dog, and an attractive young woman strolling by on her way to the swimming pool in the backyard took turns distracting the crew. As you can see, the crew is nowhere in sight.

their story lines—the where, when, who, and why of the work.

To be useful, narrative records—I will be offering many examples in the next chapter—must tell their story concisely. You don't want them to run more than one page. You want the information on the page to be sparse and uncluttered so that you can read it easily and decide efficiently which record is the best one to use for estimating the cost of a particular item or assembly in a new project.

The "item or assembly" description for each record must be exact as well as concise. Thus, the record you will find on page 220 for wall framing states that the lumber is KD (kiln-dried) because KD lumber tends to be straighter and to require less culling and fussing. A record also makes clear which items are included in the assembly and which are not included. Thus, in the illustrated record, sheathing is called out as not included (NIC). If the record were for a wall assembly that did include sheathing, it would specify the type of sheathing—say Struc 1 plywood vs. OSB (Oriented Strand Board). Because OSB is much heavier, it tires crews out more rapidly over the course of a day, and therefore lessens productivity.

A separate line provides a place to record the quantity of the item or assembly. That line can also be used to note the degree of difficulty of installation. For example, for a record of cabinet installation, you might wish to note the number of cases installed and their configuration. Why? Because a straight 24 foot run consisting of six cases will require substantially fewer labor hours than an L-shaped run of the same length incorporating twelve cases.

As the illustrated records show, the project description, time of construction, client, and designer are then recorded on the third through sixth lines. That order is important. It indicates that item and quantity are the key concerns and underscores the difference between a labor productivity record and a job cost record. When you access labor productivity records, you are looking up productivity by a specific item or assembly rather than rummaging around for a broadly similar job.

Nevertheless, project description does matter because project type relates to productivity on items. Framing walls for a second-story addition will take more time than framing the same quantity of walls for a first-story addition. Framing for remodels typically requires substantially more hours of labor per unit of production than framing for new construction. Small quantities of work will typically use far more labor per unit than large quantities. For example, installation of three sheets of sheathing at a remodeling tie-in could take three hours including layout and setup. Installation of large quantities of sheathing for a new four-unit apartment building, on the other hand, can be done at a rate of three sheets or more in a single hour. In other words, the remodeling rate for sheathing installation is a sheet per hour, while the rate for new work could reach three or more sheets per hour.

Sunny, mild weather enhances productivity. Other weather conditions may retard it. At this project in Maine, siding installation began each morning with the scraping away of ice that encrusted the scaffolding.

Clients likewise can have substantial impact on productivity. That was vividly demonstrated to me during construction of a residential addition for a client who was an executive of an international construction corporation. When home from his then-current project, rebuilding Kuwait after the first US-Iraq war, he liked to hang out with my crew and me and regale us with stories of his negotiations with the Prince of Kuwait about the fittings for the Prince's personal washroom at his personal airport terminal. (My client was against solid gold faucets because he feared hot water might cause the gold to melt; the Prince was for solid gold. They compromised on gold-plated faucets.)

When the owner was in Kuwait, his dog took his turn at distracting my crew by carrying away their tools in hopes of a game of fetch. And when the dog was not around, the client's beautiful daughter would stroll slowly out from the house and along the edge of the work site, bestowing smiles on her favorite carpenter en route to the backyard pool. All delightful stuff, but really bad for productivity.

Designs and designers likewise are part of the story behind a given productivity rate. As I have stressed before, highly refined, "minimalist" or "modernist" designs require exacting layout and installation that greatly increase labor per unit of production. Some designers respond promptly to requests for information that your crew must have in order to keep working steadily. Others dawdle, or agonize over simple issues. As the AGC observes dryly, "Some A/Es [architectural and engineering firms] have a history of poor administration, making it difficult to obtain timely responses or clarifications." But others are responsive, and their consideration greatly facilitates productivity.

Access gets a line of its own in the narrative record because it is often the single most impactful element of the story. Productivity will vary greatly depending on whether the work is up high or close to the ground, whether it is near the staging and delivery area or not, and whether it must be approached through congested existing work or is easy to get to. Right behind access is weather. As Sal Alfano commented in the *Journal of Light Construction*, "Only sailors and airplane pilots are more interested in weather than…contractors." Building in California's benign climate is one thing. Fumbling around on iced-up scaffolding in a Maine winter is another.

Beyond access and weather, many additional factors can influence productivity. They include introduction of new tools that may immediately increase productivity or initially slow it as a crew learns how to use the tool. Likewise, unfamiliar products, especially structural and finish hardware, may have unexpected consequences for productivity. Some years back, I contracted to extend a deck as part of building a larger project. The specifications provided for the new deck boards to be fastened down with bronze boat nails that matched those used for the original decking. Fortunately, I tested the nails during estimating and discovered that if driven with the usual hammer stroke, they would deform like warm wax candles and that, to avoid this, my crew would have to predrill for each nail and tap it gently into place. In my estimate, I therefore allowed for much lower productivity than I would have with standard fasteners.

Compiling your own set of records that reflect productivity in your own company may well be the best investment of time you will ever make as a builder.

The most important characters in any labor productivity story are the crew, and the single most important character is the crew lead. He is the protagonist, the person who goes toe to toe with the dark forces of inertia and makes things happen…or doesn't. I once watched the reconstruction of a large house and adjacent cottage move sluggishly forward over the course of three years before reaching completion. The crew was competent enough, but the lead had weak building and organizational skills and was unable to deploy his workers efficiently—even at simple tasks. Had one of my leads run the job with the same crew, they would have done the job in 18 months at the outside; and with his own skilled crew, my records suggest, in 12 months or less.

Productivity can vary greatly even between leads with equivalent levels of experience. The estimator for a company that builds major private and public projects says that his labor productivity records indicate his top leads get work done in 20% less time than other leads even though they, too, are highly experienced. The top leads, I'd venture to guess, are just better organized and better at organizing and deploying their crew.

Any lead, whatever his capacity for efficient organization, will benefit from having a stable crew whose members are in sync with each other. If new people are regularly being integrated into the crew, it will lose efficiency. For that reason, a good narrative labor productivity record describes both crew capability and the length of time the crew has been together.

At its bottom line, the labor productivity record should state costs in hours, not in dollars. That's critical. Because of inflation, labor records that give costs in dollars are rapidly outdated. Over the course of just half a dozen years, per hour labor costs can increase by a third or more. Thus the dollar cost of a given item can dramatically rise even if the productivity for the item remains steady.

Perhaps you are now sold on the benefits of narrative labor productivity records and even see possibilities in the format I have presented. Nevertheless, you may resist compiling records. That has been true for at least some of my readers. "I don't have time," says one. "After I get off the job, I have to go home and be Dad," another responds. "They would take a lot of persistence and discipline," a third says with a grimace.

He is right, of course. Building a file of narrative labor productivity records does take persistence. So do all the other responsibilities of running a successful construction company. So does rummaging through job cost records in search of numbers for new estimates, coming up with only uneasy approximations, and then diving down the "rabbit hole" in search of something better. In my experience, labor productivity records can be compiled with modest—if steady—effort. Over the course of years of estimating, the narrative records will save you far more time than it took to create them, even as they enable you to really nail your numbers.

Compiling your own set of records that reflect productivity in your own company may well be the best investment of time you will ever make as a builder. To help you on your way I have created a sampler of basic records. They are presented in the next chapter.

A Note to Readers:

You have just completed the first of a pair of chapters that are, to my way of thinking, the most valuable in the book.

As I have stressed, historical labor productivity records, while recognized as essential in the world of larger volume commercial and industrial construction, are almost unheard of in the world of light frame general contractors -- my world and likely yours.

I hope that I have managed to provide a clear description of the benefits of productivity records. If you are at all confused about them, please go over the chapter again, and in particular review the sample productivity record on page 185.

With that done, you should be ready to move on to the next chapter. It includes 18 more sample productivity records, each one accompanied by a discussion of the record.

My hope is that by the time you are finished with both chapters you will be determined to create historical labor productivity records for your own company. I believe that more than any other single practice suggested in this book, developing such records will enhance the satisfaction, both emotional and financial (for the two are linked), that you gain from your work as builder.

CHAPTER 12

A Sampler of Labor Productivity Records

Creating your records

In the previous few chapters we've focused on two essential building blocks of reliable estimating—namely, accurate time cards and the labor productivity records developed from those time cards. As we discussed in the last section of Chapter Eleven, labor productivity records can and should include much more information than just the hours needed to install an item or assembly. That's one of their great advantages over job cost records as a source of estimating data.

For all the richness of information they provide, actually producing labor productivity records is quite straightforward. So, you might be wondering, when should I start creating my labor productivity records? The only answer I can think of is "Now."

Fernando Pagés points out that if you are still working in someone else's company but planning to go on your own, "now" still applies. Make note of the hours it takes you and the crew you are working with to install various items or assemblies. Create a few labor productivity records at the end of each workweek. Soon you will have a basic file that you can draw on to reliably estimate labor hours for your own first projects. What a huge advantage that will be to you when you when you are just starting up your own company!

If you are already running your own company, well, as the old saying goes, "Today is the first day of the rest of your life." So use part of it to set up a system for creating and organizing your narrative labor productivity records. And do that even if you are still wearing your bags and dividing your time between producing work at your job sites and handling sales calls and office work. Do it even if you are the only person in your company and do not yet

COMING UP

- Creating your records
- The stopwatch option
- Organizing your records
- Using your records
- Constructing unknown productivity rates from known rates
- Tips from veterans
- A sampler of labor productivity records
- A summary of principles and tactics

have any employees. The records you create of your own labor productivity will be valuable to you when you do begin hiring workers. You can compare their productivity to your own. If they are much slower, maybe you will need to let them go; if much faster and still producing good work, then they may be people you want to try and keep in your company.

To begin compiling a set of records, every time you build a project determine which items and assemblies you want to create records for. Focus on creating records for a limited number of items or assemblies, say, four to six from a single project. In other words, don't bite off more than you can comfortably chew. At just half a dozen records per job, you will quickly build up a useful starter file.

When you are just beginning to create records, focus on the *big stuff*, the items and assemblies that consume 90% of your crew's time. If your crew does only carpentry work, you will create records for items such as footings, floor and wall framing, setting trusses, sheathing, hanging doors, casing and baseboard, and cabinet installation. If the crew does major items of work typically performed by subcontractors such as installing electrical circuits, drywall, and plumbing fixtures then create records for that work also.

When a project is complete, open up your time card folder and crank up your adding machine. Then total the hours spent on each of the tasks you want to create productivity records for. Once you have totaled the hours for a task, retrieve the relevant quantity of work from your takeoff sheets for the project. Then figure the labor per item of work and place the figure in your productivity record. For example, if the takeoff sheets show 78 sheets of subflooring in the job, and the time cards tell you that your crew spent 28 hours installing it, then their productivity rate was .35 hours per sheet of subflooring.

Once you have figured the productivity rate, you can record it on a labor productivity form such as the one I use. (There's a blank copy shown in Chapter Nine, a filled-out copy displayed toward the end of Chapter Eleven, and eighteen more completed forms displayed a little later in this chapter.) Then provide

Creating narrative labor productivity records

- Aim to create only an achievable number of records from each project.
- Tailor your time cards so that they clearly report hours for the desired records.
- When you begin creating records, focus on the big stuff, the major items and assemblies you build over and over.
- Later move to less common items.
- To create records, gather up time cards and total hours for each item you want to create a productivity record for.
- Transfer the hours to a labor productivity form. Figure your productivity rate per unit of the item and record it on the form.
- Concisely fill out the lines that provide the story behind the numbers

Once you have the hang of it, you will find you can create half a dozen productivity records from time cards in a few hours. After just a few jobs, you will have compiled a basic but powerful set of records.

The stopwatch option

Only rarely have I run into a small-volume light-frame builder who compiled the sort of task-specific, item-by-item labor productivity records insisted on in the world of big-time construction (and in this book). Among the few is Fernando Pagés, a homebuilder as well as the author of *Building the Affordable House* and the *Fine Homebuilding* business blog.

Fernando worked to get his time cards right and to build labor productivity records from them. But he found it a struggle, and he wanted more precise numbers than he was getting via the cards. He took to going out to his job sites with a stopwatch in hand and recording the exact time it took his crews to execute various tasks. For example, to get reliable numbers for setup time, he would show up at a project before work started and time his guys from the moment they arrived to the moment they began pounding nails.

Fernando emphasizes that "you need time-and-motion studies that are for small units of work." You can't just record framing costs in their entirety, as job cost records often will, because "that's too big an assembly."

Fernando also understands that a good productivity record gives you more than a number. It gives you the story of "who's doing the work, the material they are installing, the exact scope, and the conditions."

"It takes a weird personality," he suggests, "to want to do this kind of lab work and testing, like a scientist or engineer." Well, maybe. Personally, I think it takes a builder with the determination to nail his estimating numbers.

the remaining information called for in the form—lead, crew, clients, conditions, and more—from your knowledge of the project or by checking with the project lead.

Once you have a minimal set of records for the big stuff, you can expand it as you do new jobs, creating multiple records for basic items like framing 8' walls and installing cabinets, and also creating records for unusual items. If necessary, tailor your time cards so that your lead knows to focus on recording hours for those unusual items. Do you want a record for framing an eyebrow dormer? Make sure that your time card includes a line for the dormer in addition to the more standard work, and ask your lead to record the hours for the dormer with special care.

Organizing and using your records

As you create your productivity records you will want to organize them so that you can find those you need when you need them. You may choose to organize your records either in a binder or in virtual folders in a computer. My choice has been to stick with paper records organized in a binder. I create my productivity form with my word processing software, and then either I fill out copies of the form on the computer and print them out, or else I print out a batch of blank forms and fill them in with a pencil. If you don't yet have word processing skills, you can write out a form, make copies at your local copy service, and fill them in by hand. That's exactly what I did when I started keeping my narrative labor productivity records. And those records are useful to this day.

One computer-savvy younger builder, to whom I mentioned the binder option, shuddered at the idea. "All that paper," he said with a scowl. He could not imagine paging through it, and was appalled that I would even raise the possibility of using a paper system two decades into the twenty-first century. However, at the risk of being dismissed as a dinosaur, I want to tell you that I see advantages to keeping productivity records printed on paper in a three-ring binder.

Properly organized, the paper system can be used with great efficiency. In the binder, using tabs to divide the records into divisions and items or assemblies, you organize your records in the same sequence—the natural order of construction—as you organize your estimating spreadsheet. As you work through each division of your spreadsheet, you page in parallel through the binder.

Say, for example, that you have just completed estimating labor for the footings and stem walls of a new project and are proceeding to the garage slab. Since you are already in the concrete section of your binder, all you need do is turn a few more pages, and there are your records for slabs. Select the one with conditions most like those for the new slab you are estimating. If need be, adjust its productivity rate to reflect differing conditions that affect the new work. Enter the number into your spreadsheet. Then move on, cruising down your spreadsheet as you cruise through the binder.

Here's my estimating system: A desktop computer open to an Excel spreadsheet and a tabbed binder of printouts of my labor productivity records.

I've been kidded about my system: "Paper! Really. In the twenty-first century. Come on, Dave."

However, it is an efficient system. Both the spread sheet and the binder are organized in the natural order of construction. As I work my way down the rows in my spreadsheet, I am paging in parallel through the binder and rapidly locating the productivity records I need. I think that this is likely to be more efficient than a fully computerized system.

I may be wrong. But at the very least, mine is a very good start-up system. It's inexpensive to create and inexpensive to operate—no $400-an-hour software consultants needed.

If you prefer to stay away from paper, you can store your productivity records on your computer. Create a master folder and name it "Labor Productivity Records." Inside the master folder set up subfolders, one for each major division of your work—General Requirements, Site Work, Concrete, Framing, and so on. As your collection of records grows, sort the records into a second level of subfolders. In the General Requirements subfolder, for example, set up sub-subfolders for Lead Time, Cleanup, and so on. For Framing, set up sub-sub folders for Floor Frames, Wall Frames, Sheathing, etc.

As you can see, this organization, too, follows the logical flow of building, so it should be easy for you to locate records as you need them. When you are estimating, you will keep the Labor Productivity Records folder open on your screen. When you come to Framing in your estimating spreadsheet, you open the Framing subfolder. And when you arrive at wall framing, you open the Wall Frames folder and look over your records for wall framing to find the one most useful for estimating the wall framing costs for your new job. But I must admit to you that I have never actually made use of digital folders for labor records. I stick with my paper system. It allows me to very quickly find records—more quickly, I am fairly certain, than would clicking through computer documents. Additionally, it allows me to take my eyes off the computer screen and let them rest a bit as I read from paper. I guess my preference for paper here is a bit like the preference some folks have for old LPs over digital recordings. I think the output is better and more pleasing to the senses.

If you have no such records, that means the new work is too far beyond your experience for you to safely bid it. You should walk away.

Whichever technology, binder or virtual computer folders, you choose for organizing your records, you will have to keep your records current with regard to changes in building technology. Building materials are evolving rapidly. While building a house a few years back, I glanced around the job site as framing was being completed and was struck by the fact that none of the lumber products used, excepting the plywood sheathing installed atop the roof trusses, even existed when I began my carpentry apprenticeship—not OSB, not TJI's, not pre-manufactured roof trusses, not KD FSC studs and plates, or any of the structural hardware other than joist hangers. Because of the evolution in material technology, your labor productivity records must evolve also. Records must be retired and others created to take their place or at least supplement them. For example, records for framing floors with solid lumber must be supplemented with records for TJI's. Records for framing roofs a stick at a time must be supplemented with records for installing roof trusses.

Peter Sanchez, a plumbing contractor to whom I described the benefits I have derived from my binder of narrative labor productivity records, mused, "If I had done that starting out as a young plumber, I'd be retired by now." Peter also raised an important question. Would records help him estimate labor for work he'd never done before, such as installing 12 gas risers on a 12-story building? What kind of records could help there?

I have thought about that question with respect to builders in general. Here's my answer: If you have built up a comprehensive file of labor productivity records, you can probably find one to serve as a basis for constructing a labor estimate for work that is substantially different from or greater in scope than anything you have done previously. If you have no such records, the new work is too far beyond your experience for you to safely bid it. You should decline to submit a bid. If it is within reach of your experience, if you do have relevant productivity records, then you can go for it. In Sanchez's case, a record for installing four gas risers in a four-story building could have provided the needed starting point. In the next section, I will present a couple of examples of how production rates for unfamiliar work can be constructed from records of similar and completed work.

A sampler of labor productivity records

To begin with: A word of caution. *Do not use the records presented on the following pages to estimate labor productivity rates for your jobs.* Adapted from my records, they are no more likely to suit your work than are the averaged numbers you find in cost catalogues and databases. My purpose in presenting the 18 records is not to provide you with numbers, but to illustrate an approach to creating labor records that I think offers great benefits. To use words suggested by my able editor, Jackie Parente, my purpose is to give "you a clear idea of the type of information you will be capturing and the level of detail that is helpful."

Please note that with one exception (#18) all the records here are from a period of my career when I was employing several crews. Unfortunately, I was not acquainted with the idea of labor productivity records during the startup period when I was still wearing my bags even while running my company. If I had been, I would be able to show you records with myself listed as project lead or carpenter (or both) and the production rates I achieved. Though I missed out on the opportunity to create such records, if you are still in your startup phase, wearing your bags and producing work at the job site, you will want to capitalize on that opportunity.

CAUTION: Please make only careful and selective use of the records presented on the following pages to estimate labor productivity rates for your jobs. Adapted from my records, they are no more likely to match up closely with your productivity than are the averaged numbers you find in cost catalogues and databases. My purpose in presenting the 18 records is not to provide you with numbers, but to illustrate an approach to creating labor records that I think offers great benefits.

Labor Productivity Record #1

ITEM OR ASSEMBLY: Lead time for remodel
Included: Reviewing plans. Communicating with client, designer, & inspectors, and myself. Ordering material. Managing subcontractors.
Not included: Layout of work for crew (that time is included in hours for items or work installed).

QUANTITY & DEGREE OF DIFFICULTY of ITEM: Fifteen week project. Straightforward work.

PROJECT: Basement Conversion to Family Room. Moraga Av., Lafayette.
TIME OF CONSTRUCTION: Spring

CLIENT: Barnfield. Fairminded. Appreciative of crew.
DESIGNER: Homer. Sketchy plans but prompt responses to questions.

ACCESS: Easy.
WEATHER: Good mostly. A few showers.
OTHER FACTORS: Pleasant place to work. Crew morale high.

CREW CAPABILITY:
Lead: Daniel M. Third year as lead. Capable.
Journeymen: NA
Apprentice: NA
Laborer: NA

HOURS PER CREW PERSON:
Lead: 120.5 hours.
Journeymen: NA
Apprentice: NA
Laborer: NA
TOTAL HOURS: 120,5

HOURS PER UNIT: 2 hrs./day

LABOR PRODUCTIVITY RECORD #1: LEAD TIME for REMODEL

The illustration on the preceding page is of a record for what I call "lead time," typically (as discussed in Chapter Six) the most costly of the General Requirements. The record illustrated here is for a straightforward remodel. The bottom line figure of two hours per day is figured by dividing the total hours the lead spent on managing his project (as opposed to working with his tools) by the 60 days the job lasted from setup through completion (120 ÷ 60 = 2).

As with my other records, the hours-per-week figure is for a four-day workweek (nine and a half hours per day). I prefer the four-day workweek. It is more efficient (one less day per week of set up and clean up). It is much preferred by many carpenters to the five day week and helps to ensure they stick around, saving on turnover costs. It's a better way of life for both my crew and me.

Labor Productivity Record #2

ITEM OR ASSEMBLY: Lead time for new construction
Included: Reviewing plans. Communicating with client, designer, inspectors, and me. Ordering material. Managing subs.
NIC: Layout of work for crew.

QUANTITY & DEGREE OF DIFFICULTY of ITEM: 54 working weeks. Complex and large project.

PROJECT: New home. Bartlett Avenue. Craftsman & modernist detailing.
TIME OF CONSTRUCTION: Fall to fall.

CLIENT: Katlin. Enthusiastic but manipulative.
DESIGNER: Capable. Honest. Irritatingly pompous.

ACCESS: Good. House close to street. Site muddy during winter rains.
WEATHER: 7 months of sun. 5 months rainy off and on.
OTHER FACTORS: In firestorm area. Congested due to much other construction going on.

CREW CAPABILITY:
Lead: Daniel M. Sixth year as lead. Outstanding.
Journeymen: NA
Apprentice: NA

HOURS PER CREW PERSON:
Lead: 625 hours
Journeymen: NA
Apprentice: NA
TOTAL HOURS: 625

HOURS PER UNIT: 2.9 hrs/day 11.6 hrs./four-day workweek

LABOR PRODUCTIVITY RECORD #2: LEAD TIME for NEW CONSTRUCTION

Here we have a second record for lead time, this one for construction of a new home rather than a remodel. Daniel is again the lead, but now is in his sixth year as lead. Even though the project was much larger than the remodel, Daniel needed not even an hour more per day to run it. In part, that is due to his increased experience. Mostly it is because new work, even construction of new architect-designed homes, generally requires less labor per unit of work, including the work of project management, than does remodeling. We'll see more of the spread in labor requirements between new and remodel work in subsequent records. Together they illustrate an important principle of estimating: Do not use records from new work to estimate labor productivity for remodels. You'll run way low if you do.

Labor Productivity Record #3

ITEM OR ASSEMBLY: Setup and cleanup for remodel.

QUANTITY & DEGREE OF DIFFICULTY of ITEM: 31 weeks. Spread out job.

PROJECT: Second-story addition and other work throughout house.
TIME OF CONSTRUCTION: Winter

CLIENT: Carter. Terrific; loved crew—appreciated clean site.
DESIGNER: Marques. Conscientious but struggling. 1st remodel ever.

ACCESS: Tight spaces. Movement impeded.
WEATHER: Rain, but not much of an issue. Job mostly inside.
OTHER FACTORS: Great new vacuum cleaner. Sucks up dirt fast!

CREW CAPABILITY:
Lead: Frank P. - Experienced and efficient.
Journeymen: Fernando - 5th year / 3 yrs. w/Frank.
Apprentice: Julia – 2nd year / 1 yr. with Frank.
Laborer: None

HOURS PER CREW PERSON:
Lead: 15
Journeymen: 14
Apprentice: 131
Laborer: NA
TOTAL HOURS: 160

HOURS PER UNIT 1.29 hrs/day 5 hrs./ four-day workweek

LABOR PRODUCTIVITY RECORD #3: SETUP AND CLEANUP for REMODEL

In my binder of labor productivity records, I have separate records for morning setup—bringing tools from the job shack and trucks to the work area, laying out power cords, setting up sawhorses and saws—and end-of-day cleanup. Here, for purposes of illustration, I have combined the setup and cleanup records from a specific job into a single record.

The time spent at my jobsites on cleanup may be toward the high end relative to other general contractors. I calculate that any extra labor expended is a good investment. A clean site is good for crew morale, it is safer, and it is very good marketing.

I do know guys who are far more fastidious than I. For example, there's the legendary remodeler who is reputed to clean his kitchen projects with a feather duster at the end of each day! I've only gone as far as to buy a powerful German-built vacuum cleaner, used for the first time on this job with great results.

Labor Productivity Record #4

ITEM OR ASSEMBLY: Kitchen demo
Included: Remove cabinets, fixtures, tile, plaster, all trim. Strip nails off studs, move debris to dumpster.
NIC: Plumbing & electric. (done by subs)

QUANTITY & DEGREE OF DIFFICULTY of ITEM: 200 s.f. kitchen. Tile floor, countertop and backsplash on mortar; removal labor intensive.

PROJECT: Kitchen gut and new.
TIME OF CONSTRUCTION: Winter

CLIENT: Sudinsky – A+ Client. Straight shooter.
DESIGNER: Bethany. Good aesthetics, minimal construction detail. Responsive.

ACCESS: Good. Straight short shot to driveway.
WEATHER: NA
OTHER FACTORS: For demo used hand tools to avoid damage to structure.

CREW CAPABILITY:
Lead: Frank P. – highly experienced and efficient.
Journeymen: NA
Apprentice: Julia –1st yr. with Frank - Good apprentice.
Laborer: George from Labor Pool – hardworking, strong guy.

HOURS PER CREW PERSON:
Lead: 13.
Apprentice: 46
Laborer: 40
TOTAL HOURS: 99

HOURS PER UNIT: 2 s.f. of floor area/hr

LABOR PRODUCTIVITY RECORD #4: KITCHEN DEMO

In remodeling projects, demolition and deconstruction can amount to a significant portion of the work. Here I have included the simplest sort of demo record, one for gutting a small rectangular kitchen. There's no deconstruction—no salvaging of existing trim, cabinets, fixtures, or structural beams. Deconstruction can be finicky and time-consuming work, and merits its own productivity records. This demo record is sufficient for similarly simple projects. However, the square feet per hour figure it shows would likely be inadequate for a more complex project or a differently configured space. Rooms with the same square footage of floor area can have very different square footages of wall and ceiling surface (as in the case of cathedral ceilings or those with extensive soffits) and, therefore, very different demo costs. A more extensive record, one recording demo costs not by square foot of floor area but by square foot of total surface, and including additional detail such as the linear feet of cabinetry, would be more helpful for more complex demos.

Labor Productivity Record #5

ITEM OR ASSEMBLY: Install concrete piers for grade beam.
Rebar cages delivered to site and ready.
Included: Setup, drilling, set rebar, pump concrete, cleanup.
NIC: NA

QUANTITY & DEGREE OF DIFFICULTY of ITEM: 21 piers @13' deep. Stable and stone-free soil. Flat lot.

PROJECT: Two-story house at 19th Street
TIME OF CONSTRUCTION: Summer

CLIENT: Gerstel. Demanding. This guy wants everything dead on!
DESIGNER: Gerstel. Plans thorough and accurate.

ACCESS: Excellent. Flat lot along broad street
WEATHER: Dry and sunny.
OTHER FACTORS: Highly competent drill rig and pump operators.

CREW CAPABILITY:
Lead: David Kendall – Concrete maestro
Journeymen: Gerstel – Focused veteran; not fast.
Apprentice: Grant R. – hardworking 4th year guy.
Laborer: Ryan S. – Promising 1st year guy.

HOURS PER CREW PERSON:
Lead: 9 hours
Journeymen: 9 hours
Apprentice: 9 hours
Laborer: 9 hours
TOTAL HOURS: 36 hours (one nine hour day for 4 person crew)

HOURS PER UNIT: 2+ piers per hour for four-person crew.

LABOR PRODUCTIVITY RECORD #5: FOUNDATION PIERS

As mentioned at the close of the previous section, a labor record for work of one level of complexity can serve as a basis for work of greater (or, for that matter, lesser) complexity. For example, while this record narrates a relatively simple concrete pier installation (flat lot, shallow piers), it could serve as a basis upon which to construct a productivity projection for a more complex and difficult installation such as 37 piers, each 22' deep, on a hillside lot. Adjustments could be made as follows:

- Assume same four-person crew and same drill rig operator.
- Note that drilling will be on hillside with rockier soil and for deeper piers. Therefore, reduce expected drilling productivity from 2+ pier holes to 1 pier hole per hour.
- Note that rebar cages are much longer and will have to be lifted and placed with the drill rig, not by hand. Note also, that pump hose will be harder to handle on steep site (and that use of a boom is not feasible due to presence of power lines). Reduce expected productivity for setting of rebar and pumping of concrete from 2+ piers to 1 pier per hour.
- Note that setup and concrete cleanup will be more difficult on steep site but that crew will do that work during downtime, waiting for rig to complete drilling of each hole.
- Conclusion: Expected productivity will be less than one half of that recorded in Labor Productivity Record #5. That is, estimated productivity for drilling, rebar setting, concrete pumping, and setup and cleanup will be 1 pier per hour rather than 2+ piers per hour.
- Estimate for new project: 1 pier per hour for four-person crew installing 37 piers. Total labor, therefore, 37 hours (or just over 4 nine-hour days) for four-person crew.
- Add half day for margin of safety for this difficult work for which you do not have a closely similar record and include 4½ days for a four-person crew in your estimate.

Labor Productivity Record #6

ITEM OR ASSEMBLY: Grade beam foundation
Included: Layout, form, set rebar cages, ABs & HDs, install concrete.
NIC: Strip & clean forms.

QUANTITY & DEGREE OF DIFFICULTY of ITEM: 132 l.f. x 12" x 16.5". Formed with straight KD 2x6 (reused for framing walls).

PROJECT: Two story house. 19th Street.
TIME OF CONSTRUCTION: Summer.

CLIENT: Gerstel.
DESIGNER: Gerstel.

ACCESS: Excellent. Flat lot on wide street.
WEATHER: Clear and sunny.
OTHER FACTORS:

CREW CAPABILITY:
Lead: David Kendall – Concrete maestro.
Journeymen: Gerstel - Focused veteran; not fast.
Apprentice: Grant R. – hardworking 4th year guy.
Laborer: Ryan S. – Promising 1st year guy.

HOURS PER CREW PERSON:
Lead: 40 hrs.
Journeymen: 48 hrs.
Apprentices: 48 hrs.
Laborer: 48
TOTAL HOURS: 184

HOURS PER UNIT: 1.4 hrs/l.f
Note: Unusually high productivity, due to remarkable lead and motivated crew. Do not expect to match this productivity with all crews.

LABOR PRODUCTIVITY RECORD #6: GRADE BEAM FOUNDATION

When you estimate from experience and intuition it is easy to make the mistake of remembering your best day or week ever and over-optimistically projecting productivity for a new job from that happy memory. Similarly, estimating on the basis of narrative records that describe a particularly stellar performance is risky. You may not be able to match the performance it records with different crews or under different conditions. That's why I have posted a warning to myself at the bottom of this record, which reports productivity achieved under easy conditions by a hard-driving crew with an exceptional lead.

Labor Productivity Record #7

ITEM OR ASSEMBLY: Foundation replacement
Included: Layout, trim trench, set rebar & ABs, install concrete, strip & clean, backfill, remove extra soil.
NIC: Shoring, Remove (E) foundation & excavate.

QUANTITY & DEGREE OF DIFFICULTY of ITEM: 175 l.f. 18" x 18" footing.
12" x 16" stem. Tough work. Low crawl space. Crew on knees much of time.

PROJECT:Replacement foundation for two-story 1908 house. College Ave.
TIME OF CONSTRUCTION: Spring

CLIENT: Schneider – Delightful. Kids named dolls after crew!
DESIGNER: Pei (engineer). Really good drawings. Responsive.

ACCESS: Low crawl space. Flat lot right on quiet street with parking.
WEATHER: Sunny. Dry.
OTHER FACTORS: Crew morale high due to weather and clients.

CREW CAPABILITY:
Lead: Frank P. – Highly experienced and efficient.
Journeyman: Daniel M -- Good -- 3 yrs. w/Frank.
Apprentice: Smitty – A bull. Worked for Frank on several earlier jobs.
Laborer: Kevin, steady worker from labor pool.

HOURS PER CREW PERSON:
Lead: 162 hrs.
Journeymen: 146 hrs.
Apprentice: 146 hrs.
Laborer: 80
TOTAL HOURS: 534

HOURS PER UNIT: 3 hrs/l.f

LABOR PRODUCTIVITY RECORD #7: FOUNDATION REPLACEMENT

A good productivity record reports not just quantities of work and hours of labor expended to do the work. It covers less tangible items—such as the length of time that crew members have worked together and, as in the case of this record, crew morale. On this tough foundation-replacement job, productivity was unexpectedly high. It was nearly identical to what we achieved building two-story foundations under much less taxing conditions. In the record, I attributed the high productivity to high morale that, in turn, was fostered by the great weather and delightful clients.

Labor Productivity Record #8

ITEM OR ASSEMBLY: Frame for deck
Included: One post with pier, one post let into house wall, ledger, flashing, double 2x12 bolted to posts, braces, hangers, and joists.
NIC: Demo and patch, decking, railing, paint.

QUANTITY & DEGREE OF DIFFICULTY of ITEM: 14' x 7' (100 s.f.) deck. Only unusual detail was integrating one post into house wall frame rather than mounting on pier.

PROJECT: Deck outside newly remodeled kitchen at Andover Ave.
TIME OF CONSTRUCTION: Summer

CLIENT: Barbara & Steve Telani – Great repeat clients.
DESIGNER: Gerstel and clients.

ACCESS: Difficult. Uphill to street. 53 stair steps.
WEATHER: Sunny and warm.
OTHER FACTORS: Overhead work required staging.

CREW CAPABILITY:
Lead: Daniel M. – Skilled carpenter. Fourth job as lead,
No other workers on job.

HOURS PER CREW PERSON:
Lead level carpenter: 47 hrs.
TOTAL HOURS: 47

HOURS PER UNIT: 50 hrs+/- to frame a 100 s.f. rectangular deck.

LABOR PRODUCTIVITY RECORD #8: FRAME for DECK

Experienced estimators know that if you estimate by overly large assemblies, such as the entire frame of a house, you will tend to underestimate. The large assembly can obscure many small and time-consuming details, and, as a result, the material and labor they consume will get left out of your estimate. Conversely, estimators know that if you break work up into too many items, you may tend to overestimate because you will not see that time used for one item also supports installation of others.

Deciding just how fine a breakdown to use for a given phase of work or for a record of that work is a judgment call each estimator must make for herself. When creating this cost record for framing a small deck, my call was to treat all the work as part of one assembly rather than creating separate records for each of its components. I felt the record would be useful for estimating future small decks, and that if I did elect to estimate a small deck frame item by item, the record of labor productivity for an entire frame would serve as a good check on my results.

Labor Productivity Record #9

ITEM OR ASSEMBLY: Subfloor install (¾ t&g ply)

QUANTITY & DEGREE OF DIFFICULTY of ITEM: 2,300 s.f. Engineer specified fastening with 16d aluminum screw nails that required predrilling & hand driving.

PROJECT: Two-story house on Cochrane Avenue.
TIME OF CONSTRUCTION: Fall.

CLIENT: A manipulator, but once crew figured her out, she was not a productivity factor.
DESIGNER: Delaney Engineering. Good. Capable and responsive.

ACCESS: Below average. House set back from street and staging area.
WEATHER: Sunny or overcast. No rain.
OTHER FACTORS: Hand nailing,

CREW CAPABILITY:
Lead: Guido R – "The Natural"
Journeymen: Whitey C. – "The Tank."
Apprentice: Angela – "The Athlete."
Laborer: None

HOURS PER CREW PERSON:
Lead: 18 hrs.
Journeymen: 20
Apprentice: 18 hrs.
TOTAL HOURS: 56

HOURS PER UNIT: 41 s.f./hr. (¾ hrs. per installed sheet)

LABOR PRODUCTIVITY RECORD #9: SUBFLOOR INSTALL

Over the years, I have created multiple records for many standard tasks. Sometimes work that is standard, but with a twist to it, comes along. I have often taken the opportunity to track hours and create narrative records for that quirky work. The productivity is likely to vary greatly from standard work and the records, therefore, provide valuable data for future quirky stuff.

With this record, the twist, as you can see, is that the engineer required the T&G plywood subfloor be nailed down with 16d aluminum screw nails. (To this day I don't understand why standard nailing would not have been sufficient, but the engineer insisted his numbers dictated the darned screw nails.) We not only had to drive the nails by hand, but predrill for them. Productivity, as a result, plunged from the usual 64 square feet (or even 96 square feet per hour) down to 41 square feet per hour. That works out to ¾ hour per installed sheet as opposed to the usual ½ hour or less per sheet.

Labor Productivity Record #10

ITEM OR ASSEMBLY: Pony wall frame
Included: 3 x 6 mudsill. 2 x 6 studs. Double top plate. A couple of headers. Plumb, line, and brace.
NIC: Sheathing.

QUANTITY & DEGREE OF DIFFICULTY of ITEM: 86 l.f.
Stepped Foundation. Average height of pony wall 4'. Green lumber.

PROJECT: Pony wall for sub area of new building on Adcock Av.
TIME OF CONSTRUCTION: Fall.

CLIENT: Emerson. A++ Client. Fair, honest, appreciative. (We were replacing the family home she had lost to the Oakland Hills firestorm. She made the crew feel really good about recreating it for her).
DESIGNER: Peggy Simpson. Clear plans. Responsive to RFIs. Respectful. A pleasure to work with. Really cared about the client, not just about her design.

ACCESS: Good parking. Close in staging area. Uncluttered site.
WEATHER: Sunny mostly. Some light rain but site dried quickly.
OTHER FACTORS: None of importance.

CREW CAPABILITY:
Lead: Guido R – "The Natural."
Journeymen: Whitey C. – solid.
Apprentice: Julia – Topnotch.
Laborer: None

HOURS PER CREW PERSON:
Lead: 26
Journeymen: 29
Apprentice: 33
TOTAL HOURS: 88

HOURS PER UNIT: 1 hr./ft.

LABOR PRODUCTIVITY RECORD #10: PONY WALL FRAME

Sometimes I allow my self to get pretty candid in assessing clients and designers in my productivity records. Why not? The records are for my private use (names used here are pseudonyms). Telling it like it is makes the routine work of creating records a bit more fun. More importantly, the candid descriptions do relate to actual productivity on a job. Arrogant, demeaning, unappreciative clients can demoralize a crew and slow production. The reverse is true, too. Here's an otherwise boring record, but for a project that came with an exceptional client whose attitude supported the crew's enthusiasm for the job and their productivity. There are clients from hell. There are others from heaven.

Labor Productivity Record #11

ITEM OR ASSEMBLY: Wall frame at new construction
Included: Layout, sill, plates, headers, studs, plumb, line, brace.
NIC: Sheathing.

QUANTITY & DEGREE OF DIFFICULTY of ITEM: 510 l.f. standard 8'-high walls. Simple new construction job; a basic box with simple floor plan. Green lumber; good stock required little culling.

PROJECT: Single-story home. Adcock Avenue.
TIME OF CONSTRUCTION: Fall

CLIENT: Emerson. A++ Client. Fair & appreciative. Nice to work with.
DESIGNER: Peggy Simpson. Great to work with. Clear plans. Responsive to RFIs. Respectful.

ACCESS: Good parking. Close in staging area. Uncluttered site.
WEATHER: Sunny mostly. Some rain at night; site dried quickly.
OTHER FACTORS: Hand nailing.

CREW CAPABILITY:
Lead: Guido R – "The Natural."
Journeymen: Whitey C. – solid.
Apprentice: Julia – Topnotch.
Laborer: None

HOURS PER CREW PERSON:
Lead: 110 hrs.
Journeymen: 121 hrs.
Apprentice: 106 hrs.
TOTAL HOURS: 337

HOURS PER UNIT: .66 hrs./ft

LABOR PRODUCTIVITY RECORD #11: WALL FRAME at NEW CONSTRUCTION

Ideally, in productivity records for wall frames, openings would be separated out. One record would show labor per foot for plates and regularly spaced studs. Another would record labor for headers and studs at openings. A third would provide a figure for corners and intersections. However, tracking labor for overall wall runs, headers, corners and intersections separately is probably not feasible, at least not on time cards filled out by even a diligent lead (maybe you could do it out at your work sites with a stopwatch).

I have found it acceptable to record hours for wall framing with headers, corners, and intersections included. If I had to estimate for walls with strikingly few (or an unusually high number) of those items, and I did not have a record for a closely similar situation, I would make an adjustment in my productivity rate. In the cost record for new construction shown here, the figures are for wall runs with usual numbers of openings for doors and windows.

Note that at "Other Factors" I have recorded "hand nailing." I limit my use of nail guns for framing. They induce overly rapid production and lousy workmanship, with many nails overdriven or missing the framing members. With extra costs for purchase, maintenance, setup, and nails factored in, they don't even save much, if any, money on the size projects that I build. Some of my builder friends, all highly capable guys, ridicule my aversion to nail guns. However, I feel that even when hand nailing does cost a little more, the investment is well worth making for the added control and quality that results.

Labor Productivity Record #12

ITEM OR ASSEMBLY: Wall frame for large remodel
Included: Layout, sill, plates, studs, headers, plumb, and line.
All lumber KD.
NIC: Sheathing.

QUANTITY & DEGREE OF DIFFICULTY of ITEM: 638 l.f. 90% stick frame between (E) floor and ceiling. 10% new addition.

PROJECT: Complete rebuild & reconfigure of three-story home.
TIME OF CONSTRUCTION: Spring/Summer.

CLIENT: Aird. A joy to work with.
DESIGNERS: Mike Thomas: Excellent architect. Great drawings. Responsive and respectful. Skilled carpenter himself!
Tek Pei: Engineer. Terrific

ACCESS: Challenging and tiring. Steep site. Congested building interior. Four stories. Crew up and down the stairs all day long.
WEATHER: Sunny and dry.
OTHER FACTORS: Old building. Somewhat out of square, level, & plumb.

CREW CAPABILITY:
Lead: Frank P. – highly experienced and efficient.
Journeymen: Fernando – 6th year / four yrs. w/Frank. Good!
Apprentice: Julia – 2nd year – Good & getting better fast!

HOURS PER CREW PERSON:
Lead: 301
Journeymen: 324 hrs.
Apprentice: 307 hrs.

TOTAL HOURS: 932

HOURS PER UNIT: 1.5 hrs./ft.

LABOR PRODUCTIVITY RECORD #12: WALL FRAME for LARGE REMODEL

The productivity for wall framing at the large remodel project recorded here is far less than that achieved for the framing of the new walls recorded in the previous record. That's typical. Remodeling projects generally require from two to four times as many labor hours per unit of production as parallel work in new projects. Once again: You want to avoid using records from new work to estimate remodeling labor, and vice versa.

Labor Productivity Record #13

ITEM OR ASSEMBLY: Roof trusses
Included: Set Trusses for 24' span at 2' o.c. Plumb up and Strap. Block at wall plates.
NIC: Staging. Cut tails.

QUANTITY & DEGREE OF DIFFICULTY of ITEM: 21 @ 4/12 trusses x 24' with standing heels & tails at house. 6 @ 4/12 trusses at garage,

PROJECT: Two-story house w/single-car garage pop out. 19th Street.
TIME OF CONSTRUCTION: Summer

CLIENT: Gerstel
DESIGNER: Gerstel

ACCESS: Good. House on wide street with little traffic.
WEATHER: Dry and warm.
OTHER FACTORS: Top notch delivery via truck w/crane & skilled operator

CREW CAPABILITY:
Lead: Gerstel – experienced at roof framing
Journeymen: Grant R. – good, but first truss job
Apprentice: Ryan S. – Good 1st year apprentice
Laborer: None

HOURS PER CREW PERSON:
Lead: 9 hrs.
Journeymen: 9 hrs.
Apprentice: 9 hrs.
TOTAL HOURS: 27

HOURS PER UNIT: 1 day/26 trusses for 3-person crew. (1 person hour per truss. 3 trusses per hour for 3-crew.)

LABOR PRODUCTIVITY RECORD #13: ROOF TRUSSES

Here's a record for truss installation. If I'd had no prior experience with and no records for truss installation, I would construct an estimate of labor productivity for that task from a record for the installation of ceiling joists.

Why? Because ceiling joists are like trusses in that they must be installed from outside plate to outside plate and require blocking. Additionally, like trusses, they must be installed by a crew balancing on the top plates or working off staging that positions them at the height of the top plates.

To begin constructing a productivity rate, I would note that the record for ceiling joists spells out the work as follows: Joists at 2' o.c. Overlapping 12' joists spanning a total of 20'. Rim joists at outside wall plates. Blocking over midspan wall. I would also note that the productivity rate for the ceiling joist installation was ½ hour per unit—i.e., two overlapping joists with their midspan blocking and rim joists.

To adjust the productivity rate for the ceiling joists so that I could use it for trusses, I would first figure that trusses would involve additional work as follows:

1. Crew standby time waiting for crane operator to lift trusses to second story
2. Plumbing up, bracing, and strapping.
3. Two rows of blocking, one row at each exterior wall plate, rather than a single midspan row. Blocks ripped to bevel.
4. I would then note that the trusses would not require rims at the plate as do the ceiling joists.
5. With those steps complete I would figure the productivity rate adjustments via the following three steps:

- Allow ¼ additional hour per run for coordination with crane operator.
- Allow ¼ additional hour per run for plumbing, bracing, strapping, and for two blocks instead of one.
- Figure that elimination of labor to install ceiling joist rims and addition of labor for beveling of truss blocks cancel each other out.

Finally, I would total the adjustments and conclude that I should add an extra half hour for trusses to the half hour required for joists. That would bring me to an estimated installation time of one truss per person hour, or three trusses per hour for a three person crew.

Labor Productivity Record #14

ITEM OR ASSEMBLY: Window Install
Included: Cut away ply sheathing. Install windows and flash.
NIC: Integrate flashing w/building wrap.

QUANTITY & DEGREE OF DIFFICULTY of ITEM: 25 fixed windows. Small (2'x 2'). Flushmounted. Installed in clerestory. Windows had to be carefully aligned for continuous sill and trim.

PROJECT: New construction. Fancy office for consulting company.
TIME OF CONSTRUCTION: Spring.

CLIENT: Journell. A real loudmouth and tantrum thrower. Disrespectful of crew. Damaged morale.

DESIGNER: Swann. Conscientious and responsive. Crew respected. But agonized over RFIs, and burned time.

ACCESS: Difficult. Installation required crew to work on knees on roof below clerestory.
WEATHER: Warm and sunny.
OTHER FACTORS: Windows small and easy to handle.

CREW CAPABILITY:
Lead: Frank P. – highly experienced and efficient.
Apprentice: Julia – outstandng third-year apprentice.

HOURS PER CREW PERSON:
Lead: 22.5 hours
Apprentice: 22.5 hours
TOTAL HOURS: 45

HOURS PER UNIT: 1.8 hrs./window

LABOR PRODUCTIVITY RECORD #14: WINDOW INSTALL

I have found that the labor productivity for window installation varies tremendously from project to project. It is determined by the size and type of windows (fixed, double-hung, casement), the location of the windows in the frame (flush or inset), the flashing system specified, and the category of the project (new construction or remodeling). The productivity record shown here is for installation of small, flush-mounted windows with basic flashing systems in a new frame.

Just prior to it in my records binder is a record for installation of large double casements inset within existing framing with the following requirements: Removal of existing windows for possible salvage. Shimming and other adjustments of existing openings that were out of square and level. Integration of flashing with aged and fragile building wrap.

The double casements took over twice as long per unit to install as the windows shown in this record for new clerestory windows, itself a difficult installation due to poor access. The difference underscores again the labor intensity of remodeling vs. new construction and the decrease in productivity brought on by complexity.

Labor Productivity Record #15

ITEM OR ASSEMBLY: Door Install
Included: Install prehung interior doors in rough openings. Install Schlage passage and locking hardware. Install stops.
NIC: Casing. Final tune-up

QUANTITY & DEGREE OF DIFFICULTY of ITEM: 18 solid-core doors. Simple work: Doors prehung and prebored. Paint grade. Basic Schlage hardware.

PROJECT: One-story new house on Sunset St.
TIME OF CONSTRUCTION: Winter

CLIENT: Meyers. Polite. Not around much.
DESIGNER: Shulien Wang. Capable. Clear and accurate plans. Responsive to RFIs.

ACCESS: Good. Single-story with easy access to street & staging area
WEATHER: Rain, but negligible effect on interior production.
OTHER FACTORS: NA.

CREW CAPABILITY:
Lead: Frank P. – highly experienced and efficient
Journeymen: Whitey C. – Solid skills, focused and efficient.

HOURS PER CREW PERSON:
Lead: 8 hours
Journeymen: 9 hrs.
TOTAL HOURS: 17 hrs.

HOURS PER UNIT: 1 hr./solid core door including hardware and stop

LABOR PRODUCTIVITY RECORD #15: DOOR INSTALL

During the time I worked as a journeyman union carpenter on a large condo complex built alongside San Francisco Bay, the finish carpenters were expected to install three prehung doors per hour. After I started my own company, neither my crews nor I ever came close to that rate of production. On the other hand, none of the projects we built ever got us into a lawsuit, whereas the condo complex was tied up in litigation for 20 years.

Here is a productivity record for one of the simplest and easiest door installation jobs my company has done. Productivity was as high as we have ever achieved.

My records indicate that productivity for door installation varies as much as for window installation. While the record shown here indicates an hour per door, another of my records shows 4½ hours per door. Why the difference in installation time? There are several factors at play. The slower time is not for prehung solid core doors. It is for paneled Douglas fir doors that were not prehung but instead hung at the job site in vertical-grain fir jambs milled at the site. The panel doors were fitted with high-end, complex passage hardware, not basic lumberyard-quality hardware. Additionally, they had to be hung with unusual precision and stability because they were to be trimmed out with drywall moldings rather than more supportive wood casings.

Labor Productivity Record #16

ITEM OR ASSEMBLY: Kitchen cabinet Install
Included: Layout. Install cases, base, inserts, doors, tune-up.
NIC: Fascia and crown at uppers.

QUANTITY & DEGREE OF DIFFICULTY of ITEM: 6 uppers. 7 lowers. High-quality cabinets made by Lon Williams. Square and strong. Tricky crown molding.

PROJECT: Kitchen Remodel.
TIME OF CONSTRUCTION: Fall.

CLIENT: Lassen. Hovering but appreciative.
DESIGNER: Gerstel and client.

ACCESS: Tight. Small kitchen. Not much room to work.
WEATHER: NA
OTHER FACTORS: NA

CREW CAPABILITY:
Lead: Frank P. – Highly skilled and efficient.
Apprentice: Julia – Capable 3rd year apprentice.

HOURS PER CREW PERSON:
Lead: 30
Apprentice: 26
TOTAL HOURS: 56

HOURS PER UNIT: 4.3 hrs. per case.

LABOR PRODUCTIVITY RECORD #16: CABINET INSTALL

For productivity records of cabinet installation, the number of cases is more important than the total length of the cabinets. Though it does not hurt to record both length and quantity, it is hours per case, as opposed to hours per foot, that is most useful for projecting productivity on future jobs.

Here the productivity rate, 4.3 hours per box, might look slow to you. It does to me, too. When I am using labor productivity records, the hours per unit of work often seem off to me, and sometimes I will think, "I could do it myself a lot faster than that." I have learned to set aside that confidence. I am trying to avoid falling into the trap of estimating on the basis of a memory of my best day. Like everybody else, I can have my very best day only once. I have to ignore the memory of that best day; it has no bearing on typical productivity.

Even as I try to avoid relying on memories of best days—and, also, on labor productivity records of exceptionally good performances—I likewise avoid relying on records of exceptionally bad performances. For example, I have one record of cabinet installation at which a top lead and a new apprentice achieved barely half the productivity shown in typical records. I have kept the record, but with a note asking, "What the heck happened here?" that warns me to use it only as a reminder that production can go off the tracks, which is another reason for dampening excess optimism.

Labor Productivity Record #17

ITEM OR ASSEMBLY: Baseboard.

QUANTITY & DEGREE OF DIFFICULTY of ITEM:
76 pieces. Demanding installation. Clear v.g. Douglas fir. Nailed to frame, not over drywall. Radiused outside corners cut with router and sanded.

PROJECT: New structure.
TIME OF CONSTRUCTION: Summer

CLIENT: McCutcheon. High standards. Friendly and fair.
DESIGNER: Swann. Capable. Responsive. Not efficient. Used a lot of lead time.

ACCESS: Difficult. Small spaces. Not much room to set up or maneuver.
WEATHER: Dry and warm.
OTHER FACTORS: Modernist design requiring precision. Clean tight frame by my crew.

CREW CAPABILITY:
Lead: Guido V. – First-rate lead and carpenter.
Journeymen: Lane & Colombo. Both very skilled.

HOURS PER CREW PERSON:
Lead: 20
Journeymen: 44
TOTAL HOURS: 64

HOURS PER UNIT: .84 hrs./piece.

LABOR PRODUCTIVITY RECORD #17: BASEBOARD INSTALL

Baseboard productivity records that rely on a count of the pieces installed are more helpful than those that take into account only the total length of the material installed. The number of pieces determines the amount of cutting and fitting, and that— not linear feet of material— largely determines productivity.

Here productivity is low, again because of complexity. An adjacent record in my binder shows a productivity rate for 120 pieces that is more than three times greater, .25 hours per piece rather than .85 hours per piece. Why the difference? The faster speed is for paint-grade material installed with standard miters over drywall rather than for clear material with router-cut outside corners installed right to the frame.

Paul Cook, the author of the classic books on estimating, whom I have mentioned earlier, stresses that the "real challenge to an estimator is to pinpoint the degree of complexity which causes the cost of one item to differ from another of the same type." In his experience, Cook reports, the more complex a job, the greater the danger of underestimating the labor necessary to build it. And, he adds, the reverse is true as well. "The simpler the project, the greater the likelihood of overestimating."

Labor Productivity Record #18

ITEM OR ASSEMBLY: Whole house construction

QUANTITY & DEGREE OF DIFFICULTY of ITEM: 1,400 s.f. house w/four bedrooms and 2½ baths.
High-quality construction including pier-and-grade-beam foundation, 2x6 KD exterior wall frame, level-5 drywall, tile, maple cabinetry, bamboo flooring.

PROJECT: New home on 19th Street
TIME OF CONSTRUCTION: Spring to summer.

CLIENT: Gerstel.
DESIGNER: Gerstel

ACCESS: Good
WEATHER: Good. Inside by time rains started.
OTHER FACTORS: Not typical labor force. Lead carpenter and apprentices worked at all trades except for subcontracted work

CREW CAPABILITY:
Leads: Gerstel plus highly experienced mechanics to lead HVAC, plumbing, and electrical work.
Established subcontractors: For roofing, stucco, insulation, drywall, tile, painting.
Apprentices: Grant – advanced, very capable 3rd-year apprentice. Ryan - 1st year (unusually talented). Both hard workers

TOTAL HOURS: 4,103

HOURS PER UNIT: 4,103 hours to build house from site setup through punch list. Approximately 3 hours per square foot.

LABOR PRODUCTIVITY RECORD #18: WHOLE HOUSE CONSTRUCTION

Here we have a productivity record for construction, utilizing an unusual workforce, of a high-quality but compact four-bedroom, two-and-a-half bath house.

I built the house as research for my book *Crafting the Considerate House.* During construction, I served as foreman, working daily with two carpenter's apprentices; hiring veteran tradesmen to lead us at HVAC, plumbing, and electrical work; and contracting with subs for the other specialty trades. All labor hours put in by all persons who worked on the project added up to 4,103.

Whole project productivity figures are valuable for creating the preliminary "survey estimates," as I call them, for a project. Such estimates can help a client determine whether a project is financially feasible before investing heavily in design work.

Summing it up

The foregoing discussion of the 18 sample productivity records covers a lot of territory and emphasizes a variety of estimating tactics and principles. In the accompanying sidebar, I have summarized the primary tactics that I've suggested. They could be boiled down to three principles:

- Use your own narrative records for estimating, not someone else's numbers.
- Rely on the records that most closely resemble the work being estimated.
- Believe in your records; don't second-guess them.

A Summary of Labor Estimating Tactics

- Rely on records of your own labor productivity, not average costs from a book or computer estimating program.
- Rely on records that narrate crew capability and job conditions from access through client attitude, as well as providing hours per unit of production.
- Use records of the most similar past work as a platform for constructing estimates for more complex or different types of work.
- Appreciate that complexity is a crucial factor, and avoid underestimating for it.
- Avoid falling into the memory-of-your-best-day trap.
- Do not rely on records of exceptional performance.
- Do not rely on records of unusually poor performance.
- Use records of new work for estimating future new construction.
- Use records from remodeling projects to estimate future remodeling projects.
- Avoid estimating for overly large assemblies to prevent underestimating.
- Avoid breaking assemblies up into too many small items to prevent overestimating.
- Believe in your records!

Regarding that final point, the Associated General Contractors, has something to add. No one in any industry, the AGC reports, achieves optimal performance. Even in manufacturing, according to numerous studies, actual performance lags behind optimal performance by about a third, due to such factors as bathroom breaks, workers waiting for instruction, and materials being out of position. In construction, the AGC states, actual lags optimal not by a third but by 50%, due to "adverse factors" not present inside manufacturing plants, such as "precipitation, mud, dust, extremes of temperature, lack of planning, crews newly put together, [etc., etc.]."

I was struck by the AGC's mention of a 50% variation. Long ago I noticed that when I and other builder friends glanced over a job, we consistently tended to underestimate the time it would take by half. We were instinctively assuming optimal performance, not anticipating the kind of actual performance that narrative labor productivity records display. As an estimator, you can't believe in your hopes. Believe instead in your records.

With reliable records based on conscientiously kept time cards you can project labor hours for the items and assemblies in a project with accuracy—not with perfection, but within a small enough margin of error that any over- or underestimate will not devastate your bid as a whole. You can satisfactorily nail your numbers. Once the labor hours have been nailed down, you can move to the final step in capturing your costs. And that is figuring the actual dollar amounts to charge for construction, the subject of our next chapter.

CHAPTER 13

Nail Down the Dollars

Labor rates

If you have taken the steps I have recommended in previous chapters, you now have most of the tools necessary for building reliable estimates:

You have a well-organized and comprehensive estimating spreadsheet.

You've put in place a systematic and efficient way of doing your takeoffs.

You've established procedures for gathering comprehensive quotes from suppliers, service providers, and subcontractors.

And, icing on the cake, you have, or at least are steadily creating, a file of historical labor productivity records.

With all that in place, you can move to the crucial step you have been preparing for all along: settling on just how much to charge in your estimate and bid for direct costs of construction. Of the various dollar figures you will need, those for labor are the most difficult to nail down. So we'll begin this chapter by looking at the dollar costs of labor. Later in the chapter, we will move to the relatively simple task of including dollar figures for subcontractors, suppliers, and service providers in the estimate spreadsheet. (Profit and overhead, those critical charges above and beyond direct costs that must be included in every bid, we will save for Part IV).

> **COMING UP**
>
> - Labor rate calculations
> - Labor burden
> - The real cost of apprentices
> - Material costs
> - Margin of safety vs. fraud
> - Verifying subcontractor quotes
> - Figuring costs for job site services
> - Equipment costs
> - Auditing (proofing) your direct costs

In order to accurately determine your charges for labor, you need to determine just how much it costs you per hour per person (or per crew) to deploy labor at a job site. You must determine your "labor rates." How much does it cost you per hour to employ a journeyman? What's the rate for an apprentice? A laborer? Unless you are still at the stage of your career where you are paying cash to unlicensed hourly workers, your costs for labor will be far higher than the wages that go directly to your workers. Labor costs also include "labor burden." And labor burden is made up of (1) the wage-related taxes and insurance costs that federal and state law require you to pay, and (2) benefits that you elect to pay in order to avoid the even more burdensome costs of employee turnover, or simply because you care about your employees and want them to enjoy those benefits.

Where should you get your labor rates? Not from hearsay about what some other guy is charging and not from cost catalogues. Your labor rates should accurately reflect the labor costs you actually experience. To pinpoint them, you must create a labor rate sheet such as the one in the illustration below.

The first six lines in the sheet are for wages and five items of labor burden employers are required to pay by law. All five—Social Security, Medicare, federal unemployment tax, state unemployment tax, and government disability insurance—are figured as a percentage of wages. For example, in the labor rate sheet, the percentage for social security, not shown but underlying the costs that are displayed, is 6.2% percent of the hourly wages. For a lead carpenter, that comes to $2.36 (38 × 6.2% = 2.36). Please note that the underlying percentages used for the illustration will likely not be current at the time you are reading this book. If you are using a payroll service such as Paychex, the service will be able to provide you with the exact percentages currently in use. If you elect to do your payroll in-house, you will need to get the rates from the relevant government agencies, such as the IRS and your state tax agency.

The next seven items listed in the labor rate sheet can be more involved than the five basic tax and

HOURLY LABOR RATES **Date: dd/mm/yyyy**

	Lead	*Carpenter*	*Apprentice*	*Laborer*	*Four Crew*	**Comment:**
Wage	38.00	32.00	22.00	16.00	108.00	
Social Security	2.36	1.98	1.36	0.99	6.69	
Medicare	0.55	0.46	0.32	0.23	1.56	
Federal Unemp	0.23	0.19	0.13	0.10	0.65	
State Unemp.	1.30	1.10	0.75	0.54	3.69	*Varies by state*
Disability	0.36	0.32	0.22	0.16	1.06	
Workers Comp	3.80	3.20	8.80	6.40	22.20	
Liability Insur	0.76	0.64	0.44	0.32	2.16	
Consumables	1.52	1.28	0.88	0.64	4.32	
Supervision	----	2.60	7.80	2.60	13.00	
Training	0.40	0.40	0.40	0.40	1.60	*Mostly safety meetings*
Dental & Vision	1.00	1.00	------	--	2.00	
Pension	2.00	1.50	------	--	3.50	
TOTAL	52.28	46.67	43.10	28.38	170.43	

The true cost of apprentices

As you can see in the labor rate sheet, labor that is often thought of as cheap may not actually be cheap. In particular, the full cost of employing an apprentice is high. With a realistic amount for supervision and for workers' compensation insurance added to the wages and the other labor burdens, the hourly cost of employing apprentices comes close to that for full carpenters.

To make matters worse, the productivity rates of apprentices, even at simple tasks, is far lower than that of full carpenters. They are slower. They tend to make more mistakes. They spend more time correcting their mistakes. The result: Their cost to you per unit of work installed actually can be much higher than if the work were done by carpenters. Cheap turns out to be expensive.

That does not mean that it is not worthwhile to employ apprentices. Their presence on a job site can boost morale. Leads and full carpenters are likely to enjoy having an apprentice around to take care of routine tasks they have already done thousands of times. They may enjoy training apprentices. But to get reasonable financial results from hiring apprentices, you must have them work behind leads who know how to efficiently deploy them. The apprentices must be willing to spend day after day doing routine tasks in order to earn the chance to do more challenging work and move up in the trade. And they must be eager to grab a broom when they have no assigned task to do.

insurance components of labor burden imposed by government.

Workers' Compensation Insurance is required in my state and more than likely in yours. It provides financial life support for workers in case of injury, and is generally the most costly of the labor burden items. Like the previous items of labor burden, it is figured as a percentage of wages, in this case because that is how the insurance companies compute the cost of your premiums.

As the labor rate sheet suggests, the percentage charged for apprentices can be much higher than that charged for more highly skilled workers. You can see that the hourly cost of workers' comp insurance for an apprentice is nearly three times that of a full carpenter ($8.80 as opposed to $3.20 per hour). And that is the case even though the apprentice's hourly wage is only about two-thirds that of a carpenter ($22 instead of $32). In percentage terms, the rate assumed here for apprentices is 300% higher than for journey level workers—a differential in line with those I have actually encountered. In fact, I have seen workers' comp for apprentices climb so high that it costs half as much per hour as wages, say $11 for every hour a $22-per-hour apprentice works!

Liability insurance can be charged for in a variety of ways, but one way is as a percentage of total wages paid out. So in the labor rate sheet, for the purposes of illustration, the underlying percentage is assumed to be 2% of wages for all levels of employees.

Consumables, alternately "Crew Supplies," are a much bigger component of labor cost than many builders realize, especially at the outset of their careers. Here's a sampling of consumables in no particular order: Saw blades. Pencils. Chalk. Dry line. Ropes. Sandpaper. Power cords. Perishable small tools including tape measures, utility knives and blades, nail punches—and any other item small enough to get closed up in a wall frame. Repair of larger tools. Replacement batteries. Brooms. Dustpans. Garbage cans. Rags. Plastic sheeting.

Tarps. Rosin paper. Duct tape. Blue tape. Stuff subs "borrow." Coffee. Donuts.

In short, consumables are all those supplies that do not become a permanent part of a building, but that crews churn through daily. The cost of consumables can be provided for as a percentage of material cost. Alternately, it can be figured as a fixed percentage of wages on the assumption that higher wage workers tend to use more supplies and more costly supplies. That is, 2% of a carpenter's wage will come to more than 2% of the wage of a laborer, which is as it should be, because the carpenter will consume more consumables.

I have always used the second alternative, occasionally verifying my percentage by figuring total crew supply costs from invoices for a job and dividing by the total wages paid out for the job. Originally, I found that 2% was about right. But because the number of tools used at job sites is forever proliferating, and because the quality has dropped, necessitating more frequent replacement and repair, I have pushed the underlying percentage up to 4% here in the illustrated labor rate sheet. Again, to determine your percentage, you should consult your own numbers. Don't use anyone else's, even mine.

Like consumables, *supervision* of workers at a project can be provided for in a couple of ways. You can lump supervision together with your charges for project management time in the General Requirements division of your estimates and let it go at that (see Chapter Six). However, project management time—communicating with the designers, subs, and suppliers and so on—is determined largely by project size and complexity. Supervision of crew tends to correlate more with the skill level of the people on the job site. Therefore, you should separate out supervision from the project management tasks charged for in General Requirements and allow for it in the labor rate sheet.

Supervision costs vary greatly depending on whom the lead is supervising. If a lead is working only with highly skilled carpenters, not much supervision is required. The lead need do no more than speak a few sentences and skilled carpenters will know what to do. Likewise with laborers doing simple tasks, not a great deal of supervision is required—though, as we shall see, as a percentage of their wages, the cost of supervision is much higher than for carpenters. With apprentices, the cost of supervision as a percentage of wages skyrockets. The following couple of paragraphs will explain the calculation in demanding (*painfully* demanding, so get ready) but important detail.

In the illustrated labor rate sheet, I am assuming, quite realistically I think, that a lead will need to spend, on average, about three minutes out of every hour of his or her time supervising a carpenter or a laborer. The hourly cost for a lead, including all labor burden and not just wages, is figured in the illustrated sheet to be just over \$52. So the cost for three minutes of supervision, whether of a laborer or carpenter, is \$2.60 (\$52 × 3/60 = \$2.60). But as a percentage of wages it is 16% for laborers (\$2.60 ÷ \$16 = 16%) vs. 8% (\$2.60 ÷ \$32 = 8%) for carpenters.

When we come to apprentices, I am assuming, again quite realistically I think, that an apprentice will need about nine minutes per hour, or a little over an hour during an eight-hour day, of supervision by a lead. The cost of that supervision will come to \$7.80 per hour that the apprentice works. That's equal to a whopping 35% of apprentice wages. (Here's the math \$52 × 9/60 = \$7.80. And \$7.80 ÷ \$22 = 35%). Yes, ouch! Please see the sidebar on the preceding page for a fuller picture of the costs of employing an apprentice.

Training, like consumables and supervision, is another labor burden that is easy to overlook. It can add costs in a variety of ways: A manufacturer's rep explaining the installation of a new product. You or your project manager explaining to workers the use of a new tool, or a new company policy, or the best way to respond to a difficult client. Subscriptions to a magazine or website for employees. The weekly meetings essential for creating the safety culture that keeps down accidents.

Training costs can vary greatly and are difficult to track exactly. It's reasonable to allow for the labor costs of training events, such as weekly safety meetings, that occur regularly, and add a rough figure for events that occur sporadically. That is not exactly nailing your numbers. It's closer to the guesstimating that I try to avoid. But so long as training remains a minor cost, even a bad rough guesstimate will not throw your overall labor rate calculation off significantly. A rough guesstimate may be the best you can reasonably produce. You want to avoid spending a hundred bucks worth of estimating time getting a ten dollar cost just right (here I will leave the math to you).

Beyond the standard labor burden items described so far, your labor rate sheet may need to include additional items. Among them:

- *Union charges* for additional benefits such as *vacation pay*.
- *Living expenses* such as overnight lodging for your employees if you do out-of-town projects.
- *Medical Insurance*. Even if you operate a small company, as it gains strength you may elect to provide medical insurance for your employees. If your company grows past a certain size, you may, depending on how the political battles play out, be required to pay taxes for government-mandated medical insurance.
- *Sales tax*. Some states levy a sales tax on employee wages. You will want to review the requirements in your state.
- *Office costs* typically are included in company overhead. But you may elect to include some, particularly the costs of payroll, in your labor rates because you incur those expenses only when you are employing workers.
- *Raises*. From time to time, you may bid a project that will be under construction just at the moment some members of crew are to receive a raise. If so, you will want to adjust your labor rate sheet to account for that raise during estimating. For example, if you are bidding a job that is going to last eight months, and you anticipate that close to the midpoint of construction your lead will be receiving a raise from $38 to $40 an hour, you would adjust to an average rate for the whole project of $39 per hour.

While here I have equated labor *rate* with labor *cost* here—that is, I have equated the per hour charge included in an estimate with the actual per hour cost of putting a worker on a job—many builders estimate using a labor rate that is higher than cost, even much higher. For example, one builder I know includes charges of approximately $85 an hour in his estimate for leads. They actually cost him about $65 an hour. Some builders who use labor rates much higher than costs are simply trying to protect themselves against their own weak estimating skills. They have learned that for a variety of reasons, such as lack of reliable labor productivity records, they chronically underestimate labor hours for a project. Using a labor rate that is twenty or thirty percent higher than they actually pay out for labor and burden makes up, in a very rough and inaccurate way, for their chronic shortfall on labor hours.

Other builders, however, pump labor rates up above cost as a way of deceiving clients and designers and inflating their profit margins. They believe that clients and designers will not accept the markups necessary to provide them with the earnings they feel they need or deserve to make. To get around the resistance, their bids show only a portion of their markups at the bottom lines of the recap (summary) sheet where overhead and profit are typically presented. The remainder they conceal in their labor rates (and in their charges for material and subs). The clients and designers thus see on the Overhead and Profit lines deflated numbers that they can accept. The builders harvest the earnings they desire from the additional markup hidden in their labor rates.

It may be reasonable to add *a percentage for safety margin* to your labor costs to arrive at your labor rates (though with a well-developed estimating system that includes sound narrative labor productivity records, the margin should be 5% to 10% at the most, not 20% or more). As Paul Cook reminds us, "Figuring the cost of labor is the most difficult and inexact portion of an estimate." A modest safety margin helps to cover the slippage that regularly occurs during a project but that does not come into view during estimating from a set of plans.

Deliberately concealing large amounts of overhead and profit in labor rates is, however, another matter. Maybe it is acceptable for competitive bids submitted without charge to a client. But such concealment—when integrated into estimates for which you are charging clients as part of a preconstruction or cost planning service—is unethical. It is fraud. That's not just my opinion. It's the opinion of attorneys with whom I have discussed the deception. (We'll discuss the ethical and legal hazards of the practice in more detail in Chapter Sixteen.)

My own preference is that my estimates tell the truth, not only to my clients, but also to me. If markups for overhead and profit are scattered through an estimate as add-ons to labor and other costs, my numbers are not as clear to me as they are when overhead and profit are expressed in one place only—the lines at the bottom of my estimate and bid summary sheet where they belong.

Once you have established reliable labor rates, figuring your charges for any item or assembly in a project is straightforward. You simply multiply the labor rate by the quantity of work and the productivity rate. For example, a new project calls for 106' of exterior wall. Using labor productivity records you have estimated that the productivity rate will be 1.1 hours per foot of wall, including all work from layout through standing and plumbing and lining, though not including sheathing. You expect to have a two-person crew with a combined burdened labor cost of $100 per hour, or $50 per man-hour, doing the work. You then figure the cost of framing the walls as $5830 (106 feet of wall × 1.1 hours/foot × $50 per hour = $5,830).

Charging for material

When you are creating an estimate, you will likely figure material costs in tandem with labor costs, entering figures for both at each item and assembly as you work your way down your estimating spreadsheet.

Good material quotes

- Always in writing
- Material specified in quote matches material called for in project plans and specs
- Delivery included
- Price of shop drawings, if any, included
- Sales tax, if any, included
- Warranty included
- Good-until date included
- Charge for returns stated
- Prompt payment discount offered
- Quotes from different suppliers organized in same sequence

As part of your material costs, you will want to make certain to allow for waste (as done in the sample takeoff sheets in Chapter Four). Additionally, the figures for material that you enter in your spreadsheet must include supplier charges for delivery, shop drawings (as, for example, in the case of roof trusses), and for sales tax, if suppliers in your locale are required to add it to their charges to you. You may also want to add a safety margin for potential inflation and, if your supplier charges a return fee, an allowance for returns.

Some builders estimate material from price sheets that are updated periodically, typically once a month. That's a dubious practice. Construction material prices are like the price of gasoline. They fluctuate violently over short time spans. A price obtained at the beginning of the month can be significantly out of date three weeks later. To nail your numbers, you should get fresh material quotes for every project.

Your yard may offer a discount for prompt or early payment. If it does, and you are in the habit of paying promptly, you may want to reduce the amount in your spreadsheet for materials by the amount of the discount, on the assumption that you will pay promptly and receive it. On the other hand, you may want to keep the prompt payment discount for yourself. I always have. I figure the discount is compensation for good company management. Some readers may smile and think to themselves, "That's a pretty fancy bit of rationalization." However, I think I will stand my ground. Given the cash flow choke-offs that occur in construction, there is never any guarantee that you will have funds on hand to make prompt payment. There's a possibility that, instead, you will have to endure interest fees for late payment while you fight for your money. You take the risk. You earn your reward if you manage it successfully.

You may want to keep the prompt payment discount for yourself. I figure the discount is compensation for good company management.

Material quotes, unless they are for very small amounts, should always be obtained in writing. I was reminded of the importance of that guideline recently. A builder friend relied on a phone quote—from an excellent yard we have both used for decades—for a large quantity of custom exterior sheet goods. He incorporated the quote in his bid, won it, and then was told when he ordered the material that the charge was three times the amount he had been quoted over the phone. With no written quote in hand he was unable to contest the upcharge.

If you are asking for written quotes for the same material from more than one yard, you can increase your efficiency by asking that the quotes be provided with items listed in the same order. You can then easily compare the quotes line by line and achieve savings—sometimes really substantial savings—by taking delivery of certain items from one supplier and other items from another.

Because material prices are so volatile, quotes must include a "good until" date. If an upward swing in price occurs during a job, it can do serious damage to your earnings if you are not protected against it. If you are doing a large job involving much costly material, and your suppliers will not guarantee their prices for the length of the job, you may want to protect yourself against volatility with one or more of the measures (such as use of an allowance) to be discussed in Chapter Fifteen. Alternately, you can speculate, incorporate the supplier's quote in your bid and hope that prices will go down rather than up and that you will be able to pocket the winnings. Not recommended. Building is risky enough without supercharging the risk with casual gambles.

Subcontractor quotes

Compared to figuring labor charges or even material costs, determining charges for subcontractor work is beautifully simple. Still, it must be done systematically. Much of the task has already been covered in Chapter Eight. You need to establish relationships with reliable subs who will get quotes to you in a timely manner and who will perform if awarded the job. To the extent possible, you want to winnow slippage out of subcontractors' quotes by alerting them to any unusual responsibilities called out in the plans and specs—and especially by requiring that they tell you in writing what is included and not included in their prices.

Additionally, you may want to add a contingency amount to sub quotes—such as the 3% that one builder of large custom homes tacks on to each quote. The contingency plays the same role for sub work as a completion percentage does for each division of work to be performed by your own crews. In

Unit pricing of subcontractor bids

Some builders do not solicit sub quotes for each project. Instead, they obtain periodically updated unit prices—so much per square foot of paint, so much per electrical device, etc.—from their subs.

With unit prices in hand, the builders figure quantities of work for the subtrades, apply the unit prices, and come up with the cost for the subs' work. For example: 4,000 s.f. of interior wall and ceiling painting at a unit price of $1.59 per s.f. for primer and two topcoats results in an estimate for painting of $6,360.

In effect, the builders are quoting the subs' work for them, determining what they will be paid.

Unit pricing of subs work is fast. The owner of one company that specializes in design/build remodeling work claims that with unit pricing it takes him three hours to put together a bid for the additions he builds. With unit pricing, you don't have to go through the process of walking subs through job sites and plans or collecting their Included/NIC statements and making sure there is no slippage or redundancy in their quotes.

Unit pricing may work well for builders doing cookie-cutter work, i.e., projects that are both simple and similar, one after another. For more complex work, unit pricing has the same drawback as cost catalogues. It relies on average prices that may be high for very simple work and seriously low for complex work. Most seriously, unit pricing sidesteps review of job conditions, which, as we have emphasized, can result in variations of several hundred percent from one job to another.

both cases, the percentage covers the small slippery items, often not visible or even implied in the plans and specs, that almost inevitably will get by you during estimating.

On occasion, you may want to have your crew handle a sub trade, rather than involving a subcontractor. Maybe you need to assign the work to your crew to ensure that they are fully employed. Maybe the volume of work is so small that a sub's minimum charge would be substantially greater than you would spend handling it in-house. Maybe you simply want to do the work in-house for the fun of it.

If you do decide to have your crew handle a trade that you normally subcontract, you can include the cost for that work in your spreadsheet in one of two ways:

1) You can include the cost just as you would any other cost for work to be done by your crew. You simply estimate your costs for labor, material, and services and enter them in your spreadsheet. If you go with this option, you may want to enter a higher amount than usual for completion. In estimating for a trade that you don't regularly handle, you are more likely to experience slippage. You'll need the margin

Services

- Scaffolding
- Sani-can
- Temporary power pole
- Heating
- Lighting
- Storage sheds
- Dumpsters
- Recycling
- Concrete installation inspections
- Welding inspections
- Other special inspections
- Site fencing
- Traffic control
- Security personnel

provided by a generous amount for completion to cover the cost of that slippage.

2) You can enter the costs of the trade as if it were being subcontracted, but with your company positioned, in effect, as a subcontractor to itself. Say, for example, that you have decided to have your crew handle the plumbing needed for a renovation project. You set up a separate spreadsheet for the work, figure your costs for labor, material, and services and then—here's the wrinkle—also add charges for General Requirements as well as figuring markups for overhead and profit (a process we will cover in Chapter Fourteen). Once you have completed your estimate for the plumbing work, you will forward it to the line for plumbing on your recap sheet rather than placing the costs in a division of your crew's work.

There are good reasons for electing the second option. It's not just a way of padding your numbers. Doing the sub trade work in-house will entail additional general requirements—extra lead time to plan and organize the work, additional pickups at a separate supplier, additional recycling and cleanup, and more. It will also create extra overhead including more estimating and project management time, vehicle use, and office paperwork for such items as job costing and workers' comp insurance filings. Most emphatically, handling a sub trade in-house creates additional need for profit. It will bring additional liability—the risk of failure of the work, callbacks, and even insurance claims. If a sub handles the work, he includes profit in his charges to cover that liability. If you substitute yourself for him, you can reasonably do the same.

Service and equipment costs

Service providers are similar to subcontractors in one way. You contract with them rather than hiring them as employees. They differ from subs in another way: The work they do—providing, install, maintaining, and removing such items as scaffolding, sani-cans, fencing, and others listed in the accompanying sidebar—does not remain on site as a permanent part of a project as does, say, the work done by your plumbing sub. Estimating spreadsheets often cover services provided by outside companies in the same column as subcontractors. I prefer, however, to provide for services in their own column. Sub and service costs are fundamentally different. With subcontracted work, you are generally dealing with "closed costs," i.e., costs based on a firm quote from a sub. Additional costs may arise if work is added during the project. But for the work agreed to prior to the project, both you and the

PROJECT: 221 Broadway Add **ESTIMATOR: Jones** **DATE: dd/mm/yyyy**

DIV,ASSEMBLY, ITEM	*Division Totals*		LAB HRS	LAB COST	MAT COST	SUB COST	SERV COST	COMMENT
GENERAL REQUIR'MTS								
Lead Time			80	$4,800				
Set up & Cleanup			40	$1,200				
Sani-Can							$140	
Total	$6,140							
SITE WORK								
Remove (E) shed			2	$100		$400		
Utility Trenching			2	$100		$1,100		
Total	$1,700							
CONCRETE								
Excavate			8	$400		$1,600		
Form & Rebar			120	$6,000	$1,300			
Install Concrete			32	$1,600	$1,800		$280	Pump
Total	$12,980							
FRAMING								
Mudsill			4	$200	$220			
Floor			20	$1,000	$1,100			
Walls			40	$2,000	$1,800			
Roof (Trusses)			20	$1,000	$1,600		$580	Crane
Sheathing			40	$2,000	$1,100			
Total	$12,600							
EXTERIOR FINISH								
Siding			72	$3,600	$4,100		$900	Scaffold
Trim			9	$450	$320		$200	Scaffold
Total	$9,570							
GRAND TOTAL	**$42,990**						*Proof:*	$42,990

Though simplified and abbreviated, this spreadsheet does provide for a separate column for service costs (for a sani-can, concrete pumping, a crane, and scaffolding) rather than mingling them with subcontractor costs.

The spreadsheet also contains a proof, a subject we will get to in the last section of this chapter. To proof the addition in a spreadsheet you simply add the numbers two different ways and see if they match up. Here, I first added the totals for the divisions and arrived at a grand total of $42,990. Then, I added up the costs of the items. Again I arrived at $42,990 (as shown in the lower right hand corner of the spreadsheet).

Had proofing resulted in two different totals, I would have known that I had an error somewhere and would have had to search it out.

sub are committed to the price in their quote. Your final cost is not time dependent. It does not vary in relation to the length of time the subcontractor spends at the project.

Service costs, in contrast, are more likely to be "open." Services are often charged by the month, week, day, or even hour. In other words, they are time dependent. For my own estimates, I figure service costs last, because as I build my estimate I will dial in the length of time needed for various phases of the project; and those calculations will tell me, in turn, for how long I will need various services.

For example, let's say that I am bidding for the renovation of a three-story building, a project that will require scaffolding. Relying on my productivity records, I calculate that the lead and the two carpenters that I plan to put on the job will need 102 hours to install items from flashing and windows through siding and trim on the exterior walls. At 38 hours per week (4 days at 9.5 hours per day) that comes to 2.68 weeks (102 ÷ 38 = 2.68). When the carpenters have finished their work, the painter will spend a week on the exterior. Realizing that there may be a few days' lag time before the painter arrives with his crew and that the scaffolding may be useful for miscellaneous other tasks, I decide to allow for five weeks of scaffolding rental. However, the estimate remains "open" because my estimate may be off to one side or the other, and if it is, my actual costs will be higher or lower than estimated.

Equipment charges are like service charges in that equipment may be rented from outside vendors rather than owned by your company. Equipment and charges for it break down into four categories:

- Small tools, which can be grouped with the consumables and charged as a component of labor burden, as discussed earlier.
- Larger power tools. The purchase, maintenance, and repair costs for larger power tools, such as table saws, compressors, and compound sliding miter saws are most efficiently and effectively recaptured as part of the markup for overhead. True, use of such tools will vary job to job. But exactly allocating the cost of, say, table saw use to a particular project gets you into the foolishness we've alluded to earlier as "spending a hundred bucks of estimating time on a ten-dollar item."
- Rented equipment. Rentals of equipment, like rentals of services, are typically time dependent. However, the length of time of the rental may be harder to forecast than for services such as scaffolding. For example, if you rent a power shovel and gas-driven compressor to dig out a basement, you may be able to produce an estimate based on an assumed length of time for the work. But the actual time and cost may vary from the assumption in accordance with subsurface conditions and the difficulty of digging out and removing earth. For such reasons, for equipment rentals it is often advisable to categorize the charge in your estimate as an allowance (We will get into allowances in Chapter Fifteen. If you need a definition now, please see the Glossary).
- Company-owned heavy equipment, such as a dump truck, an excavator, or a concrete pump. Here estimating becomes more complex. If you want to sidestep the complexity you can, of course, simply aim to recapture the costs of such equipment with overhead markup as you do for power tools. However, here we are talking about much more substantial costs than those for a power tool like a table saw. Just lumping those costs into overhead, appealing though that shortcut may seem, can significantly skew your understanding of your costs and your estimates.

Here's why: Including the costs of heavy equipment in overhead will inflate your Overhead charge significantly. As a result, you will be marking up jobs by larger amounts to recapture your overhead costs. In some cases, you may be marking up excessively. For example, for interior renovations where you probably will not use your heavy equipment at all, you will—since the equipment cost is included in Overhead—be charging

Estimating the costs of owned heavy equipment

Estimating the cost of using heavy equipment owned by your company on a specific project is an involved, multi-step process.

To begin with, you must figure out the full cost of ownership, and not only as a lump sum, but on a per-month basis. If you don't buy the equipment outright, cash on the barrelhead, you must total all payments—down payment and monthly payments with interest—to determine your purchase cost. You can then amortize the cost of the equipment over its expected useful life to determine your monthly cost of ownership.

For example, say that you purchased a dump truck and Bobcat for $100K out of your cash reserves. You expect to get 10 years' service out of the machines, after which they will be obsolete or nearly worn out or both and not likely to bring more than $10K when you put them up for sale. In that case, your cost of ownership will be $90K (100K - 10K = 90K). That comes to $9K a year, or $750 a month, for 10 years.

However, ownership is not your only monthly cost. You will have to insure, store, and perform periodic maintenance of the equipment. To keep our numbers simple, let's say those additional items will cost you another $250 a month. That pushes your ownership costs to $1,000 a month (750 + 250 = 1,000).

The costs of deploying the equipment to a specific project will run higher still. To arrive at the full cost of deployment, you must make still more calculations. First, you figure annual costs of ownership. They come to $12,000 (12 × 1,000 = 12,000). Second, you determine how many months of the year you are likely to make use of the equipment. That's tricky because workloads can vary so much month-to- month and from year to year. But let's say that you figure that you will be using the equipment about half of each month, or a total of six months out of a year. That means that ownership cost per month for the months you actually use the equipment is $2,000 (12,000 ÷ 6 = 2,000). To that figure add costs for pickup and delivery to a site, for fuel, and for repairs in the field. Now, finally, you have arrived at the full monthly cost of the heavy equipment (though that cost is for the equipment only. Labor cost for the operator is a whole other issue).

Of course, if you don't own your own equipment and do not have a skilled operator on your crew, you will have to hire a sub who owns and operates his own equipment. But before investing in your own equipment you will want to take a close look at whether or not the costs of subcontracting will come anywhere close to the costs of ownership. There's a good chance they will not. Regularly, I see builders who buy impressive pieces of equipment only to leave them standing idle the great majority of the time. Those builders would have been much better off renting equipment or hiring subs as needed.

That's especially the case because there is yet another cost to owning equipment, one often left out in the kinds of calculations we are making here, namely opportunity cost—the amount you could have earned with your money had you invested it rather than spending it on underutilized equipment. I won't torment you with all the calculations here. But if your cost for ownership comes to $12K a year and you save half of that by renting or subbing, that's $6K a year (or somewhat less if you sock it away after taxes) you can put into a retirement account. Over the course of a career that investment can grow to a truly impressive amount.

Checking your estimate

- Let the estimate sit for a while before taking a fresh look.
- Determine a systematic pattern for reviewing the estimate and stick to it.
- Check plans, specifications, and addenda to make sure all work has been covered in the estimate.
- Eliminate any double coverage from the estimate.
- Look out for misplaced decimal points.
- Focus especially on the high-cost items in the estimate.
- Provide for buffers against the inevitable oversights that occur during estimating, by including amounts for:
  - Completion
  - Subcontractor slippage
  - Margin of safety in labor rates
- Compare total labor hours in estimate to expected length of project.
- Check General Requirements as percentage of direct costs.
- Check subcontractor charges as percentages of direct costs.
- Proof math underlying direct cost totals.
- Make sure totals for divisions have been accurately transferred to the summary sheet.

too much Overhead. Conversely, you may be marking up less than you should be on those projects—say, construction of a new building—where the equipment is heavily used.

Neither excess nor inadequate markup is desirable. You want the projects for which you actually use your heavy equipment to carry the burden of your ownership of that equipment. Likewise, you want the projects where it is not used to be free of the burden.

Accomplishing that goal and accurately charging for heavy equipment in your estimate requires some serious number crunching. If you do not own or plan to own heavy equipment, but prefer to hire an operator with his own equipment when one of your projects requires it, you will get little benefit from reviewing the calculations. For that reason I have included the review in the sidebar on the preceding page. You can take it on or just move on the next section, as you prefer.

Auditing your direct costs

With labor, material, sub, service, and equipment costs all calculated and entered in your spreadsheet—and not before—-you can total your direct costs. A word of caution here: Avoid peeking at total costs, either for divisions of work or the project as a whole, before you have entered all costs for all items. As one successful builder puts it, "You want to keep your mind free" during estimating. But if you see the totals for divisions and the project mounting up, your mind may instead freeze with concern that your price is running too high to be competitive.

The concern may nudge you into using unrealistically optimistic labor productivity rates, ignoring complexities, disregarding access problems, or downplaying completion costs. All in all, you may undershoot significantly on direct costs. That is not what you want. You want your direct cost numbers to reflect not your hopes of building the project and your concerns about losing it, but its financial realities.

Once you have figured and totaled all your direct costs, you should perform an audit, i.e., a systematic review, of your estimate. Ideally, a second skilled estimator performs the audit. The owner of a large construction company says, "One nice thing about having two people look at estimates is that you are less likely to overlook something huge… We are always missing stuff." But the typical, smaller construction company is fortunate if it has just one person skilled at estimating, namely the owner.

If you are the only estimator in your company, after completing an estimate, set it aside for a day or two. Then give it a fresh look, taking at least these steps:

- Go through the plans and specifications, including all addenda, and make sure that all major items and assemblies are covered, either in your estimate for work to be done by your crew or in a subcontractor's bid.
- Make sure that for the high-cost items—those that involve large quantities of work, unusual complexity, or expensive materials—your figures look reasonable.
- Check your multiplication, especially for big numbers.
- Be alert for any misplaced decimal points and, if you are using an automated program, for any erroneous default figures. (More about that in Chapter Nineteen when we discuss the use of Excel).

Horrible errors occur regularly both in estimates being composed on a computer and those being done manually. To make matters worse, the damage does not stop with the initial error. It is amplified when you figure overhead and profit. For example, if you miss an $8,000 item in your estimate, and are marking up your estimate by 10% for overhead, that's another $800 of shortfall: ($8,000 × 10% = $800). Bottom line: review your estimates not only for omissions but also for mistaken entries and miscalculations.

Make sure as well to embed in your estimate any necessary buffers against the oversights that inevitably occur when you figure the cost of a construction job.

Sage advice

The Associated General Contractors' manual on estimating recounts the story of a contractor who left the HVAC work—which would have amounted to a quarter of his bid—completely out of his estimate.

"That seems ridiculous to prudent contractors," says the AGC, "but it actually has happened many times. Every bid is potentially a successful or disastrous venture. There is no substitute for looking at the detail of the estimate during quiet times before the bid."

Paul Cook, the estimator whose superb books are described in Resources, notes that "mistakes, errors, and slips crop up in every estimate just as they do in championship archery or other target sports." Therefore, he advises, review all estimates thoroughly before submitting your bid.

You will catch some of your errors and be able to correct for them. Estimating does enjoy this advantage over archery: If an estimate misses its target, you can pull that estimate and take another shot.

I have suggested such buffers at several points in our discussion. They include: An appropriate percentage for completion at every division of work done by your crew. Amounts to cover slippage in subcontractor quotes. For complex or unusual jobs, a modest margin of safety in labor rates.

Now you can run a few vital final checks:

- *Determine the percentage of total direct costs allocated to General Requirements.* For example, for a project with $40,000 in direct costs, if

$4,000 is allocated to General Requirements, that amounts to 10% of direct costs: (4,000 ÷ 40,000 = 1/10 = .1 = 10%). Make sure the result is in line with your usual percentage for General Requirements for projects of a similar type. If not, review the General Requirement items to see where you might be running off.

- *Especially if you have only one subcontractor bid for a given trade, check it against the total costs of the project.* For example, say that the electrician you have been working with regularly has bid $9K on a job with $54K in direct costs. That's close to 17%. But your records show that for the type of job you are bidding, electrical usually comes in at around 11%. (You don't keep such records? Paul Eldrenkamp has written about their value in one of his outstanding articles. You will find them at his website, which is listed in Resources.). You will want to give your sub a call. Maybe he made an honest mistake or maybe there are unusual complexities to the work that justify his high price. But it's just possible that he is opportunistically pumping up his bid because you have been giving him every job that comes along. If so, you will want to rein him in.
- *Especially important: Compare the total labor hours provided for in the estimate to your sense of how long the project should take.* For example, say that you are counting on using a three-person crew for the job. You expect them to put in 40 hour weeks. For all three, that's 120 hours per week. Your estimate indicates you have allowed for a total of 1,200 hours. That means you are allowing 10 weeks for your three-person crew to get the job done: (1,200 ÷ 120 = 10). Does 10 weeks seem reasonable for the size and type of job you are bidding? Is it in line with the length of time needed for similar projects in the past? If so, you are at least assured that the most treacherous part of your estimate (of any estimate), the calculation of labor hours, is not horribly off.
- *Run a "proof" of your spreadsheet math* by adding your figures two ways—as suggested in the spreadsheet illustrated on page 244 of this chapter—and checking whether or not the sums match up. For example, total all the individual item costs and compare the result to a grand total of the subtotals for all the divisions. If they are equal, you have proofed your math.
- *Check that the totals for divisions of work have been accurately transferred to your recap (summary) sheet.* Total the figures on the summary sheet and compare that figure to the total for all divisions on your main spreadsheet. If the totals agree, you've successfully completed your audit.

If you are doing large jobs and proofing your estimates for them with an adding machine, you will find the work tedious. The tedium may prompt you to move to estimating with an electronic spreadsheet such as Excel. It will convert proofing from a grueling chore into one that requires merely a click or two with your mouse. But before turning to Excel in Part V, there is another phase in the process of nailing your numbers that we must discuss. That's the crucial business of expanding an *estimate* of direct costs into a *bid.* To do so you will add a charge for the overhead costs of operating your company. Then you will add a further charge for profit, a judgment call that demands recognition of the challenging financial realities we all face in the construction business.

PART FOUR

TAKE COMMAND

CHAPTER 14

Overhead, Truth, and Profit

Beginning to see a problem

After working his way through college with summer construction jobs, Chuck Tringali decided to turn away from the white-collar world and learn carpentry. A few years later, he acquired his general contractor's license and opened a one-person company specializing in residential remodeling.

Chuck attracted plenty of work, first from friends, then from friends of friends, then from their friends. You know how it goes: You do good work. You are reliable. Your customers perceive you as fair and are eager to tell other people about the good builder they have found.

Chuck thought of himself not as a guy out to rack up big profits but as simply wanting to "make things and help people." He felt amply rewarded when a building inspector took photos of a deck Chuck had built so that he could show other builders in the area how the work should be done. He felt even more rewarded when, for a tight budget, he renovated a "crappy little cut-up Victorian" into a pleasant home for a young family.

COMING UP

- The inescapable reality of overhead
- A traditional way to categorize overhead costs
- A frugal builder's view of overhead costs
- "Profit costs"
- Why you must earn a profit
- The fixed percentage markup method of recovering overhead
- The capacity/time-based method of recovering overhead
- The gross profit margin method of recovering overhead and realizing a profit
- Profit as a policy decision
- Overhead and profit goals, hopes, and realities
- A not-for-profit business?

But even as his satisfaction mounted, Chuck said, he was "beginning to see a problem." Though he felt good about being fair and even generous to his clients, he increasingly felt, also, that he was not treating *himself* fairly, as least when it came to pay. Chuck did think his hourly compensation for work at his job sites was okay. The problem, as he saw it, was the complete lack of financial compensation for all his costs and work *off* the job sites. Chuck was not getting paid for the weekend and evening work of keeping his operation going—not for taking and making phone calls, keeping the books, maintaining his truck and tools. He was not getting paid for the demanding work of creating estimates and bids for his projects. In fact, he was not even charging for the time he spent designing the projects. As for the risks that go along with building—all the costs for everything from minor callbacks to major failures that can arrive long after a project is completed—forget it.

Though he had begun to see that he did have a problem, Chuck was uncertain as to what he could do about it. He was having trouble adjusting his prices upward from the "friends and neighbors" rate that he had established when he opened his company. "I'm not sure how to move forward, how to pivot my hat," he said.

Do you recognize something of your own beginnings in Chuck's story? I do. When I started out as a general contractor, I charged only the carpenter's wage I was accustomed to getting for my work on the job site—nothing more. When I bought my first orbital jigsaw and experienced its smooth cutting action, I was so pleased that I instinctively thought of my acquisition as part of my pay. I had no idea that its purchase was part of something called "overhead." I'd never heard the word. Later when I was introduced to the concept, it felt intimidatingly sophisticated and ominous, implying something hanging over me that could come crashing down at any moment. That, at least, was not such a bad take. Mismanaged, overhead *can* take you down.

The problem I had and that Chuck was recognizing is commonplace among craftsmen who attempt the transition from working for wages to running their own companies. One builder put it this way: "I got into the building business because I loved working with my hands. If I could have afforded it, I would have paid people to let me work on their houses." Eventually, he realized that was exactly what he was doing. By leaving his legitimate and necessary off-site work charges in his customers' bank accounts even while taking on long-term risk, he was, in effect, subsidizing his clients' projects. He was gifting them his money for the privilege of working on their homes.

What overhead?

When start-up builders are pushed to recognize the reality of off-job-site costs, they often respond dismissively. "I don't have any overhead, or none to speak of," they will say.

"Really?" a more experienced and savvy builder might answer. "How about your truck and your tools?"

"Oh, I would have those anyhow. I'm a truck guy, man. I like tools."

"How about the time you spend visiting potential clients and estimating and bidding?"

"I get that done at nights and on the weekends so I can still get in my hours on the job. Estimates don't take me long. I know how much most of the work is going to cost, framing, subs, and so on. Other stuff, like windows, I can get it pretty close. I work T&M anyhow."

"Have any clients decided to not pay you because they think you took too long? Never mind that. Back to 'no overhead.' How about answering the phone and paying the bills and keeping the books?'

"My wife takes care of that."

"Oh, so she isn't paid for her work? She's a slave? What happens if she decides she's tired of being a slave and decides to focus on a career of her own? Or leaves? Sounds like you would be up the creek without a paddle."

"Oh, no. That's not going to happen. We've been married twelve years."

When you are starting out, you may have to eat many of your overhead costs. You are building a business. Your uncompensated outlay of time and money is part of your initial investment in that business. But you do not want to let the fact that you are not yet recovering those overhead costs lull you into telling yourself, "I don't have any overhead." That's a habit that, once started, can be hard to stop. The California contractors' license board reports, "The number of contractors who are not aware of their [actual] overhead is amazing."

Selling, General, and Administrative expenses (SGA)

SELLING EXPENSES

- Printed business cards
- Digital business cards
- Job site signs
- Truck signs
- Mailings
- Web sites
- Social media
- Networking group dues
- Contractor listings
- Advertisements
- Award application fees
- Home shows
- Charitable donations
- Entertainment
- Handouts for prospective clients
- Meetings with prospective clients
- Sales commissions

GENERAL EXPENSES

- Yard purchase or rent
- Yard maintenance
- Shop purchase or rent
- Shop maintenance
- Office purchase or rent
- Office maintenance
- Utilities
- Office furnishings and fixtures
- Property insurance
- Property taxes
- Technology
- Computers
- Printer/copier/scanner/fax
- Accounting software
- Estimating software
- General business software
- Landlines
- Smartphones
- Internet service
- Vehicle purchase and depreciation
- Vehicle operation & repairs
- Vehicle insurance
- Construction equipment depreciation
- Construction equipment maintenance and repairs
- Construction equipment insurance
- Contractor's license
- Business licenses
- License bond
- Special certifications
- Theft insurance
- Health insurance
- Liability insurance
- Warranty work
- Goodwill work
- Bad debts
- Interest on debt
- Cost of capital

ADMINISTRATIVE EXPENSES

- Project management
- Office management
- Design and specification
- Estimating
- Bookkeeping
- Human resources (hiring & firing)
- Payroll service
- Office supplies
- Postage
- Bank charges
- Research
- Education: Magazines, books, classes, conventions
- Computer training
- Business training
- Consultants
- Builder's association fees
- Travel including bridges, tolls, parking
- Accounting and legal services
- Phone time
- Owner's draw

The inescapable truth is that if you are operating as a builder—whether as one person with a pickup truck and an office in the corner of your garage, or as a larger company employing a dozen office staff and fifty guys in the field—you *are* burdened with overhead. You may pay with dollars for some of the numerous overhead items listed on the previous page. Others you may pay for with time. You may manage the costs tightly and minimize them. You may let them drift upward (as many builders, unfortunately, do). But pay them you will. We all do. They are our costs of being in business. If we don't or can't pay them, we are out of business.

Even as they become steadily better equipped to improve their clients' properties, their own space becomes steadily less habitable.

One useful way of comprehending the full extent of overhead is to adopt formal accounting language and think of overhead as SGA—selling, general, and administrative expenses. As the sidebar illustrates, all the many items of builders' overhead settle nicely into those categories.

Many of the SGA expenses listed in the sidebar are self-explanatory. A few deserve explanation and comment. Let's begin with the first category: selling expenses. Now, in the twenty-first century, you are almost certainly going to want to invest in an Internet presence. Potential clients expect to find you on the web. When past clients want your services again, they will look for you on the web. If you are not visible there, you risk losing those clients to someone who is—someone the clients come across while they are looking for you. Creating and maintaining web sites and other types of web connectivity will cost you money or time or both. (Time, by the way, as the old saying goes, is money; and after you have invested a lot of it without compensation, you will have lost a lot of money.)

Other selling costs may be optional. But builders do generate substantial amounts of work with, among other methods, imaginative truck and job site signs, booths at home shows, mailers, and listings that purport to screen the builders they present to the public. Some well-established remodeling firms employ salespeople, paying them salaries that they account for as overhead. You may decide to minimize your cash outlay for marketing (as I have). But even so, you will have one inescapable and substantial sales expense—the many hours you will spend (as I have) talking with prospective clients on the phone and at their property.

With regard to the next category on our SGA list, general expenses, you do not have as much freedom to pick and choose as you do with selling. No matter the size of your business, you will almost certainly experience, in some measure, most of the general expenses listed in the SGA sidebar.

- *Yard, shop, and office.* Builders who say, "I don't have any overhead" are overlooking the fact that there is a cost to operating out of their homes. Their yards, garages, and living space fill up with salvaged material, tools, and office equipment. A computer ends up on the dining room table. A laser, planer, and chainsaw are stored in the entry hall. Even as they become steadily better equipped to improve their clients' properties, their own space becomes steadily less habitable. Because they don't recapture, via an overhead charge, the costs of using their homes as their place of business, only they and their families are burdened with the costs. If they eventually move their business out of their homes, the previously ignored overhead costs suddenly become out-of-pocket costs. They must now write checks monthly for everything from rent or mortgage through utilities. Since their customers are not accustomed to contributing to any of those newly visible overhead costs, they may be put off by the abrupt increase in charges necessary to cover them. Now instead of losing habitable space, the builders are losing money.
- *Technology.* Just a few decades ago, the only office equipment a construction company really

needed was a desk or two, chairs, a typewriter, a corded phone, a stapler, pens and pencils, and miscellaneous office supplies. Now, even a compact and frugally run construction company requires all the costly items listed under Technology in the General Expenses section of the sidebar.

- *Equipment and vehicles.* Builders must own many more tools now than in the past. The tools are not as durable—plastic has replaced steel, manufacture has moved from the United States to Asia—and they need more maintenance and more frequent replacement. The tools can enable increased productivity. They can reduce the direct on-site costs of construction. But they also add to overhead.
- *Warranty work.* Some builders elect to charge for warranty work, aka "callbacks," as a component of overhead, allocating a percentage of direct costs on each job to cover it. My own preference is to view it as a hit to my retained profits and to allow for it as part of profit markup—an additional charge beyond even direct costs and overhead and a crucial subject we will get to shortly.
- *Goodwill*: Closely related to warranty work are defects and problems that you take care of even though, strictly speaking, you may not be obliged to. "That scuff mark the customer had her housekeeper scrub away," explains one contractor, "We know we probably were not responsible for it, but we sent her a check anyhow. We want to enjoy her goodwill."
- Finally, we come to a major yet often overlooked general expense item: The *cost of capital.* If you are in the construction business, you will have invested capital (i.e., both money and the monetary value of your time) in your operation in three ways: (1) In the form of those uncompensated overhead costs that you absorbed to get your company going. (2) In the equipment you added once you were under way, and in your yard, shop, and office—whether they are at your home or elsewhere. (3) As the funds you have (I hope) socked away in a financial institution as an Operating Capital Reserve Account (OCRA). In other words, you've got money in the bank, ready to be put to work for your company when it is needed.

An OCRA is essential tool of business. John M., a builder of custom homes, now recovered from the bankruptcy he experienced earlier in his career, advises, "Every builder should build up cash in his or her business. When I finally had cash in the business, my life became so much easier." That cash—which he prudently keeps at 8% to 10% of revenue—has made it possible for him to ride out recessions, to comfortably cover unexpected expenses, and to pay his bills on time, always. "You want to keep your reputation pure," he emphasizes.

All that money in your OCRA, as well as your other capital investments, entails a capital cost, or "opportunity cost." The OCRA must be readily available for use in your business when needed. It must therefore be kept in an account such as a money market account that you can draw from on a moment's notice. The other capital, meanwhile, is tied up in the business assets you make use of every day, which means you are losing the opportunity to invest that capital elsewhere (hence the term "opportunity cost").

There are a variety of ways to recover capital costs. But for the typical light-frame construction outfit, the most straightforward way is to view them as opportunity costs and to charge for them is as part of a markup for overhead. You want to recapture the opportunity cost, minimally at 7% per year of your total capital investment, for historically, over the long term, that is roughly the average return on a prudent stock and bond portfolio.

For example, say that you have \$100K tied up in your business: \$20K to get your company going, \$45K worth of equipment including your pickup, and \$35K in an OCRA (20 + 45 + 35=100). Your opportunity cost can be figured as \$7K (\$100K × 7% = \$7K) annually, or \$583 a month that you need to recapture as part of your charges for overhead.

Note that recovering opportunity cost is different from recovering the money you laid out for a tool. Say you spend $1000 on a saw that is gong to last five years. You recover the $1000 by "depreciating" the saw over, say, the five years of its life at $200 a year, including that $200 in your annual overhead, and recovering a portion of the overhead on each job. Meanwhile, you will also want to be recovering the opportunity cost associated with the $1000 you have tied up in the saw at $70 a year ($1000 × 7% = $70). In other words, the saw is

Overhead drift

The SGA divisions are one useful format for viewing overhead costs. The bar chart below suggests another way of categorizing them, one that relates to choices about overhead management—which in turn relate to your ability to bid for projects, especially during hard times.

Inescapable out-of-pocket overhead is the cash you absolutely must lay out—for computers, insurance, and to maintain equipment. It can be kept to remarkably low amounts. I've managed to get down to 2%.

Erasable overhead is additional SGA expenditure that you determine will strengthen your company—but that you take on in such a way as to be able to ditch it in a hurry when recessions storm ashore. Carl Esposito observed that his chief competitor's 30 vehicles, each displaying the name of his company, gave the competitor a powerful market presence. But ownership of the trucks also burdened him with ineradicable overhead; he would be paying for those trucks regardless of how good or bad business was. Carl achieved market presence with a more flexible expenditure. He paid his employees $200 a month to display his company sign on their vehicles. When the market goes south, Carl erases that overhead, explaining to his employees that the company needs to go into survival mode, tightening its bids to preserve their jobs, and will not be able to pay for the truck signs until the recession recedes.

Owner's draw is both essential and erasable. You have to pay yourself enough to live on. But the lower you keep your essential personal expenses, the more painlessly you will be able to minimize draw and tighten your bids when times get tough. Builders who have taken loans to purchase "contractor's toys" (luxury trucks, fancy offices, ego-scale equipment) can't reduce their draw—and, therefore, can't tighten their bids—without losing their toys. They do lose them. You see the toys for sale at auctions during every recession.

Business sages have been preaching attentive overhead management for a long time. Mark Kerson, a successful general contractor and author of *Elements of Building*, quotes one sage, Ben Franklin: "Beware of little expenses. A small leak will sink a big ship."

Inescapable Overhead	Erasable Overhead	Owner's Draw

costing you money every year in two ways: By wearing out. And by denying you the chance to invest the $1000 in something that gains value rather than losing it. Moving down the SGA list, you come to the third group of expenses: administrative. For the most part they are straightforward. If you employ project supervisors or managers, however, you have a judgment call to make. You may want to recover their wages or salaries, together with related labor burden, project by project as part of direct costs. However, if they are sometimes running jobs but at other times attending to office tasks like estimating, you may find it better to include their cost in overhead. Alternately and perhaps more accurately, though it requires more book keeping, you can split their cost between direct and overhead costs.

The cost of other staff is clearly an overhead cost. And don't overlook human resources costs, namely the cost of hiring and firing. It's time-consuming and expensive work, whether you or an employee is doing it—and it is especially costly if your operating style or other factors create significant turnover.

The remaining administrative costs are largely self-explanatory. Sometimes they are steep. If you do not have a substantial OCRA, then bank charges for an emergency line of credit can pile up. If you must call on lawyers, if you have to hire an accountant to unscramble badly organized books and straighten out your taxes, if you need a computer consultant to support your use of software or fix a computer glitch, or a business consultant to help you steer your company, you will spend a bundle. During my career, I have laid out little cash for such services. But that's only because I have spent a great deal of time at self-education, rather than hiring other people to educate me. In other words, I endure the costs, too. I just tend to spend more time and less cash. You may choose to spend more cash and less time so that your time is freed up for other tasks you prefer, such as marketing your company's services or working at your job sites.

One way or the other, if you are to succeed as a builder, you will have to invest some combination of money and time in research, education, and training—not only for yourself but also for your employees. The construction business is changing and is likely to continue changing rapidly due to an onslaught of requirements for energy efficiency, evolving building codes, changes in material technology, and advancing information technology (not to mention huge changes in the global economy). Whether you run a one-person or a fifty-person operation, you will forever be investing in the acquisition of strategies and practices to deal with and capitalize on those changes.

For most light-frame companies, the largest of the SGA expenses is, or at least should be once the company is well-established, the last one in the list: owner's draw. As owner of your construction company, you will take on management tasks from sales calls, to maintenance of your shop and office, through accounting, and—of course—estimating and bidding. You must bring in pay for that work, and to be perfectly clear, that's pay above and beyond what you pay yourself for producing work on your construction sites, whether as a laborer or lead carpenter or something in between.

The pay you give yourself for on site work should be figured as hourly wages, just as the pay provided any other crew member is hourly wages. Those wages, whether paid to a crew member, or to yourself, are part of the direct costs of on-site production. On the other hand, the draw you take for running your company is more akin to salary that will be figured on a weekly or even monthly basis. And that draw, or salary, is part of overhead cost (i.e, your SGA expenses). Therefore, you should provide for the draw as part of your overhead charges for taking on projects. How much of an owner's draw should you take for running your company? Perhaps just a token amount when you start out and are still earning most of your income from working with your bags on. You may have very little idea what you are doing as a businessperson, as was certainly true in my case at the beginning of my career. Maybe, to begin with, your owner's draw should be a mere 50 bucks a month, just enough to get you in the habit of providing for a draw. Later, when you know what you are doing, the draw can grow to a substantial amount, as we will discuss in the later sections of this chapter.

Why profit?

A few years ago, I got a call from a friend who had put together an estimate of direct cost for the complete renovation and remodel of a three-story San Francisco Victorian. A superb craftsman, my friend intended to be both lead carpenter and project manager himself. He had estimated the cost of his own labor for the project at a fair market rate for a top lead carpenter, incorporating not only wages, but also all the labor burden he would have experienced had he hired someone else to work in his place. As he packaged the estimate into a bid, he had added an additional amount for overhead, even including an owner's draw. But when it came to adding something more for profit, he balked.

"I am charging so much already," he told me, pointing out that with just direct costs and overhead figured, his clients were already looking at a price tag in excess of a million dollars. "Is it really right to charge more?" he wondered.

Yes, I insisted. The dictionary definition of profit is simply income from a product or service above and beyond what it costs you to produce it. But given the realities of construction, it's wiser for us, I suggested to my builder friend, to view it as coverage for additional costs that we are likely to experience—if not immediately on any given job, then over time. With luck and his usual relentless attention to detail, I added, he might emerge from the Victorian project with a good part of any profit he included in his price to his client still in his bank account. But chances were that at least some of it would get gobbled up by unpredictable events, if not during the reconstruction of the Victorian then afterward, during the ten years he would remain legally responsible for the job. In short, in the real world of construction, profit is not gravy. Profit, is not, as the dictionary definition suggests, income that you can count on. It's a charge for costs you are likely to experience in one way or another and at one time or another.

My friend did understand that there was an awful lot that could go less than optimally on a complex project like his Victorian rebuild. Still, he was uneasy. "Do I deserve a profit on top of everything else?" he wondered again. Back and forth we went. Finally, he agreed to add a little something in the way of profit to his bid.

If you have read my book on running a construction company, you may recognize this story. Here, however, is the conclusion of the story, which had not yet occurred when I first published it. My friend celebrated with his very pleased client what he assumed to be completion of the project, then loaded up his tools and moved to New England to build his dream homestead on the wooded acreage he owned there. A year later, he got a call from the now-worried owner of the Victorian. A siding product and detail—installed in exact accordance with the architect's drawings and specs—had begun to fail. The events that followed are complex. Here, however, it is enough to report that after having paid another builder for what turned out to be a pricey but unsuccessful attempt to repair the work, then finding himself with an increasingly anxious and angry client who was muttering about a lawsuit, my friend had to load his tools on his truck again, drive back across country, and work for a month and a half without pay to rebuild the faulty design.

Did the minimal profit he had hesitantly agreed to during our conversation cover the cost of his giant callback? We never penciled it out. I doubt it.

Such severe hits to profit occur regularly in construction. In fact, they occur so frequently that I have come to think of "profit" not, as is conventional, as extra earnings or as a reward for taking on the responsibility of running a business. Instead, I view it as a third division of the *cost* of building. In other words, if you are in the building business, you experience three kinds of costs: Direct costs. Overhead costs. Profit costs.

Even if you are a conscientious builder and capable manager, over time at least some of what you include in your bids as profit will be sucked back out of your accounts. It will be consumed by "profit costs," those costs arising from future events that cannot be defined or predicted with certainty when you are compiling your estimate and bid. These profit costs overlap somewhat with ordinary contingencies, those incidental items you can provide for in your estimate with lines for completion or with a safety margin of a few percent in your labor rate. But they can also be much more severe. And they can come from sources having little or nothing to do with on-site production. Moreover, as the accompanying sidebar suggests, they can hit you at any time—before, during, and after (even many years after) you undertake a seemingly profitable project.

Sources of Profit Costs

Before Construction Begins

- Item(s) overlooked during estimating
- Underestimating of items
- Bids leading to jobs beyond your capability
- Bids resulting in a volume of work beyond your capacity
- Delays caused by building departments
- Delays caused by designers
- Client delay or cancellation of a project
- Under-pricing jobs to stay busy

During Construction

- Abrupt inflation of material costs during a job
- Contradictory or incomplete plans or specs
- Change-order conflict
- Project halted due to inadequacies in design
- Projects halted due to client indecision
- Freebies to placate upset clients
- Layout errors
- Weather delays and disruptions
- Labor shortages or strikes
- Bankruptcy of subs
- Suppliers failing to deliver material on time
- Defective material

After Construction

- Warranty and goodwill work
- Failures of completed work
- Bad debt and bandit clients

Anytime

- A client damages your reputation via social media
- Key personnel leave your company
- You or other key personnel become ill
- Litigation
- Theft
- Vandalism
- Accidental equipment or vehicle destruction
- Computer and software breakdown
- Embezzlement
- Tax and insurance audit penalties
- Recession
- Bankruptcy

Before construction of a project begins, despite your best efforts, you can miss items during an estimate or you can underestimate the costs of building them. If that happens, you still must install them correctly, because in the long run, hasty work will increase profit costs, compromise your integrity and reputation, and betray the clients who have put their trust in you.

Delay of projects can be even more damaging than estimating errors. The planned start date for projects can be set back by building departments slow to process permit applications and by clients beset with unexpected events of their own. In fact, clients sometimes cancel projects after accepting a bid, and occasionally even after signing a contract for construction. A recession strikes. Clients see the value of their assets plunge. They panic. You get a phone call or e-mail saying "we just can't go ahead" (with that project you had planned to start on Monday).

Yes, profit is essential. It strengthens a company in two ways. It provides protection. It opens up possibility.

When projects are delayed or canceled, you must scramble for work to keep yourself or a crew employed. To get work, you may estimate and bid too tightly—then, when you begin to lose money on those underpriced jobs, you are in a deeper hole. For such reasons, the consultant and author Michael Stone warns against bidding without adequate profit just to keep your guys working. If there are no jobs that you can get with a profit included in your price, he says, then everyone "just goes home. That's the nature of construction."

True enough. But don't think that by just sending your guys home you will evade profit costs. If you don't keep them working, you undermine their trust in you. At best they will be upset with you. You may lose them altogether as they accept a job offer from a competitor known for making sure his or her people have work all the time. Replacing employees is expensive. Finding and hiring new workers is labor-intensive work. Once you have hired someone, it takes the new person a while to get up to speed in your operation. So either way, send your guys home or keep them going on a tightly priced job, you are vulnerable to serious profit cost when a project is delayed or canceled.

During construction, a whole other assortment of profit costs can occur: You can discover that plans, which appeared coherent during estimating, contain glitches that require labor-intensive adjustments. You may not be able to get paid for making those adjustments, but instead, as one builder puts it, have to placate clients or designers and hand out freebies, just "write a check and pay for things you should not have to pay for."

Your workforce, or a key part of it, can suddenly become unavailable or lets you down. A lead gets seriously ill, is hurt, or abruptly demands vacation time or family leave. A subcontractor goes bankrupt. A supplier fails to deliver special order material on schedule. As a result, the whole project is bottlenecked, with one delay dominoing into the next. For example, the built-to-order windows are held up for four weeks. As a result, exterior siding and trim cannot be installed. That puts the thermal boundary, drywall, and all interior finish work on hold. By the time the windows are in, the drywaller is tied up with other commitments. Work stops for two more weeks while you wait on him or her. Meanwhile you must pay the ongoing expenses of running your company. That eats into your planned profit on the job; profit costs are mounting.

Profit can be impaired even after construction is over. When you build, you are, in effect, making short-term loans, using material and labor that your clients will pay for in full only after the work is done. Like any other loans, these can go unpaid, leaving you with a so-called "bad debt." Some builders have experienced bad debt so often they include an amount for it in their price for every job. Now and again, a builder is held up by a bandit client,

Selling price

Even successful builders tell stories of struggling during the early years of their careers with the need to charge for more than on-site production costs. One veteran recalls that to become more businesslike he had to absorb repeated "shock treatments." Two final jolts brought him face-to-face with the truth of his situation. First, his accountant informed him that if he were to properly allow for such costs as the maintenance of his truck and the use of space in his home, he would see that he was actually losing money on every job he did. "Ridiculous," he sniffed. But then he tallied up his earnings on a large remodel project that had taken him many months to complete. He saw that he had made about a dollar an hour for his work at the project and nothing to cover his off-site costs. He was, in actuality, losing money.

Likewise, Paul Winans, who eventually rose to become head of the National Association of the Remodeling Industry, struggled with the financial realities in his early years. When the legendary industry consultant Walter Stoeppelwerth told Paul that he needed to double his markups if he wanted to build a sound business, Paul became frightened. "No way," he thought to himself. "I will never get another job." That did not turn out to be true. Paul more than doubled his markups, did hundreds of more projects, and built a respected company. Looking back over his career, he would reflect, "The hardest person for me to sell [on adequate markups] was not my clients, it was the person living inside my head."

Both veterans long ago came to terms with the truth that Chuck Tringali was just beginning to see when I spoke with him. Direct construction costs cannot—if you want to prosper as a builder—equal selling price. You cannot sell a job to a client for your costs of labor, material, and subs at the job site—even if a lot of the labor is your own. You must also charge for your off-site work and costs. And you must charge for the unpredictable costs, often arriving long after projects are complete, that inevitably result from the rewarding, but financially hazardous, enterprise of operating as a builder. In short, you must charge for overhead, and you must charge a profit.

A SOUND BID

Direct costs
- Labor
- Material
- Subcontractors
- Services

+ Charges for off-site costs
- Sales
- General
- Administrative

+ Charge for business risk and development

Selling Price (Bid)

someone who has repeatedly bilked construction contractors, but who has not been screened out by the builder during project qualification. There are ways to minimize your exposure to bandits, to cancellation or delay of projects, and to other profit costs. We will discuss some of them in the next chapter.

Among the profit costs that can strike at any time there are some, however, against which there is little or no defense. You can become sick. You can get swept into litigation for building failures for which you have no responsibility, because you happened to be working on the site at the same time as the contractors who did cause the failure. You can be hit by theft—on a job site or off it. Thieves have hit my job sites or pillaged my pickup half a dozen times. One event cost me twelve grand. Other builders I know have been hit harder.

Among the other "anytime" items listed at the beginning of this section at a few deserve special emphasis:

- *Embezzlement*. Recently, a contractor friend was telling me enthusiastically about the new bookkeeper he had just hired. The next week, several cops walked into his office and arrested her—for embezzling a million dollars from her previous employer, who like my friend, had been fooled by phony references into employing her after she got out of jail for still another embezzlement. Many builders, because they just don't want to be bothered with accounting and bookkeeping, never learn how to spot and prevent embezzlement (it's easy; just do a web search on "embezzlement prevention" to learn how in half an hour). An embezzler sucks up their profits and takes them home—for good. The huge sums embezzled by my friend's bookkeeper were never recovered. Fortunately for him, the cops got her before she grabbed his company's cash.
- *Recession*: If a builder is unable to reduce overhead quickly enough and attempts to keep going with tight bids, recession can siphon away years of profit. Carl Esposito, a custom homebuilder I mentioned earlier, had built up his operating capital reserve account to $750K. Then in 2007–2008 the "Great Recession" struck. By the time it was over, notwithstanding Carl's attentiveness to overhead management, his reserves had dropped to roughly $100K.
- *Going Broke*: Carl Esposito is an experienced and capable builder. He survived the recession and is prospering again, replenishing his capital reserve. Many builders do not survive. As their profit costs escalate they are forced into bankruptcy, which itself imposes further costs—many hours of uncompensated work, legal expenses, confiscation of company assets, and along with the financial costs, humiliation, shame, and anxiety.

Building failures, law suits, embezzlement, bankruptcy, recession, and other brutal costs that eat up hoped-for profit (and often more) strike even able builders with painful frequency. The better news is that some builders do manage to retain profit via relentlessly thorough attention to company and project management. They insistently include profit in their bids along with their estimated costs of on-site construction and overhead. They understand that they must have it not only to cover unpredictable costs but also to strengthen their operations. Iris Harrell, the founder of an unusually profitable Silicon Valley design/build firm, insists with evangelical zeal on the need for including profit in selling price. "It's like putting corn in the barn," she preached to her audience at a builder's conference. "It's the seed for future crops. You need it to upgrade the services that allow you to take better care of your clients, to upgrade computers, to buy construction equipment, to buy a building to keep it in," and to secure a life "beyond work life," i.e., to provide for your retirement and that of your people.

Yes, profit is essential. It strengthens a company in two ways. It provides protection. It opens up possibility.

Recapturing overhead

By now, hopefully you are convinced of the need to package a charge for overhead and another for profit into each of your bids along with charges for direct construction costs. Questions remain: How? And how much? Let's begin with overhead and then move on to profit.

You may have already encountered percentage-based formulas. They instruct you to mechanically tack on a certain percentage of direct costs for a job as your charge for overhead. There are several such formulas floating around our industry. I do not use or recommend them. Coming up with an appropriate overhead charge for a particular bid requires more than use of such simplistic, if appealing, formulas. It's a bit more challenging.

It's not that the actual procedures for determining overhead charges for a job are difficult. They involve only a bit of simple math called "marking up"—i.e., adding that charge for overhead to your direct costs. The challenge derives from that fact that your company overhead is a moving target. Even as you are trying to nail it down and to then nail an appropriate portion of it to each job, it fluctuates as costs for all the many items that make up overhead ebb and flow over time.

Even more problematic, the portion of your total company overhead that should be allocated to a particular project can prove elusive. Ask yourself, is it likely that overhead will have the same mathematical relationship to a bathroom remodel in one neighborhood, an extensive structural repair in another, an addition in yet another, and a new building over yonder? Will the identical percentage prove equally appropriate in each case? No, it will not. Coming up with an appropriate overhead markup for such a range of projects requires judgment, discernment, and thoughtful analysis—though, really, not all that much of it, just more than many builders seem willing to do.

Even experienced builders, apparently uneasy over the need for analysis, sometimes avoid it by strictly limiting the range of work they undertake. Industry consultant Wayne DelPico observes, "The need to predict...[overhead] costs explains why so many contractors develop a market niche." Because they are unable to calculate and appropriately charge for overhead on a wide range of projects, they cannot "compete in a wide variety of construction markets." They decide to specialize in kitchens and baths, or apartment and condo construction, or restaurant build-outs, or custom homes, or structural repair.

Along with specializing, builders avoid the challenges of determining appropriate overhead markups a project at a time by adopting one or another of the percentage-based formulas and applying them across the board. I am going to march through the two most commonly used formulas before describing what I view as a better approach to determining overhead for a project. Because these formulas are so widely embraced and are pushed by a variety of industry consultants—in other words, they are out there, waiting to ensnare the innocent—I feel it may be helpful to you to be aware of them and of their appeals and shortcomings. You may feel, however, that you can somehow manage to get through the rest of your life without reading my description and takedown of what I view as inadequate approaches to handling overhead. If so, you can quite reasonably skip over the next few pages and move on to the next section, "Capacity based markup."

The most simplistic (and, therefore, most tempting) of the percentage formulas involves simply adding the same fixed percentage of direct costs for each and every job as a markup for overhead. For example, if you estimate your direct costs at $10,000 for a repair project, you add 10% of those costs (i.e., $1,000) into your selling price to cover overhead ($10,000 × 10% = $1,000). For a kitchen upgrade with $20,000 in direct costs, you'd add $2,000.

Where do builders who use such a formula to mark up direct costs for overhead get the percentage? Not, unfortunately, in too many cases, from any overhead calculations of their own. Rather, they have heard the figure, which very often is 10%, at the lumberyard or a favorite pub, from other builders who are using it because they, too, when they realized that they had to charge for more than direct costs, picked up 10% somewhere around town. In fact, today, while I was preparing to write this chapter, I got an e-mail from a young builder. He had determined that he was now well-enough established to begin charging for overhead. How much would he charge? "10%," he told me. Why? "Well, isn't that what you are supposed to charge?"

Even using that arbitrary 10%, he will have taken a big step up from forgoing overhead charges altogether. And he is a smart guy who understands that he must steadily strengthen his understanding of his numbers. Likely, he will soon make the next move favored by builders who start off by marking up for overhead using a constant percentage. He will:

1. Total his overhead costs for the past year.
2. Total his direct costs for the past year.
3. Divide the overhead total by the direct costs total to arrive at a percentage.
4. Use that percentage to figure overhead charges for future jobs.

This procedure, detailed a bit more in the sidebar on the next page, has the appeal of simplicity. Therefore, it is used not only by builders just coming to terms with the necessity of charging for overhead. It is a procedure that builders settle into and use for their entire careers.

This fixed percentage markup method may, at least for a time, work out okay for builders confining themselves to jobs that are consistently of similar type and scope (say residential kitchen and bath remodels, or small additions, or tenant improvements) and whose annual overhead costs and volume of work remains fairly stable. In that case, overhead will amount to a fairly consistent proportion of direct costs from job to job. It can be recaptured with reasonable reliability by marking up direct job costs by a consistent percentage. Even then, however, the fixed percentage markup method does have two major pitfalls:

First, *figuring overhead on the basis of a past year's total* <u>*revenue*</u> *and applying the resulting percentage to* <u>*direct costs*</u> *for the next year*. Making that mistake—really, nothing more than a simple math error, but one that is widespread—you will substantially undercharge for overhead. Here's an example:

For year *x* you bring in $100K in revenue, of which $80K is direct costs and $20K is overhead.

Then you make the mistake. To figure overhead markup for the next year (*y*), you divide $20K by $100K (i.e., total revenue rather than the $80K in direct costs). That gives you 20% (20 ÷ 100 = .2 = 20%) as your overhead markup.

Then, in year *y* you have the same revenue, direct costs, and overhead costs. You mark up direct costs by 20%. As a result, you charge for only $16K in overhead (80 × 20% =16). You fail to recapture the other $4K.

To fully recapture your overhead you should have initially divided $20K by $80K. That would have given you a 25% rather than a 20% markup. Using the 25% for year *y* you would have recaptured your entire overhead because $80K × 25% = $20K.

Second, *using a percentage figured on the basis of numbers from an especially good year for a more typical or bad year*. Here the problem is that the percentage is being used robotically and without understanding how it should vary with year-to-year changes in financial conditions, i.e., to what economists refer to as the "business cycle." Example:

You have a good year with direct costs of $800K and $100K in overhead including a draw for yourself. Using those figures, you calculate that to recapture your overhead for the next year you must mark up direct costs by 12.5% (100 ÷ 800 = .125 = 12.5%).

The following year, a recession sets in. Your volume of work falls by half so that direct costs come to only $400K but your overhead—because you are locked into most of it and cannot ratchet it down in proportion to the drop in revenue—drops to only $80K.

Sticking with your percentage based on the previous year's performance, you mark up the $400K in direct costs by 12.5%, thereby collecting $50K in overhead charges (400 × 12.5% = 50). You are short of recapturing overhead by $30K (80 – 50 = 30).

To actually recapture your full $80K in overhead you would have had to use a markup based on a projection, made a year in advance, of that $400K in revenue. Such a projection is almost impossible to make with assurance. How do you know at the beginning of a year what your revenue for the year will be? You don't. For that reason, builders who rely on the fixed percentage for overhead markup based on a past year's revenue often experience the kind of shortfall described in the above example—or the reverse happens. They overshoot their target when they use figures from a slow year for a boom year. Either way, their markup method is haphazard.

If you've got a handle on both the procedures and the potential pitfalls of marking up direct costs for overhead via the fixed percentage method, good! A great many builders do not. A *Journal of Light Construction* (JLC) editor attending a seminar on overhead markup found that over half the audience was prone to making the errors described above. And it's no wonder that they are. A widely sold book about estimating and bidding offers an example of marking up which makes *exactly* the second of these errors—using a percentage figured on the basis of numbers from a good year for a bad year.

Builders seeking to step up from the percentage markup method of calculating overhead often move to the so-called Gross Profit Margin (GPM) method. I am even less of a fan of the GPM method than of the fixed percentage method. It seems to have been brought into construction estimating and bidding by guys with MBAs trained in financial analysis. However, while GPM may be useful for comparing the gross operating results of corporations, it is, in my view, an awkward and imprecise bidding tool.

If you are interested in learning more about its use and faults, and are willing to battle through some tricky logic, please go the sidebar on the next page that is titled "Gross Profit Margin Method." There you will find the method described and critiqued. Or, you can just move directly to an alternate and in my

Percentage markup method of recapturing overhead

Step One:

Total your overhead costs, including an appropriate owner's draw, for the previous year.

Step Two:

Total your direct costs—including all costs for labor, material, subs, and services—for all on-site construction for the previous year.

Step Three:

Divide overhead total by direct costs total to arrive at a percentage.

Example:

Overhead total for 2017: $44K

Direct construction costs for 2017: $336K

Overhead calculation:

44 ÷ 336 = .13095 = 13.1% (rounded off)

Overhead markup on direct cost of each job for 2018 will be 13.1%.

Note: I do not recommend this method (See text for reasons).

Gross profit margin method

The first thing to understand here is that "gross profit" does not actually refer to profit. "Gross profit" refers to overhead and profit combined.

When you use the GPM method, you actually figure for overhead and profit simultaneously with the formula Selling Price = Direct Costs ÷ (100% – GPM). Here's how:

1) Project your hoped for revenue and hoped for gross profit for a year. Next, calculate the gross profit margin as a percentage of revenue – for example, for the year you hope to enjoy revenue of $600K with $180K for overhead and profit and, therefore, a gross profit margin of 30% (180 ÷ 600 = .3 = 30%).

2) For a particular job, first estimate direct costs, $40K for example.

3) To figure your selling price for the job, plug $40K and (100% - 30%) into the formula: Selling price = $40K ÷ (100% - 30%).

4) Calculate as follows: 100% - 30% = 70% = .7; $40K ÷ .7 = $57K = selling price.

Builders who favor the GPM method apparently do so because with GPM they can use a smaller percentage figure than with fixed percentage markup and yet arrive at the same dollar amount. (I will leave the math to you). They find the smaller percentage easier to sell to clients.

The greatest problem with the GPM method is this: Overhead and profit are very different creatures. By determining them with one computation you can blur the distinctions and fuzz understanding of financial performance.

That is exactly what does happen. An industry consultant informed an audience of builders that if they just made sure to include a generous gross profit margin in their bids they would do fine. But the fact is, as one builder in her audience pointed out, you could have a high GPM but still have a shaky business if your GPM is made up mostly of overhead and includes little profit. No one in the audience indicated any understanding or support for his entirely valid point. And the consultant breezily dismissed it and continued merrily on her way.

Derivation of Gross Profit Margin Formula

The strange-looking GPM formula is derived algebraically from a more straightforward formula as follows:

1. Note that Selling Price × (100% - GPM%) = Project Direct Costs.
2. Divide both sides of the formula by (100% - GPM%).
3. Cancel out (100% - GPM%) over (100% - GPM%) on the left side of the equation.
4. Arrive at Selling Price = Project Direct Costs ÷ (100% - 70%).

In other words, what you are doing is setting up a proportion—selling price is to 100% as direct costs are to 100% minus your GPM%. {Selling Price/100% = Direct Costs/(100% - GPM%)}.

With the proportion set up, you can solve for your selling price. Fun, eh?

view much better, even if less widely used, method, coming up next.

Capacity-based markup for overhead

Using capacity-based markup, you do not rely on mathematically derived percentages (as both the fixed percentage markup and GPM methods do) to recapture overhead. Instead you include overhead in your price for a job on the basis of (1) your capacity to do work, with that capacity being determined by the number of project leads (or, alternatively, the total number of workers) you employ, and (2) the amount of your capacity the job uses.

Let's get right to a few concrete examples:

You are in your start-up phase—or simply prefer a very compact operation—and you are serving as both manager of your company and project lead so your capacity is one job at a time (Yes, you might be wrapping up a few items on a previous job while concentrating on a current job. But to keep the example simple, let's just assume you complete one job before moving to the next). Let's assume that your overhead (including a draw for running your company) for your one-job-at-a-time operation is $40,000 a year, which works out to $770 a week (40,000 ÷ 52 = 770 rounded up).

To put it a little differently, your capacity is one job at a time, and you experience overhead costs of $770 weekly to support that job. Thus, for a three-week repair project, you would include $2,310 (3 × 770 = 2,310) for overhead in your selling price for the job. For a 12-week remodel, to recapture your overhead, you would include $9,240 (12 × 770 = 9,240).

For our next example, assume that you have decided to grow your company. Its capacity is now two jobs at a time with a lead running each job while you concentrate largely on marketing, estimating and bidding, and otherwise managing the company. Your overhead has now grown to $100,000 (including the enlarged draw you will need because you will be working almost full-time running the company).

That works out to $1,923 per week for your company as a whole, or $962 a week for each of your two leads and projects: (100,000 ÷ 52 = 1,923; and 1,923 ÷ 2 = 962 rounded up). Therefore, if you bid a 10-week project for one lead, you will aim to recapture $9,620 (10 × 962 = 9,620) to cover your overhead for that lead's job. If you bid a 16-week project for the other lead, you will need to include $15,392 (16 × 962 = 15,392) for overhead in the bid.

Capacity-based markup can be adapted to larger and more complex companies as well. For example, say that you have an operation with two project managers (aka supervisors), overlooking eight projects between them. Now your overhead (including the supervisors' salaries, your draw, yard and office, and office staff) could easily run to $500K a year, or $9,615 a week. In that case, you would need to recapture $1201 per week for each of your eight projects: (9,615 ÷ 8 = 1,201).

Some builders who use capacity-based markup prefer to base it not on their number of leads or project managers but on the total number of crew members they employ. They allocate overhead using a three-step process.

First, they figure their overhead for a selected period.

Second, they calculate the total hours they expect the guys on their crews to work during that period.

Third, they recapture overhead at an hourly amount per worker.

For example, a builder might:

1. Figure overhead as $8,000 per four weeks.
2. Calculate that his crew of six will put in a total of 960 hours for a four-week period: (4 weeks × 40 hrs/week × 6 = 960 hrs).
3. For each hour of crew work, include $8.33 in his bid: (8,000 ÷ 960 = 8.33). If he figured a project was going to take five weeks for a three-person crew working 40 hours a week, he would include $4,998 in his bid for overhead (3 × 40 × 5 × 8.33 = 4,998).

Capacity-based markup and the time factor

The core idea behind capacity-based markup for overhead is this: Overhead does not vary primarily in proportion to the dollar cost of a job but in relation to the time the job takes. Here's why: Your capacity to take on jobs is determined by your ability to run those jobs properly. For a job to be run properly one of your leads, if not yourself, must be managing it. Therefore, the longer a job takes, the longer it will tie up a lead and the more of your company's total capacity for running projects will be used by the job. The more of your capacity the job uses, the more of your total overhead it also uses.

A somewhat different and helpful way of explaining the centrality of time to overhead calculations is offered by industry consultant Martin King. Mr. King's key point is this: the costs of construction fall into two categories—closed costs and open costs. He defines closed costs as fixed. He defines open costs as time-sensitive. The cost of a box of nails for a job is a fixed or closed cost. The cost of labor by your own employees is an open, time-sensitive cost because the more time an on-site task requires, the more it costs.

Overhead, King maintains, is also an open cost. "The amount of overhead a project uses depends on the length of time it runs, not its dollar amount" because the longer the project takes the longer it ties up a portion of the assets—equipment, office capabilities, and so on—that support your work in the field. And the longer those assets are tied up, the more they cost.

King goes on to suggest that one of the worst mistakes an estimator can make is to treat an open cost as a closed cost. The percentage-based methods of marking up for overhead, so widely used in the world of light-frame construction, are exactly that kind of mistake, he says. Why? Because with percentage-based markup you are figuring overhead partially or even largely on the basis of closed costs—those for material and subs.

As a result of calculating overhead as a closed rather than an open cost, concludes King, you will overcharge for it on some jobs and make unexpected profits, while on other jobs the reverse happens. You may get lucky and come out okay, but that's not good enough. You want your financial results to be based on accurate estimating and bidding, not chance.

The crew-based approach to capacity-based markup seems to me to be less useful than the lead-based approach. Let me explain why via an example: Suppose that you figure overhead on a per-worker basis and have two leads and seven other workers. One lead is running a job with only one helper because work at his project is being installed largely by subcontractors. The other lead is running her job using only four of your other workers, while the remaining two are kept busy cleaning up your yard.

Since you are charging overhead only for workers out at the job sites, the result will be that you will fall short of recapturing your overhead even though you are at full capacity. If you had allocated overhead to the leads and their projects rather than per worker, you would recapture all your overhead.

Capacity-based overhead markup, as the above examples indicate and the sidebar on the preceding page amplifies, delivers the advantage of factoring in the time a job takes as a basis for calculating overhead charges. But it is also true that it is not a perfect solution to the tough challenge of charging accurately for overhead. (There are no perfect solutions; reliably figuring and allocating overhead is a huge challenge even for large corporations armed with battalions of sophisticated financial analysts.) First difficulty: The size of a job, and not only time and capacity, does influence overhead. If you have one lead working with a helper and using just a few subs at a small addition and another putting up a new structure with a crew of five and numerous subs, the two leads and two jobs are not going to be using exactly the same amount of overhead per week. The larger job will be burdening your company with more weekly office work and heavier use of tools and equipment.

How do you solve the problem and make an adjustment? One approach: Start with your capacity-based figure for overhead, then reduce it appropriately for the small project and increase it appropriately for the larger job. Of course, it's easy to say "Adjust appropriately." But there are no easy adjustment formulas. Adjusting will require looking at your specific SGA costs.

However, if you do calculate an adjustment, you will likely be surprised at how small the difference in weekly overhead burden is for a small project vs. a large one. An adjustment of roughly 10% to 15%, upward for the large job and downward for the small job, is likely to be all that is needed. For example, if your base rate was $900 a week, you might bump it to $1050 a week for big projects and push it down to $800 for smaller ones. In other words, a kitchen remodel can burden you with nearly as much overhead per week as the construction of a new home, and a rebuild of the bathroom at your local Y with not a great deal less per week than the build-out of a new restaurant.

Another difficulty: If you allocate overhead to a project on the basis of capacity and time, and the job runs longer than expected, your overhead allocation will be proportionately inadequate. For example, say you figure the job will run seven weeks. Instead it runs nine. The result: You will be short two weeks' worth of overhead. On the other hand, if the project takes less time than estimated, overhead markup will be proportionately excessive.

Here the solution is to improve your estimating. Develop a file of labor productivity records (as discussed in Chapters Eleven and Twelve). Rely on them to accurately estimate the time projects will take, along with the costs they will incur. Use that accurate estimate of time to more reliably determine markup.

No matter what method of markup you use, you must *first* nail your estimates of direct costs. If you underestimate or overestimate direct costs, your markup will be off to the same degree. For example, if you were using the percentage markup method and estimated direct costs at $20K but they ran to $24K, a 20% markup would be $800 low. (20K × 20% = 4,000; 24K × 20% = 4,800; 4,800 - 4,000 = 800). No overhead markup method will save you from bad estimating of direct costs, and bad estimating of direct costs will throw off your markups.

A third difficulty: Capacity-based markup can go astray if it is based on an inaccurate assumption of utilization of capacity. That's a mouthful. Let me again explain by example: Say you figure your overhead at $770 a week for a 52-week year, but in actuality—because of bad weather, vacations, dead time between projects, and so on—you are busy only 47 weeks a year. You will be losing out on five weeks' worth of overhead markup. The solution? Divide your anticipated overhead for a year not by 52 but by a more realistic number of weeks derived from your experience.

Robert Criner, a nationally known remodeler based in Virginia, suggests a final difficulty. Criner

Profit Markup Considerations

Market conditions
- Your need for work
- Competitive position
- Other opportunities

Project
- Type
- Size
- Location
- Schedule
- Client
- Designers
- Plans & specs
- Specialness

Your values

fears that capacity/time-based markup can render you noncompetitive on smaller projects by pushing your price for those projects too high. He's got a point. If you allocate the same weekly overhead to a kitchen renovation as to a large addition, your price may prove noncompetitive for the kitchen.

On the other hand, the reality is that if the kitchen is taking one of your leads and the addition the other, the two jobs are, in fact, using roughly the same amount of your capacity per week. And they are, therefore, incurring roughly the same amount of weekly overhead costs. A solution to Criner's concern is, therefore, not to turn a blind eye to the reality of your overhead costs for the smaller job. Rather, you take it on with an appropriate allocation of overhead and, if necessary to win the job, whittle down your profit markup, accepting the risk of having little to buffer yourself against future profit costs which might arise from the job. Failing to recognize and recapture the ongoing week-to-week overhead costs of running your company is just not acceptable. You can't pretend the costs aren't there. Knowingly going without profit on a job may occasionally be necessary or even worthwhile. You do so to keep a crew busy, to establish yourself in a new market, or for the joy of building a special project. But you don't want to do it often.

Realizing profit

Marking up a bid to recapture overhead is one thing. Adding profit is quite another, for overhead and profit are very different in nature.

Overhead is ongoing. It's knowable. With a well-organized accounting system and reliable bookkeeping, you can see just how overhead is mounting up as the year passes. You can reliably figure what it will be in the near term because it is largely made up of a series of costs that you have incurred in the past on the same regular schedule that you will incur them in the future. Yes, for sure, those costs can fluctuate. Insurance, gas, phone and Internet fees, consultant charges, office supplies, and other overhead expenses can suddenly jump up (and now and then slide downward). But they are not likely to all move at once. Unless you deliberately add a big new chunk of expenses, such as ownership of new trucks, your overall overhead will probably not move abruptly over the near term or even for a year and more. Therefore, you can systematically recapture it job by job.

Profit, as I have suggested earlier, can, like overhead, realistically be viewed as coverage against costs. But unlike overhead costs, profit costs are barely knowable in advance. Minor profit costs may be somewhat predictable; you know you will occasionally provide clients with modest freebies or suffer the loss of a few tools from a job site. The serious ones are not. For a long period, they may not occur at all. Then they may come in a surge—as a lawsuit, a serious failure of completed work, and an injury to a lead in charge of a major project—that engulfs you over a short span of time.

One builder, who has kept his company alive and growing over four decades, likes to sum up the difference between overhead and profit this way: The one "is a fact," the other "a policy." In other words, overhead is a fact that you can know and manage as surely as you can know the direct production costs at the job site, i.e., not perfectly but closely. But when you add profit to arrive at selling prices for your projects, you are making

Predictable and pernicious profit costs

For my own projects, I have, thanks to the abilities of my leads and subs, had to do little warranty work. Because I have had to deal with it only infrequently and irregularly, I think of warranty work as a "profit cost." Though I never know exactly when or how much, I know that now and then a callback will take a small bite out of my retained earnings.

Other builders, however, who have tracked their costs for warranty work, have found that at least the routine items (think sticking doors, leaking faucets, loose deck boards, window problems) come at a fairly regular rate. They elect to charge for them as overhead, as a normal cost of operating their business, rather than as profit loss. In fact, two builders I interviewed for this book, both of whom do sizeable volumes of work, report that they consistently have found that adding .5% (half a percent) to their overhead reliably covers their usual warranty work.

In his *Fine Homebuilding* business blog Fernando Pagés describes a particularly useful way of tracking (and of managing) such routine costs. For each year, he accounts for his warranty costs, not by the projects on which they actually occurred, but in a separate general ledger account—as if they were all part of a single job. Makes sense! Attending annually to that profit-diminishing work is a separate project—one that you could name, for example, "Profit Costs 2020."

Fernando's approach has a benefit beyond showing you the total impact of your profit costs. If you track the costs as he does—namely by trade—you will see where you need to improve your work. Window installation? Plumbing? Coatings? Something else? Maybe client selection!

Beware, however: The routine and trackable profit costs are not the only costs that you need to cover with a profit markup. The big bears—major company breakdowns, serious project failures, lawsuits, recessions—lurk in deeper, darker caves and come out less often. But they are the profit costs that do, by far, the most damage. To be prepared for them, you must earn profits and sock them away in reserve accounts.

policy decisions, i.e., judgment calls. The judgments will vary greatly from company to company. You can readily find ten knowledgeable builders who will tell you that marking up for profit is easy: you just add on a percentage of the costs. However, if you ask them to give you the percentage, you

Buying a job

Now and again a builder will "buy a job," agree to build it without profit or sufficient markup for overhead, and sometimes even for less than the cost of construction. Builders buy jobs, among other reasons, to keep a crew busy, to keep their doors open through recessions, to win a high-profile project that will boost their companies' market presence, to build a project that somehow touches their hearts, or to break into a new market.

Seeking to rapidly expand into new markets, Mike Tolbert bought several jobs over the course of just a few years. To begin with, he contracted to build a custom kitchen for a third of what he knew it would cost him to complete. Mike figured photos of the project, posted on his website, would catapult him into the high-end remodeling market. They did. He considers his $40K loss on the kitchen project money well invested in growth.

Encouraged by his roll of the dice with the kitchen, he took on the remodel of an old commercial structure at a price with a built-in $200K loss. And yes, that project enabled him to break into a booming sector of the commercial market. Soon enough, his company was awarded the contract for renovating a seven-story office building. He took it on at a price steeply discounted from market levels, with only slim profit to buffer against unexpected shocks. Last time I talked with him, he was still building the office tower and was gradually working his way out from under the debt heaped up by the earlier jobs on which he had deliberately taken a loss. Who knows to what new level the office project will lift him…or drop him.

Having come into construction from the high-tech start-up world where failure is extolled, Mike thinks like a tech entrepreneur: Failures are the stepping stones to success. Burn money now to make it huge in the future. Mike happens to be working in a very, very hot construction market. Maybe his strategy will work for a time longer.

But construction is not like high tech, where if you do create a successful product, you can mint and sell a million copies with little additional investment. In construction, failure tends to increase vulnerability to future failure and successes are stepping stones to future success. If Mike's tech world like stepping stones take him up a mountain rather than over a cliff, he will have traveled a path most unusual in the world of construction. I hope to keep you posted on his journey.

may well get a different answer from each one. They are making different policy decisions. Numerous factors figure into those calls, but they fall into the three broad categories:Market conditions. The project. Your values.

"Profit?" a builder who employs 30 people said to me. "It's what the market will bear." Because his company had just (barely) survived the Great Recession, he was keenly aware that market conditions shape profit possibility. He had seen that competition around price can become so intense at times that you must either bid work with no profit included in your selling price or go without work and lay off your people, something he could not bear to do. So he placed bids without adequate markups for profit—and quite likely without enough even to cover his overhead—and lost so much money he had to take out a half-million-dollar mortgage on his home to keep his company afloat.

His story has a happy ending. When the recession ended and the recovery accelerated in his area, his phone began ringing with calls from clients whose first question was not "How cheap can you do it?" but more like "What do I need to do to make sure you can fit me into your schedule?" Because he had not laid off his people, he had in place the skilled and reliable workforce he needed to take on the large projects that again came his way, now with healthy amounts of profit included in his bids. Different market conditions, different profit levels allowed.

Even in relatively strong markets, builders will sometimes go without profit to "buy a job." One builder told me he will consider "selling at a discount" to keep a crew employed during a gap in his schedule. Sometimes he will buy a job to give a deserving carpenter a chance to step into a lead position on an appropriately sized project.

In good times more than bad (when you may feel compelled to go for any opportunity), the project itself, and the client and designers that come with it, will influence your judgment call regarding markup. Is the project of a type and size you can readily handle? Is the schedule reasonable? Are the clients reliable? Can you count on their fulfilling their contractual obligations as well as expecting you to fulfill yours? Are they the sort of folks you will be content to have in your life not only for the duration of the project but for the years afterward during which you are legally responsible for it? If the answer to some of those questions is "No," you may still go for it, but with a high markup for profit, in essence thinking to yourself, "We will take on this project, but with extra coverage for the risk," or "If this character wants to work with us, okay, but we will be going into the project with protection, and the protection is going to cost him."

On the other hand, perhaps the clients have a small budget but really need a better home, and you want the satisfaction of making a difference in their lives, profit be damned. I once rebuilt a home for a single mom with a shoestring budget, i.e., my profit on the job would not have bought me a pair of boot laces. But years later my client spotted me in a coffee shop. She came over to my table and told me, "I used to hate my house, but now I love coming home to it." I am glad to this day that I bought that job.

As for the designer and design: Perhaps the designer has a special gift that allows her, as one architect expresses it, to create those forms and details that catch our eye and make us wish, when we see her work around town, that we could pull over, open the front door, and take a look around inside. Maybe you want to get established with that designer, so much so that you are willing to go nearly profitless (one time, anyhow) in order to gain the opportunity to build more of her designs.

If the designer's work is not all that special, are her or his plans and specifications at least satisfactory? Will your crew be able to work from them efficiently? Is the designer okay personally? Responsive? Fair? Respectful? Or, oblivious to builders' needs, narcissistic, and obsessive—bound to drive up your direct costs and severely impair your chances of preserving a decent profit? If the answer to the last series of questions is "Yes," you may bid the job, but only with higher than usual profit protection built into your selling price.

Along with market conditions and the project, its owners, and designers, your values can shape your decisions around profit markup. Personally, I have wanted my prices to be what I think of as fair to everyone, including myself. "Fair" here means that clients are paying a price that is solidly competitive for the quality of performance they will receive. I would not be comfortable working with the knowledge that a client had made a financial error in contracting with me, that they could have done much better by going elsewhere. At the same time, I want the charges to be sufficient so that my employees, subs, and I can earn a good living while doing work that will make us proud.

For every builder I have met who shares my sensibility, I have met another who embraces a different or even opposite set of values, who believes that maximizing profit is the moral thing to do. Thus, one builder I know consciously charges his "poor" clients less than his more affluent ones. Another, with equal resolution, goes after the same unusually high overhead and profit markups (amounting to about 50% of selling price) for projects of both less and more affluent folks. He argues that otherwise the more affluent would be subsidizing the less well-off clients and that it is not his place to play Robin Hood and allocate his clients' resources for them. Still another responds angrily to any suggestion of practices that might restrain his profit margins, calling them virtually un-American. Another, who also ardently harvests 50% gross profit margins, quotes scripture—"The worker is worthy of his wages" (Timothy 5:18)—to justify her charges. She insists that her markups are indicative not of avarice but of superior emotional health.

How much?

For overhead, the answer is: you must know your overhead expenses and you must mark up enough to recapture them. For profit, the answer is a matter of judgment, since profit is a buffer against unknown future costs coupled with a hope that you will be able to hold on to at least part of it. Given the decisions and judgment calls, it's natural for a builder to wonder, are the markups I am using or have in mind okay? Reasonable? How do they stack up against my competitors?

Answers to those questions are promoted by various industry educators. You will find the answers in books and on web sites. You will hear them at builders' meetings and industry conventions. The answers, however, are all over the map. To try and make some sense of the topography, I will first review the *range of markups* (all figured on a percentage and not a capacity basis, by the way) urged by the industry savants. Following the illustration on the next page, my review will begin at the level of the most dismal markup goals and climb the ladder from there. That done, I will take a look at the markups that builders *actually achieve.* As you will see, there is a gulf between the recommendations and the actual results.

The very most anemic recommendation for markup that I have ever run across—10% for overhead and profit combined—comes, horrifyingly, from a widely sold book on bidding for residential remodeling work. There is no indication in the book as to how the authors came up with 10%. I've wondered if it was a misprint, for we can project the awful results that would come from using it with a simple example: A builder keeps two crews going. They complete 10 remodels per year. He handles all off-site work from sales through accounting.

The builder's direct costs for the ten jobs average around $40K each, for a total of $400K. A 10% markup provides him with $40K to pay all overhead expenses—truck, equipment, computer, phones, insurance, and on and on. What would he have left to live on? Pathetically little at best. And how much profit would he be able to put away as a buffer against those unpredictable costs that inevitably strike construction companies? Zero. If you care to run the numbers yourself, you will see that even with higher-end projects, such as you find in prosperous metro areas, and twice the revenue, a builder using

10% for overhead and profit combined would be better off financially if he were to work instead as a lead carpenter or project manager, letting his boss take the risks of contracting.

Moving up the ladder a step, you come to a markup of 14% to 16%. That may seem like a small bump up from the 10% at the bottom rung. But while you may be going up only a few percentage points, the difference in dollars brought in for overhead and profit is huge; for here the figures come from an entirely different stratum of the construction world than remodeling. They are the range of markups used by a builder of huge "estate" homes with price tags running from four to nine million dollars and occasionally even higher. In good years, with his 14% to 16% markups the builder aims to take a draw of $50K a month (not a typo, that's each month), generously bonus his employees, and also build up his operating reserves.

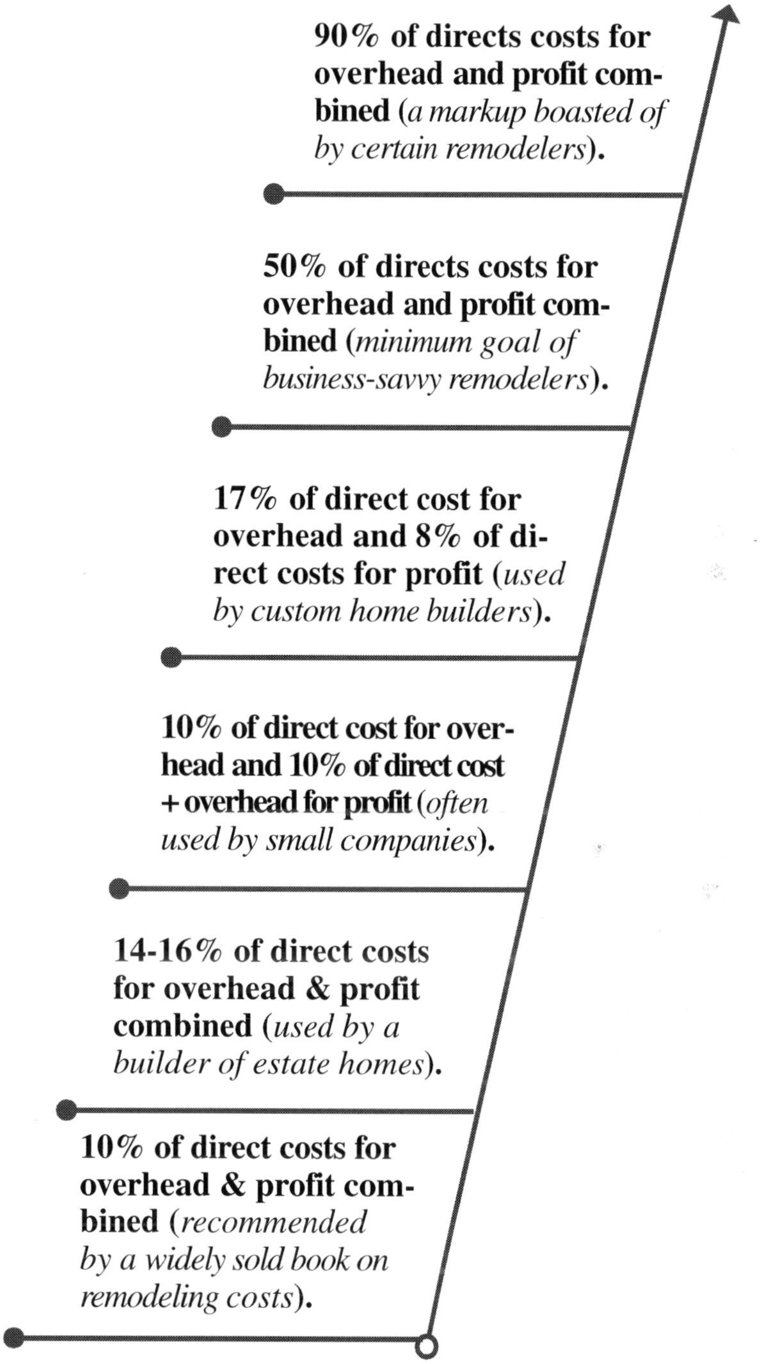

The next step up the ladder brings us to 10% for overhead and 10% more for profit, markup goals that turn up as commonly as F-150s do at job sites. I have no idea where the figures originated. But they turn up regularly in construction publications or discussions. They are, for example, presented without explanation in a recently published estimating and bidding manual, as if they were as much a fact of construction life as the costs of labor and material. As mentioned earlier, 10% for overhead is regularly adopted arbitrarily and on faith by startup builders, who just as readily grab 10% as their profit markup. They cast about for numbers, are told to go with 10 & 10 by cousin Joe who, in turn, picked the numbers up from second cousin Ernie, owner of "Top Doggie Construction," and decide to accept that hearsay as gospel. The numbers often remain embedded for years in the bidding belief system of builders who have never troubled to acquire a serious business education and so are never introduced to alternatives and less

A Range of Markups

90% of directs costs for overhead and profit combined (*a markup boasted of by certain remodelers*).

50% of directs costs for overhead and profit combined (*minimum goal of business-savvy remodelers*).

17% of direct cost for overhead and 8% of direct costs for profit (*used by custom home builders*).

10% of direct cost for overhead and 10% of direct cost + overhead for profit (*often used by small companies*).

14-16% of direct costs for overhead & profit combined (*used by a builder of estate homes*).

10% of direct costs for overhead & profit combined (*recommended by a widely sold book on remodeling costs*).

formulaic thinking. It's 10% for overhead (regardless of what their overhead actually is) and 10% for profit because some higher power apparently declared that is what it is supposed to be, forever.

Among financially more assertive builders and industry advisors—the sort of people who show up at contractors' conventions and take the business end of construction very seriously—you encounter markup goals at the higher end of our ladder. For custom home construction (not the estate-sized places described above but more typical new houses) markups of at least 17% of direct costs for overhead and 8% for profit are urged. "Custom homes" here includes not only new homes, but whole house rebuilds. Such projects are like remodels in name only. When done, they are actually new homes that are made up only in small part—some foundation, some framing, often not much else—of their original structures.

At the next step up the ladder, you arrive at the much higher markups, 50% for overhead and profit combined, recommended by business-savvy remodelers who specialize in smaller projects—baths, kitchens, additions. Why such a big jump from custom homes? The answer derives from the capacity/time factor discussed earlier. With essentially the same overhead costs and the same number of lead people, you can do a much higher dollar volume of custom home construction than of smaller remodels in the same period of time. As a result, even using much lower markups for the custom homes, you can reap as much or more revenue for overhead and profit.

For example: If you charge $300K in direct costs for a custom home and mark up the costs by 25%, you reap $75K (300 × 25% = 75). However if you charge $36K in direct costs and mark up 50% for a modest-sized remodel you reap only $18K (36 × 50% = 18). Therefore, even at twice the combined markup, you would have to do four of the remodel projects in the time consumed by the custom home project to bring in close to the same dollars for overhead and profit: (4 × $18 = $72). And in fact, completing four modest-sized remodels in the time it takes to build a custom home would be a tall order even for a very capable lead. To match the overhead and profit taken in at 25% on the custom home, you likely would have to use a markup even higher than 50% on the remodels.

For that reason, the 50% markup included in the ladder is actually the *minimum* that certain business-savvy remodelers and their advisors recommend. That markup, it should be noted, translates into a 33% gross profit margin (GPM)—the way of looking at and figuring overhead and profit preferred by many builders specializing in remodeling. (Here's the math: $100K direct costs × 50% = $50K markup. $100K + $50K = $150K selling price. $50 ÷ $150 = 33% gross profit margin).

Some industry educators—most notable among them, Walter Stoeppelwerth, who passed away in 2013, have advocated GPMs of 33% and more for remodelers. Walt insisted that a construction business that handles mostly remodeling jobs "is actually a retail service business that requires a tremendous amount of customer communication and coordination" and, therefore, has high overhead.

To small-volume, light-frame builders who do largely remodeling, and who have accepted the 10/10 markups (which translate to a GPM of only 16%) as sacrosanct, the higher ranges advocated by Walt Stoeppelwerth may seem startling. They may seem unreachable or even unethical. But around the country many well-established remodelers, especially those with design/build operations, do aim for combined overhead and profit markups of 50% (i.e., GPMs of 33%) and higher.

In fact, in certain highly affluent markets, you can find companies including markups of 90% to 100% (i.e., GPMs of 45% to 50%) in their selling price. In other words, their charge for a project is double their estimate of direct costs for its construction. When a customer pays them a dollar, their hope is that only half of it will go for the work actually done at the job site. The other half is intended for overhead and profit.

Goals and recommendations are one thing. Results are something quite different. One of the industry's most ardent apostles of profit laments that few builders ever develop the "business maturity" to realize even the 8% profit he advocates as a bare minimum. I am not sure, however, that the problem is lack of maturity. It may instead be that the world of construction presents such serious financial obstacles it is not possible for many builders to sustain more than minimal profitability. In fact, profits often fall so far short of the markup goals touted by industry educators that I wonder from time to time whether, for a great many companies and for the industry as a whole, construction is in actuality a not-for-profit business. (Not nonprofit; that's when profit is not allowed. But not-for-profit, meaning you want to make a solid profit but can't.) I have explored that idea in the sidebar at the end of this chapter.

No matter the odds against sustaining long-term profitability, you do need to attempt to realize a profit and build up your operating reserves with every project you do—because over the long haul you will almost certainly need that buffer in place. A few tips and techniques may help you stay the course.

- If you mark up for overhead and profit separately (as I have recommended), you may want to *apply your profit markup to all your costs*—not only the direct construction costs but also the overhead costs allocated to a project. For example, say the direct costs of a project are $60K, and that you have marked up those costs $9K (i.e. 15%) for overhead, and hope for a profit of 8%. If you apply the 8% only to the direct costs you will create a profit markup of $4,800 (60K × 8% = 4,800). Figuring the profit on direct costs plus overhead, i.e., $69K, will give you a $5,520 markup (69K × 8% = 5,520). While not huge on any one project, over the course of many jobs, the additional profit adds up.

- *Winning too high a percentage of your bids* may signify that you are undercharging for overhead and profit. One remodeler in my area who has a one-man remodeling operation contracts and builds eight out of ten jobs he bids. It is not hard to figure out why. He submits estimates and bids for free and does beautiful work for about half of what his peers charge, pricing his labor on the job sites at under market and including only anemic amounts for overhead and profit in his bids. He would be better off financially bidding higher and winning a lower percentage of his bids. What might be a reasonable percentage of "wins?" Paul Cook suggests the following rule of thumb: Winning 1 of 10 bids = seriously uncompetitive. Winning 1 of 4 = too competitive. Winning 1 of 6 = about right. Cook worked in the world of larger scale commercial construction. Builders that I am acquainted with in other niches aim to win 1 out of 3 of the jobs for which they submit bids. A different ratio may be appropriate for you. What is clear is that at some point—and 80% is certainly past that point—a high winning percentage can suggest your markups are unnecessarily low.

- If you determine that you should or must improve your financial performance, you don't have to jump up your markups in one huge leap. You can *increase markups incrementally, even in baby steps*. For example, one veteran remodeler suggests raising markups "1% a month until your rate of closing deals begins to drop."

Choosing the method of markup that will work best for you, zeroing in on the markups necessary to recapture overhead, and making your policy call on profit markups demands disciplined management and thoughtful judgment. There is another truth, too, voiced by Mark Kerson, in his insightful book, *The Elements of Building*. "Every one of the companies from whom you buy includes overhead and profit in their prices. The ones that didn't are no longer in business."

In construction, however, unlike the vendors Kerson refers to, you are not offering a product for a set price with overhead and profit included. Instead, you set the price and then produce the product. So you not only have to include overhead and profit in your price; you have to defend your markups from erosion during production. In the next chapter, we'll talk about ways to do that.

Construction: A Not-For-Profit Business?

Recently I was invited to speak to a group of experienced builders. I said to them, "I've been wondering, is construction a not-for-profit business—not nonprofit, that's when you don't want to make a profit, but not-for-profit?" I did not mean the question as a joke. The audience, however, responded with a knowing laugh. It seemed they had already come to answers of their own.

For years, I had heard ambitious profit goals touted at builders' events. Just as regularly I had run into established, capable builders whose actual profits were far lower. One of my first mentors told me that her typical bid included about 11.5% above her direct and overhead costs, but she expected to retain over time only about 1.5% as profit. The other 10% would be siphoned right back out of her reserves by what I have come to think of as "profit costs."

A builder, whose remodeling company enjoys national recognition, reported that while he hoped for a 6% annual profit, he regularly realized only around 4% and during a recent year only 2.5%.

A builder of large custom homes in an affluent metropolitan area says that his retained profits run about 2%. Likewise, the founder of a light-frame company that at times employs close to 100 people and focuses on remodeling of everything from retail shops and cafes to whole houses and bathrooms also reports retained profits at the 2% level.

Both the custom homebuilder and the remodeler participate in "executive roundtables," groups of business owners who gather to share business ideas. They report that their profit experience seems about typical for others in their groups. One says bluntly that the higher profit levels preached by the self-appointed construction industry gurus "are for the birds. No one is getting that!"

Of course, these are just anecdotes. What may be more telling are surveys. One, from the National Association of Home Builders, says that the "best remodelers" average about 4% profit—which suggests that for the more typical firm it is likely to be much lower. In fact, a consultant who has made his living by running roundtable groups for remodelers promotes 14% profit markup as a businesslike goal when he is recruiting new clients. But he has released summaries of his clients' actual results over the course of many years. Their slippage from the dream he dazzles them with is huge. They have realized, on average, not 14% but 2.5%.

Other surveys I have dug up report similar figures. But even those numbers may be unduly encouraging, for the surveys cover only companies that have survived the minefield that is the construction business. That's problematic. You can't count only survivors to arrive at the overall results for a whole industry, any more than you can to tally the total impacts of a war. The dead must be accounted for as well.

And here is another dark truth about even the modest profit levels mentioned so far. They were achieved during relatively good years, the decades before the Great Recession that began in 2007. With that experience included in the story, the anecdotes get worse: One builder takes on half a million in personal debt to keep his company going. Another borrows a million and then has to fold his business. A third goes without a draw for several years. A fourth gets lucky; he sells his company and retires just before the recession strikes. Two years later the company folds.

Severe recessions comparable to the Great Recession happen every few decades. Shallower but still harsh recessions occur far more frequently. Recessions are not anomalies. They are a regular feature of the economic landscape you have to traverse as a builder. When they strike, whatever "profit pumps" construction companies have managed to install can go into reverse. Accumulated capital flows out the door. One example: Pulte, one of the largest homebuilders in the country. According to its financial reports over the two decades including the Great Recession, Pulte's profits averaged a few tenths of a percent per year. During good times, Pulte netted hundreds of millions annually. During the hard times it lost just as much.

Of course, even when recession is not in play, profitability is always under attack. As Barry LePatner puts it in *Broken Buildings, Busted Budgets*, construction "bears a risk disproportionate to the return on investment... Years after a building is completed each of the team members remain liable for problems that may and often do arise." That's as true for a small remodeling company as for the commercial builders LePatner deals with in his Manhattan law practice.

If construction is, overall, a not-for-profit business—not deliberately nonprofit, but not-for-profit despite every effort—that is all the more reason to put profit into your bids. If you don't, you will have no buffer against recession and the other drivers of profit costs. The costs will come out of your paycheck, the salary you take for the work of running your company, or your personal savings.

That would be a shame. Because even though it is hard to retain much profit over time, you can earn a good living running a small-to medium-sized light-frame construction company. Though not likely to be a path to great riches, it's a damned good life. If you hone your estimating and bidding and building skills you can, over time, produce decent financial results as well as good structures.

During good years you can enjoy a substantial draw even as you reap a lot of other satisfaction—command over your own time, satisfying relationships, and pride of work. If you choose to live and operate frugally, then invest a sizeable portion of your draw and whatever profit you can retain, you can even obtain financial independence at an early age, then go on to do what you darned well please and find meaningful.

CHAPTER 15

Support Your Bid

Hold that line

You are going after a project. Relying on your labor productivity records and on quotes from reliable subs and suppliers, you have nailed down the direct costs for the project. Using either capacity-based markup or one of the percentage markup methods, you have figured overhead. Sizing up the market, the client, the designers, and the project itself, you've made your judgment call and settled on an additional markup for profit. Adding the direct costs, overhead, and profit together, you've arrived at your selling price for the project and are ready to present your bid.

If you are bidding in the world of public or larger-scale commercial construction, you may be required to present your bid in accordance with a strict protocol—by a precise time, with bonds in place, and with your numbers broken down in a specific format. Your bid will be opened at an appointed hour along with all the others, and the winner, the low bidder, will be announced. His bid will be regarded as a promise. He will contract for the job for the bid amount or face financial penalties.

In much of the light-frame world, however, the bid process is more relaxed. Your bid may be more proposal than promise (though that can depend, in part, on how you word it and on the laws in your area). If you back out of the proposal, the client and designer may be angered, but the price you pay will likely be long-term, in the form of a hit to your reputation, rather than an immediate financial penalty.

On the other hand, if after making your initial proposal you are intent on winning a bid, you may be able to push it as the best bid even if it is not the lowest. Along with the bid he submits to his clients, one veteran builder regularly sends a letter describing the strengths of his company, then follows up with a phone call to answer any questions. As you can see from his letter, excerpted in the sidebar, on the next page, his company has

COMING UP

- Following up on a bid
- Standing by your price
- Creating protective contract clauses
- Stating assumptions
- Using allowances
- Setting up a change order process
- Charging for change orders
- Integrity

exceptional strengths. Nevertheless, the letter can serve as a model even for a builder who has been in business for a much shorter time. If you have reliably served clients and built with skill and care, you too can point to your demonstrated commitment to quality, mention your satisfied customers, and note your personal involvement in your projects. Likewise, you can follow up with a phone call, or better yet in some cases, an offer to stop by and go over the bid in detail.

Sal Alfano, who was a general contractor before becoming editor of the *Journal of Light Construction* describes meeting with potential clients for a small addition. He spent two hours with them, suggesting ways to correct flaws in their design, going over construction details, answering questions, and explaining his estimate division by division. "I was awarded the contract," he says, "not because I had the lowest price, for I was high by nearly 10%, but because I had presented the most thoughtful and thorough estimate."

In my work, I have also found that sitting down with clients, explaining just how I will build their project and how my charges relate to the steps in the process, strengthens their confidence in me. People want assurance when they are contracting for construction. They have heard an abundance of horror stories. While price always matters, *it is not all that matters to clients.* With follow-up, you can improve your chances of contracting with them for construction of their project on the basis of more than price.

Some builders will put their charges up for negotiation when they visit potential clients to review a bid. That may be unavoidable during recessions. During those hard times, it may be that the only clients available are the bargain hunters. On occasion, even during good markets, giving a bit of ground on charges may make sense. For certain projects, I have made cuts in my charges, typically for a small amount of profit markup, and asked each of my subs to do the same. But I am talking about adjustments amounting to one percent or even less of the total price for the project. And they have been made as goodwill gestures toward clients with constrained budgets who had already accepted reductions in the scope of their project so that we could move forward to construction together.

Other builders, with justification, raise an eyebrow at even such minor flexibility as I practice.

Bid support letter

Ward Construction is a 50-year-old structural repair company in Richmond, CA. Their letter to potential clients, excerpted below, can serve as a model even for builders with far less experience.

Dear ______,

Attached is our estimate to correct your problem.

I'm confident we have designed a workable, cost-effective solution. Our reputation has been built on years of quality work... We've had hundreds of repeat customers.

Our foremen are skilled and experienced... Your job will be personally supervised by me, the owner of Ward Construction.

We are available to address any of your concerns regarding our work at any time in the future. If you have a problem five or ten years or more after completion, we will be there. We will call you in a few days to review our proposal and answer any questions.

Best Regards,
Art Ward

One likens making cuts in a bid to voluntarily accepting a bad grade in school. It takes "a lot of good ones to make up for" a single subpar mark, he points out. If you do attempt to lift your "grade point average" back up after giving a client a break, he says, you will have to raise your charges to those clients who have not cajoled you into giving them a discount. These financially more desirable clients end up paying the freight for those he views as less desirable clients.

The consultant and author Michael Stone likewise advises against discounting, especially in response to a client who tempts you with the promise of other work. It should be explained to the client, Stone urges, that "the price of every job is based on its costs and that each job must stand alone." If you cave in to the demand for lower prices, he says, the demands will continue; and once you have shown that your prices are negotiable, you will be forced to continue negotiating, not only with that particular client, but also with the friends he recommends you to. Stone has a point. You don't want your references to go out wrapped in sentences like, "He was a little pricey to start off, but it wasn't that hard to talk him into giving us a really good deal. He'll work with you."

Even if you choose to be less firm with your pricing than Stone urges, you will want to resist what one builder calls "grinders," the clients who relentlessly try to force your bid down with a combination of threats (to give the job to someone else) and manipulation ("I have a guy who will do it for $10K less; and I like him, but my partner thinks you are the right guy for the job"), or by attacking your bid line by line. If your Foundation and HVAC costs are as low or lower than other bids, the grinder will leave those alone and attack your numbers for the Frame or other divisions where you are higher. One builder with thousands of projects under his belt reports he got so tired of being worked over by clients who went after his bid line by line that he stopped producing itemized bids altogether. Now he simply provides clients with a proposal that includes a single number, his price for the entire job. They can take it or leave it.

If clients genuinely want to work with you but price is a problem, you have available better options than discounting. You can suggest that the clients postpone the job until they have the necessary financial resources in place. You can offer to work with them to make reductions in scope of work that will result in a lowering of direct costs and perhaps even of overhead costs and profit. You can combine the two efforts, offering to do the bulk of the job now and showing the client portions that can be reasonably postponed and completed later when money is available.

All in all, there's a lot you can do to win a bid after first submitting it: Follow up your bids with letters and phone calls. Sit down with clients to hear their concerns and answer their questions about your bid. Show clients that you have thoroughly thought through their project and your charges.

At the same time, as builder Robert Hanbury put it in a *JLC* article, you must ignore that voice that tells you, "Be nice, you are charging too much." In other words, stand by your carefully figured bid. Resist any manipulations intended to force you down. Hold that line.

A protective contract

You are not home free after a client has accepted your bid. It will need protection from the erosion that can carry away your potential earnings during the time leading up to construction, all through it, and afterward. Among the most important protections you can put in place is a thorough construction agreement that includes:

- A statement of assumptions regarding construction scope and quality.
- Allowances, rather than firm charges, for items that are not precisely specified in the plans and specifications.
- A clear description of a legally sound change order procedure.

A thorough contract sets boundaries. It makes clear your obligations to the client and the client's obligations to you. Obvious and necessary as that may seem, some builders resist creating comprehensive contracts. They offer a variety of rationalizations such as "My clients would not put up with such a long contract." Then there's the notion, surprisingly common, that it is somehow morally superior to forgo a contract—to work on the basis of a handshake and trust, like in the old and supposedly better days. One builder I know maintains that because his clients all "love" him, the legalistic mumbo jumbo of a contract, relied upon by smaller men, is not necessary for him. Or at least he felt that way until recently, when one of his clients apparently stopped loving him, and for lack of a proper contract he found himself defenseless against her aggressive lawsuit.

A handshake is great. It is a fine gesture with which to end a meeting after you and your clients have carefully reviewed and signed a thorough written contract.

My own take is that a handshake is great. It is a fine gesture with which to end a meeting, but only after you and your clients have carefully reviewed and signed a thorough *written contract*. Thereby, you have elevated mutual trust from a sentiment—the feeling that you are working with good people—to an understanding solidly anchored in law. If, during construction, you have a difference of opinion, you can go back to the contract. It will help you to resolve your differences and get back on course.

A thorough contract will run to multiple pages; my own is nine single-spaced pages. I have never had a client take exception to its length. If I were tempted to cut it, I would probably be brought up short by Michael Stone's blunt warning that you should leave out of your contract only those things you are willing to pay for yourself. I have, however, made my contract client-friendly, with a table of contents, boldfaced headings, and the use of color. In the condensed version of the contents page (illustrated on p. 285), I have highlighted certain areas that are especially important for the support and protection of a bid.

The *project description* must enumerate the design documents that were the basis of your bid, including: All pages of the plans, whether that's just a one-page schematic for a bathroom remodel or "pages A-1 through A-14" of architectural plans plus all pages of the engineering plans for a new structure. All pages of thc specifications. And all addenda, whether produced by the designer or by you.

Schedules and obligations regarding timely performance should be spelled out. You, as the builder, must commit to a start and finish date. One builder advises underpromising and overdelivering. He might allow for seven months from start to finish in his contract, let the owner know he hopes to do better, and aim for six months. I have found that to be a helpful practice.

At the same time, you should spell out that you are not responsible for delays beyond your control, such as bad weather, natural disasters, or an injury to your project lead. You want to clarify to the clients, in a helpful way, that they, too, have an obligation to perform in timely fashion. Provide them with a schedule listing actions they must take and decisions they must make, such as selection of fixtures.

Though you would first want to consult with an attorney, you may even wish to spell out penalties in your contract for client failure to act in a timely fashion. Mercifully, I have not had the experience, but I have heard builders describe the severe emotional and financial burdens of working with indecisive clients. One described construction of a house, initially scheduled to take nine months, that stretched out to three years as he waited, over and over, for his clients to make up their minds. A penalty for client failure to act in a timely fashion may be an especially fair balancing requirement if you are subject to penalties for taking longer than specified for construction.

Your contract should protect your bid by requiring clients to make *timely payments.* Paul Winans, the former head of NARI (National Association of the Remodeling Industry), insists, "Never get behind. If you do, you will never get paid." I have sat in meetings with builders who did not follow Paul's advice and were bilked for tens of thousands of dollars—in one case for $100K—on a single job. A builder I know is now nervously waiting to see if he'll get his last $50K.

These guys are not big-time operators, but general contractors with small- to medium-sized operations who do remodeling or build new homes and small commercial spaces for their local communities. How do you forestall financial hits such as they have taken? One builder, after getting burned for $80K on each of two separate projects, realized it's easier to collect small amounts because owners gain relatively little leverage by holding them back. He has avoided further bad debt by billing frequently for relatively small amounts and collecting within five days.

For my own projects, I have been ultravigilant about payment. I bill weekly for whatever percentage of work has been completed since the last payment. I require payment within three days. I allow only small amounts to be held back till completion of any punch list, for as Barry LePatner laconically notes in *Broken Buildings, Busted Budgets.* "Owners have been known to work over contractors on occasion by withholding final payments." My clients have accepted my payment terms. I have enforced them as necessary, telling one client who tried to hold back a final payment of a few thousand dollars for the construction of her new house that I would check her mailbox on the way to the courthouse, and that if my check was there, I would take it straight to the bank rather than filing a lien on her home. The check was in the mailbox. It cleared. As a result of my vigilance I have only once experienced a bad debt. That hit was for $1,000 and it came early in my career. And it came, of course, from folks who were neighbors and friends, who had pleaded with me to take on their project and with whom I had only a handshake agreement.

This brief discussion does not cover every possible contract clause you may need to protect bids for your type of work or for your area. For projects involving bank loans or public funding, you may want to set up some special procedures to provide for prompt payment. For even small residential projects, you may need to provide multiple legal notifications beyond the mechanic's lien warning noted in my condensed table of contents. They can be essential to protecting your bid. For if they are not in place, you may not have a legal leg to stand on if you do have a dispute with a client. To fully develop your contract, you may wish to purchase a legal manual (contact Builders Booksource via buildersbooksource.com). And because the law governing construction contractors varies from state to state, have any contract you develop reviewed by an attorney really expert in construction law (your local divorce and jack-of-all-specialties guy will not do).

Unfortunately, for certain classes of projects you may not be able to use a contract of your own making, but be required to use AIA (American Institute of Architects) contracts. I am inclined to agree with the harsh warnings expressed by other builder/authors who have variously described these contracts as "troublingly complex," "disconcerting," and putting builders "under the thumb" of a project's designers. I am not much mollified by hearing from my own construction attorney, who, in recognition of the AIA's recent efforts to address builders' concerns, says that he has moved from "never sign an AIA contract" to feeling that "the current editions are markedly improved," so that, if "substantial modifications" are added, he will "reluctantly" accede to using one.

AIA contracts remain hazardous, and your best chance to protect your bids remains via a contract you create together with a good construction lawyer. Bear in mind, though, as you develop your contract, that your purpose is not to build a fortress around yourself while stripping a client of protection. A contract should be balanced and

Contract for Construction Services Cover Page

Company Name
Contract for Building Services

Highlighted on this contract cover page are items that are especially important for the protection of a bid.

Hardison Remodel -- 187 Crescent Circle, Raleigh, N.C.

Assumptions, Allowances, Schedule of Owner's Decisions

ASSUMPTIONS

Not Included: Exterior and interior painting (owner will contract directly with painter of their choice)
Final cleaning (by owner)

Roofing: 40-year composition shingle

Decking: 2x6 select-grade redwood (some small, tight knots)

Cleanup: The project will be left broom clean at the end of each workday and at completion

ALLOWANCES

Exterior door: $700

Tile: $600

Door hardware: $200

Note: These allowances are for midlevel and readily available products. For more costly or complex products and those requiring long lead times for ordering, the builder reserves the right to increase labor, overhead, and profit charges. Likewise, he may decrease them when appropriate for less costly and simpler products.

SCHEDULE FOR OWNER'S DECSIONS:

Exterior door and hardware specification: September 10

Tile choice: October 1

For a small project, you may be able to include your assumptions, allowances, and even the schedule of the owner's decisions on a single page. More typically, your statement of assumptions will be longer.

fair to all parties—for legal as well as ethical reasons. It should not be adversarial but instead should serve to strengthen the trust that helps bring construction projects to successful conclusions. Along with the incorporated documents (plans, specs, assumptions, other addenda), it will reduce the potential for surprises to the client and thereby avoid disagreements during construction. It's a lot easier to establish a clear understanding with clients *before* you start work at the job site than it is to change perceptions afterward.

Assumptions

Earlier I suggested that as you build your estimate you keep a list of assumptions—i.e., additional specifications and clarifications such as those in the accompanying sidebar—to fill in gaps in the documents provided by the project designers. Even if you have a design/build operation, it's wise to compile assumptions as you are imagining the construction of the project and estimating its cost. No set of plans, even those you draw yourself, is ever quite complete, and you will become aware of additional details as you fill out your spreadsheet.

You can create assumptions efficiently, dropping them into a preformatted Word document as you create your estimate. Assumptions cover three basic categories: (1) items of work you are specifically excluding from your proposal, (2) the quality of the items you are including, and (3) the scope of the items you are including.

Items that are not included (NIC) must be clearly defined. The statement on the previous page of assumptions, allowances, and schedule of owner's decisions includes a couple of excluded items that are quite straightforward. But NIC items can be trickier, especially around interfaces of your own work and that of others. Say, for example, that you are building a restaurant in a new commercial complex and are contracting with the owner of the restaurant for that work. In your estimate, you provide for installation of an electrical panel with circuit breakers, but assume that the developer of the complex will be bringing power to the panel. Meanwhile, the developer believes that his lease with the restaurant owner holds them responsible for bringing over the power. Fail to clearly state your assumption (and make sure the restaurant owner has read it and signed off on it), and you may find yourself, as one builder actually did, facing an angry client who, not unreasonably, thinks that providing power was part of what you were supposed to do to get their restaurant ready for operation.
In general, when your work interfaces with that of another contractor, you must clearly state where your work ends and their work begins.

Assumptions prevent surprise and disappointment, and you do not want surprised, disappointed clients. You do not want to find your client, who has just arrived home from her 10-hour shift tending her patients in the critical care ward at the hospital, staring at the redwood decking you have meticulously installed and feeling terribly disappointed because in her mind's eye she was imagining ipe. You do not want clients startled to find an orange peel texture on the walls of their renovated offices when they expected a level 5 smooth finish. You do not want clients to come home to a completed job that you have carefully swept up, but that they expected to find scrubbed and polished by a professional cleaning service. To prevent such surprises, the assumptions accompanying your bid must state exactly what you will and will not be providing in the way of finishes and services if they are not already called out in other construction documents.

To fully protect your bid, you should also specify that your assumptions take precedence over any specifications provided by project designers. There are several particularly treacherous specifications that crop up in design documents that you will want to overrule. They are "as needed," "or equal," "highest quality," and "match existing." The first three are just too ambiguous and leave you exposed to indefinite demands. Who is to decide what is "needed," what is "equal," and what is "highest quality"? You, the designer, the owner, or someone else?

Work that is needed but is not called out explicitly in the design documents should be defined in the assumptions. If the design documents require you to excavate "as needed," for example, your assumptions should specify a defined depth or volume of excavation, thereby overriding the vague requirement. Likewise, "or equal" should be tossed out and replaced with a fixed specification for a particular product. "Highest quality," too, must be overridden and replaced with reference to a specific product. Of course, sometimes quantity or quality cannot be defined with precision before construction starts, due to unknown conditions or time and availability constraints. Then you can call out a specific item as a placeholder, describing quantity and quality exactly but noting that during construction the specification may have to be modified. Or you can simply provide for flexibility with an allowance or change order procedure—topics we will get to in the coming sections.

As for "match existing"? Now there's a phrase that opens the door to naïve expectations and serious disagreement. Does it mean "roughly correspond to," or "closely resemble," or "look exactly like"? An experience of my own, narrated in the sidebar on the next page, illustrates the potential for conflict and costs. For your client's sake and your own, when it comes to matching of existing finishes, you need to clarify what is possible and what you will provide, rather than leaving the client free to imagine an unrealistic outcome.

Allowances

If an item of work eludes exact specification prior to construction, often the best way to provide for its cost in your estimate is with an allowance. For example, if you are building a new kitchen and an appliance has not been specified, you can provide for it with a cost allowance such as $900 or $3,900. Allowances are useful when:

- Design documents leave specification of an item, typically a finish item or fixture, to an owner, and the owner does not have time to make a choice prior to the completion of estimating and bidding.
- A quantity of work, such as a volume of excavated soil to be off-hauled, cannot be reliably predicted.
- Construction of a project is so far down the road from estimating that costs may shift substantially before building begins.

Match existing

A few years back, as I was completing a residential remodel, my client confronted me angrily. "I hate it," she said, pointing to the new oak flooring in her kitchen. I was startled. The installer, "Bill," I will call him, is a topnotch hardwood flooring guy. To my eye his work looked impeccable.

"If he is the best you know, that's pathetic," my client answered, "because this doesn't match my existing floors. It's ugly."

She demanded I have Bill tear out his work and install new flooring. I told her that wasn't happening, that her demand was unfair. In turn, she threatened me. I was "supposed to take care of her," she said, so that she would refer me to her friends. In other words, no free new flooring, no references.

From there, our exchange got harsher still (I don't take well to being threatened, and this client had already proved herself to be a grinder). Eventually, the client, Bill, and I were able to agree that she would mark half a dozen boards that she particularly disliked and that Bill and I would split the cost of replacing them.

Reviewing the incident, I realized that I could have forestalled it with more clearly stated assumptions in my contract. I should have made clear to my client, for her sake as well as mine, that any new material would look very different from her 80-year-old floors, darkened by time, worn by countless footsteps, and clumsily refinished.

Matches are never perfect and often not even close. My run-in reminded me that the limitations need to be spelled out, especially when a designer's specs say "match existing."

Postscript: Ten years after I completed the job, the client called me for a reference to a good plumber. I asked her about the flooring. She loved it, she said. It had held up wonderfully. However, she proved a woman of her word and made good on her threat. She never did refer me to anyone else.

"The beauty of an allowance," says the founder of a Southern California design/build company, "is that it partially releases the builder from the pressure of having to predict the future and guess at unknowns." Yes, true, but "partially" needs emphasis. Like other opportunities for bid protection, allowances entail obligations. You should not lowball allowances to fool clients into thinking the costs of their project are less than they will actually be. If you do lowball, you may (and deserve to) experience blowback. The client may rightly feel ripped off, and resist paying all of the difference between the actual steep cost of an item and your puny allowance for it. You will certainly diminish any chance for future references.

Unfortunately, there are builders who do lowball allowances to make their bids look better. If you suspect you may be bidding against one of these characters, point out to the clients that your allowances are realistic. Ask that they compare them to the allowances used by other bidders. If you are forced to use an allowance for items whose cost is particularly volatile, and it turns out the allowance will be significantly exceeded, let the client know ASAP and in writing. As with contract stipulations in general, you are trying to avoid surprising your client. If the cost of copper needed for flashing explodes upward by 30%, as can easily happen, better the clients find out earlier rather than the day you hand them their bill.

You may wish to even further protect your bid by providing not only for increases in material cost but also for additional labor, overhead, and profit charges in the event that an owner selects items that are substantially more costly or complex than assumed in your allowance. If an owner decides to upgrade, for example, to a stove that is of far higher cost and quality than the midrange product your allowance covers, it's possible that the related labor, overhead, and profit costs may go up as well. The higher-end stove may involve a more elaborate installation with higher labor costs, add to overhead by requiring you to set up an account with a new supplier, and burden you with liability for a more complex and costly item. A client of mine once upgraded during construction from a standard to an ultrafancy refrigerator. It failed repeatedly. I spent many hours getting the supplier out to fix it and, finally, to virtually rebuild it on site. Because I had not charged additional markups for the upgrade, I was not compensated for the value of those lost hours.

In the event that you do provide for increases in labor, overhead, and profit charges, you may also want to provide for lowering them when an allowance item is less costly or complex than anticipated. If your experience is like mine, you will find that clients rarely choose products significantly less costly than those provided for in a realistic allowance. Nevertheless, the balancing clause produces benefits. It emphasizes to your client that while you do need to protect your bid from erosion, you seek to do so in a way that is fair all around.

Change orders

As construction projects move along, changes are regularly made to the scope of work specified in the contracts. Changes can have benefits. They can result in improvements in a project. "They are a chance to make your clients happy, to give them what they want," says builder Dee Bailey. They provide earnings opportunities for the builder, and sometimes satisfying work.

Changes and the costs that result from them, however, are also a potential minefield. Owners are afraid of the costs. Designers worry that they will be blamed for the additional costs by clients who feel that the extra work should have been anticipated in the plans and specs rather than cropping up during construction. Changes and the change orders they lead to can blow up a relationship between a builder, his clients, and the project designer. In order to realize the benefits while minimizing damage, you must handle changes with a well-put-together, fully legal, and fair-minded change order process. Repeat: *must.*

"Change orders" are exactly what the term implies. They are orders for a change in the work agreed to in the contract. While occasionally they

Causes of Change Orders

Hidden conditions

- Boulders, roots, streams, buried debris, collapsing soil, and other obstacles to excavation below ground surface
- Concealed decay and other failure in existing structures
- Covered-over layers of plaster or other finishes
- Substandard, failing, or abandoned electrical, plumbing, and mechanical components beneath finish surfaces

Owner's requests

- For additional work
- For upgrades

Designer's requests

- For additional work due to omissions or errors in plans
- For enhancements to design

Building department requirements for work not called for in design documents

Factors, such as storms, that are beyond a builders control and that result in delay of construction

may call for deletions, change orders are usually for additional work. Even though the client or designer may ask for the work, it may well fall to you as builder to write up the change order. That is the case even if you are working under a contract stipulating that the project designer is responsible for creating the change order. Designers, in my experience, are oblivious to any requirement, even those included in contracts drawn up by their own professional associations, that they write the change orders. In my own work, I have never had a designer or any other party create a change order. I write them up and present them to the owners for their signatures.

To prepare clients for the change orders that will almost inevitably be called for during construction, you must educate them in advance. Otherwise, they won't be prepared—emotionally or financially—for the extra charges or for the delay in the completion of their project imposed by the change. Even clients who have contracted for construction before may be in the dark about the change order process. Their previous builders may have neglected to produce change orders even when justified.

When clients are not provided with a clear change order process, and are surprised by charges for extra work, they may be so angered that they become unwilling to pay for it, insisting that the work should be covered by the initial contract price. In a case for which I served as an expert witness, the client had initially been told his project would cost $40K. As the project neared completion, the contractor abruptly presented the client with a $20K invoice for additional work. None of it had been documented in written change orders. The client, claiming he had been ambushed by the extra charges and had no idea they were coming, refused to pay. Both the client and the contractor "lawyered up" and went to arbitration. The arbitrator allowed the builder only a small portion of his extras—likely less than he paid out in legal fees.

I have repeatedly heard variations on that scenario. To avoid it, provide your clients with advance notice of the likelihood of change orders, of their possible sources, and of your process for handling them. This can all be accomplished by

A Sample Change Order

SAMPLE CHANGE ORDER

OWNERS:__________________PROJECT:____________________________
DATE:________
We hereby agree to make the following changes in the work:

Description of changes inserted here.

INFORMATION FOR THE OWNERS REGARDING CHANGE ORDERS:
Please note that prior to executing the changes, the builder requires written authorization via a copy of this change order form for any additions to the work or deletions from the work agreed to in the construction contract. Causes of changes include but are not limited to:

- Hidden conditions such as boulders or debris beneath the surface of the soil.
- Hidden conditions within existing structures such as hazardous material, decay, and abandoned electrical, plumbing, and mechanical work or hidden layers of plaster or drywall.
- Requests by the owner for additional items of work or for deletion of items.
- Additional requirements by building departments or project designer beyond those shown in the documents referenced in the construction contract.

The owners are strongly encouraged to set aside contingency funds to cover the costs of change orders, for they are a normal part of any construction project. When a change order is required, the builder will write it up using this form and provide it to the owners for their signatures.

Cost of this change: ____________________

Payment for this change order due at: ______

Cost of change: ________________________

Change in schedule from this change: ___(days)
Total change in schedule to date:_______(days)
Adjusted completion date for project:_______

Builder's signature: ____

Owners' signatures: ____

NOTE: This illustration is intended only as an example of a form useful for educating clients. You should create your own form with support from a person knowledgeable about construction law in your area.

including a sample change order and a description of change order basics in your contract.

During a project, if you feel you are dealing with a trustworthy client, you might, for small items, relax the requirement that change orders be signed before any extra work is done. For example, on a remodel project you uncover an abandoned flue that must be removed immediately for framing to go forward. Rather than holding up your crew, you might simply get a voice, text, or e-mail approval to go ahead with the flue removal, agreeing with your client to process a change order later in the day.

But for substantial change orders, never—excuse me for getting insistent here—repeat, *never, ever* execute them without *prior written authorization.* If you fail to adhere strictly to that practice, you will eventually find yourself embroiled in a change order dispute such as the one for which I served as an expert witness. Or, as happened to a builder in my community, you may find yourselves facing clients (two attorneys in his case) who deliberately go along with verbal change orders, happy in the knowledge that they will never have to pay for all the nice extra stuff provided for by the change orders because they understand that without written ones you will have no chance to collect for the work.

Your change orders must include a description or statement of:

- The extra or deleted work
- The charges or credits
- Payment terms ("due upon completion of work" for smaller items; a series of payments such as "due at completion of framing" for larger items involving several divisions of work)
- The total cost of all change orders to date, perhaps including an adjusted total contract price, i.e., the original contract price plus (or minus) all change order amounts
- The extra time needed for construction as a result of the change
- An adjusted completion date for the project

For years, I neglected to provide clients with the adjusted completion date. I assumed that they would see my total of extra days and adjust the completion date for themselves. Wrong! I found out how wrong when a potential new client called ten of my past clients. My references were strong with only one blemish, she told me. Past clients had told her their projects took longer to complete than expected. At first I was puzzled, as I knew that we had completed all our projects within the time provided for in the contract plus the additional time stipulated for change orders. Then it hit me: clients had not made the schedule adjustments for themselves. If a contract set September 30 as the completion date, that was the date that stayed in the clients' heads. They did not adjust to the third week of October for the 15 additional workdays spelled out in the change orders. In their minds, we ran over schedule.

Charging for change orders

Successful builders generally agree on several basic principles when it comes to change order work: You do the work only after its cost has been agreed to in writing. You collect for the work promptly. You charge for all direct costs for the change including general requirements from project management through recycling. However, when it comes to charging for the cost of producing the change orders and for overhead and profit, opinions diverge, not only among builders, but also among clients.

In my experience, clients generally have little problem accepting charges for the direct costs, but they may balk at additional charges. One of my favorite clients wrote me a letter requesting that I drop my markups for overhead and profit for an entire deck she added to her project as we neared completion of the work covered in our original contract. Direct costs were fine, she said. But she felt, as she put it, that "my administrative costs had already been incurred" and that I had no need to charge for them again.

I explained to her why I did need my markups: All the administrative costs from estimating through payroll processing that I had experienced for her initial project would recur for the deck. The project would also burden me with the other usual overhead outlays for insurance through office cost, since it would tie up one of my crews for a couple of more weeks. She thought all that over and accepted my explanation.

Clients may also question the reasonableness of marking up change orders for additional profit. In fact, one well-known construction attorney advises owners that they should resist profit markups on change orders. He does not explain his position, and I find it incomprehensible. Profit markups on change order work are not only reasonable but also essential. Change order work puts you as much at risk for the whole range of profit costs I discussed earlier as does work specified under the original contract. Profit is your buffer against those costs—or it is your compensation for such excellent management of your client's project that profit costs are minimized and you actually get to retain the profit.

Beyond the normal direct, overhead, and profit costs, change orders burden you with their own special costs: Specifying and designing the work to be done. Estimating its cost. Explaining the change orders to the clients. Working to make sure they are signed in a timely fashion. Communicating the work that they require to a project lead or manager. Efficiently fitting the changes called for into the flow of work at the job site.

That work will be done by the most skilled people in your company—you, or, if you run a larger operation, one of your top people. Change orders are way too important to be handled by less capable folks.

For my own projects, because change orders are an extension of estimating and I have long declined to do free estimating, my contracts provide for a charge for change order production—but with the proviso that a set number of hours of the work will be done gratis. For a small project, that might be just a few hours. For a nine-month job it might be 20 hours. In practice, I rarely levy the charge if I exceed the allotted hours by only a modest amount. But I keep the possibility of the charge alive as a protection against unusually finicky and indecisive clients—and for the occasional project, like the one described in the sidebar on page 295, that has spawned a torrent of change orders and an enormous amount of work to prepare them.

Other builders, however, prefer not to even suggest the possibility of charging for processing change orders or even for the specification and estimating work that they entail. One sees change orders as a profit center and does not want to risk losing them by charging for the work of preparing the change order. Another has calculated that 8% to 10% of her company's revenue is generated by change order work. She sees it as work she does not have to compete for. "It's ours once the project is underway," she says. For those reasons, she, too,

Change Order Cost Items

- Design of extra work
- General requirements
- Demolition and recycling
- Site work
- Structural work
- Electrical, mechanical, and other infrastructure work
- Finish work
- Subcontractor work
- Disruption of work flow
- All other direct costs
- Overhead
- Profit
- Estimating cost of change
- Writing up and processing of change order

avoids measures that might discourage clients from initiating extras.

Builders also take different approaches to handling overhead and profit charges for *deletions* of work by change order. One prominent builder takes the position that all overhead costs have already been incurred by the time the client initiates a request for the deletion. Therefore, he concludes, no refund of markups is appropriate. In my view, his position is valid for overhead. It is true that you have absorbed costs—from takeoff through estimating to drawing up material orders—for the work covered by the deletion. And you will have additional costs for processing the change order and transmitting it to your lead person.

Retaining profit markup on deletions is, however, another matter. If you are deleting work from a project, you are also thereby deleting the risks and potential profit costs of building it. Clients might well argue that you have no right to a profit on work you are not even going to produce, and that by refusing to credit back profit markup on a deletion, you are taking undue advantage. Even if you can think of a counter to that argument (I cannot), and persuade yourself that holding on to the profit is ethical, there is a practical reason for not doing so, namely risk to your good name.

Take undue advantage of the monopolist's power and you may pay a price. You may be tagged as a "change order artist."

"Change orders," one builder emphasizes, "have a bad reputation." A powerful dynamic underlies that bad reputation. When a project is in the bidding phase, the clients are in the driver's seat, choosing among builders who are operating in a competitive marketplace. They can compare prices submitted by several builders. They may even have the possibility, especially for larger commercial or high-end residential projects, of obtaining reliable data from professional estimators who work as consultants. That's true even when clients decide to negotiate with a single builder or hire a design/build firm, for during the preconstruction phase of a project the clients can always decide to ditch the builder and move on.

Once a contract is signed and construction is underway, however, the power relationship suddenly flips. Though technically the clients can terminate their builder, it's difficult. The clients need the builder to see the job through to completion. They do not want to be stuck with a partially completed project, especially given the challenges of finding a reliable builder who wants to pick up where someone who has been fired has left off. The upshot: The builder has gone from competing for the project to enjoying the position of a virtual monopolist. The client has little choice but to continue to do business with him.

When a change order is presented to the now relatively helpless client, he feels pressured. If he does not sign, the job may be stalled, or something he wants may not get built. On top of that, he is not in a position to knowledgeably challenge the builder's charges. He's not an estimator (and probably his designer is not either). The builder is.

Therefore, once construction is underway the builder is not only in the position of a monopolist, but sits on the power side of what economists describe as a transaction characterized by "information asymmetry." The builder has tons of information about costs; the client has little, if any. And critically, because of time pressures—the need to process the change order so that work continues—it is probably no longer practical for the clients to obtain alternate estimates from other builders or from a pro estimator, as they might have during preconstruction. They are now in a position similar to the one a builder is in when he goes to see his dentist. The dentist has lots of information about the builder's teeth. The builder knows only that his bicuspid hurts like hell. He is not equipped to effectively counter either the dentist's insistence that he needs a root canal or the $3000 charge.

Take undue advantage of information asymmetry and the monopolist's power and you may pay a price. You may be tagged as a "change order artist," someone

The power of no-charge extras

When I signed the contract for the "Monroe" project, as I will call it here, I was delighted. It involved reconstruction and remodeling of a large old house built with magnificent materials including two-foot-wide, vertical-grain redwood planks, some nearly 30 feet in length, used for paneling of the living and dining rooms. The project called for the kind of complex problem solving and exacting craftsmanship my crew and I had come to love.

Going into the project, I felt that I had in place a balanced contract that would protect my bid yet be fair to the client. Because the house had been badly maintained for decades prior to the Monroes' buying it, and because they wished to move forward on a "fast track" basis, building from a sketchy set of plans, I took special care to explain the change order process. I cautioned them that they would need to have in reserve at least 20% of the contract price for hidden conditions and additional work they would likely order up during construction.

Work began. Hidden conditions and requests for extras came at us thick and fast. Soon it was apparent that Mr. Monroe's deep-seated optimism had led him to disregard my caution about reserve funds. He became so agitated by the change orders that he refused to review them. He could not handle the stress, he said. We were able to proceed, but only by putting into place a written agreement that he was giving his wife sole authority to approve change orders. She did approve them, all 100 or so pages' worth.

After fourteen months, the house was beautifully rebuilt. I had been paid and had realized my intended markups. I was left wondering how I could resolve the one blemish on the project, the residual bitterness that the change order process had left behind. I hit on an idea: I excerpted from the change orders a complete list of the no-charge extras we had done and mailed all 10 pages to the clients.

Immediately, I heard back from Mrs. Monroe. She and her husband now felt much better, she said. Until they had received the list, they had felt I had taken advantage of them; now they could see they had been treated fairly.

I am not sure Mr. Monroe actually did feel all that much better at that point. But four years later he sent me a note, thanking me and my crew for the wonderful home we had created for him and his family. Accompanying the note was an $800 gift certificate for dinner for all of us at one of the San Francisco Bay Area's best restaurants. "Class act," I thought, though at $800 he really got a pretty good deal on all those extras. But I am more than okay with it.

adept at milking excess profit from projects via change orders. By exercising some restraint with change orders, you can reassure clients they are not in the grip of a monopolist out to exploit his position.

One way to reassure clients, as I have suggested, is to credit back profit on deletions (and, if you can't credit back overhead, to explain why). Deva Rajan, a legendary Northern California builder, experimented with another option. He offered clients an exchange. For an additional markup to his bid, he agreed to absorb the cost of work that would normally be handled via change orders. If you adopt such a practice, you should, of course, offer to take responsibility only for correction of hidden conditions, and perhaps for minor design omissions and errors as well as minor additional building department requirements. Other extras would still require a change order. While you might be willing to take on the risk of repairing hidden structural decay, you would not want to open the door to the client upgrading from carpet to solid oak flooring or adding a bath, and expecting you to provide these without charge.

To use Rajan's innovation, you have to be a bit of a gambler. In effect, you are estimating the cost of work you cannot see during estimating. You might lose big. You might win big.

I am more into managing the huge risks inherent in building than into adding to the risk level, so my preferred method of reassuring clients is to issue "courtesy extras." Particularly on remodels, but even in new construction, you regularly pick up on the need for odds and ends of work—removal of a boulder or large root for a foundation trench, a flashing cap for an exposed beam end, an extra duplex outlet—that are not called for in the contract documents. Such items can be written up as "no-charge extras" to soften the impact of the more substantial items for which you must charge.

Those no-charge extras can add up. If you make use of them, you do want to keep tabs on what they are costing you and take care to not go overboard. That said, I have found that no-charge extras do comfort clients and are a good investment in our relationship, and therefore, in marketing. They also are a good investment in my own mental health. Like other builders I have talked with about change orders, I do not like issuing them. They are usually bad news for clients, and delivering bad news is not the most gratifying part of being a builder. Delivering a little good news along with the bad helps.

Integrity

All the thorough documentation and careful communication discussed above is necessary to protect a bid and emerge from a project financially whole and with satisfied clients. But something else is necessary, too. I call it integrity.

To my way of thinking, integrity in building means investing care to assure that your work is sound—plumb, level, and square with all fasteners properly set; with all wiring, pipes, and ducts properly supported and connected; with the thermal boundary and water management systems conscientiously installed; with finishes skillfully crafted and installed.

Integrity is not a characteristic. It's not like being tall or blond. It's a job. It's something you have to work at day by day.

Integrity means that if an item of your company's work fails, as inevitably some items will, you go back and take care of it immediately. Integrity means that on-site safety is vigilantly maintained and that your employees and subs are fairly compensated. It means that in communication with your clients, employees, subs, and suppliers, you try to be clear and candid. You practice transparency.

As a builder friend of mine and I once agreed, maintaining integrity is a constant challenge, for "integrity is not a characteristic. It's not like being tall or blond. It's a job. It's something you have to work at day by day, relationship by relationship, and nail by nail." You have to push yourself. Maintaining integrity takes work. Sometimes it's just a chore. Sometimes it involves judgment calls.

The work pays off, and not only spiritually. It protects your bids. It prevents the profit costs brought on by failing work and frustrated clients who refuse to pay, or sue, or stab a hole in your reputation via social media. When you build a project with integrity, you leave behind a sturdily put together structure that will serve your clients reliably and that will not drain down your retained profits with costly call-backs. You've created a relationship with the clients that causes them to recommend you enthusiastically, because you proved trustworthy. You were clear, candid, and fair. You rarely, if ever, surprised them. You did what you said you were going to do. You can be counted on. They tell their friends that both you and your work have integrity. That's money in the bank—that stays there.

CHAPTER 16

Move Beyond Bidding for Free

You'll still be competing

A few decades ago, I volunteered to take a turn hosting the Splinter Group, an association of builders who work in the cities across the bay from San Francisco. The turnout for the meeting was unprecedented. So many guys signed up that we had to hold it twice to accommodate them all in our 100-seat venue. I was not the draw. I had not yet written *Running a Successful Construction Company* and few in the building community had ever heard my name. The draw was my subject: "Beyond Competitive Bidding."

At the time, light-frame builders in my area largely traveled one of three paths to acquiring customers. They built "spec" houses. They operated Main Street boutiques that offered one-stop shopping, design through turnkey construction, for bath and kitchen remodels. Or they survived via "competitive bidding"—and disliked, even deeply resented, having to invest untold hours of unpaid estimating for projects they might never build, as competitive bidding forced them to do. They were eager for an alternative. I had invented one for myself, which I eventually came to call "cost planning."

In a nutshell, cost planning is this: The builder teams up with the owner before design begins, or even before the designer is selected, rather than after design is completed. He or she provides technical input coupled with detailed cost information and value engineering as design proceeds. The

COMING UP

- You'll still be competing
- Free bidding: A bad system with benefits and costs
- Why builders hate free bidding for free
- Alternatives created by owners, designers, and builders
- The author's alternative: cost planning
- Presentation of a cost planning program to clients
- Cost planning in detail
- Value engineering
- Finding your own path beyond bidding for free
- Typical charges for estimates and bids
- Reasonable expectations
- Rewards
- Ethical obligations

intention here is to foster cost-effective use of time, money, and design effort. Ergo, the name "cost planning." The builder gets paid for the cost planning work and performs it with the understanding that he or she will handle the project through construction all the way to completion.

Since the night of that Splinter Group meeting, cost planning has become a standard approach to getting jobs in my area. One of our most respected builders responds to clients who call him up for a free bid with "Well, you know, the better builders in our area have moved beyond that. We've found that so-called 'competitive bidding' does not work very well for our clients or us. We think we have a better way. I would be happy to get together with you to discuss your project and our preconstruction program if you like."

"The better builders in our area have moved beyond that. We've found that so-called 'competitive bidding' does not work very well for our clients or us. We think we have a better way."

Cost planning has not only taken root in my area, but has spread across the country. It is now widely embraced, not only by builders, but also by designers. They have learned that it is in their interest to team up with a builder during design rather than battling with the winner of a competitive bid after construction begins.

In the last half of this chapter I am going to go over cost planning in detail. I will lay out all the steps the process can include. I will even provide you with a sample cost planning agreement.

First, however, I would like to offer a realistic perspective on cost planning, for it's not without its own challenges; to discuss competitive bidding, especially its benefits; and to go over the variety of cost planning programs (often going by different names) that have been initiated by others. If you prefer, you can jump ahead to cost planning and skip the next few sections of this chapter. However, they do provide information that may be very helpful as you fashion your own bidding practices.

Even as cost planning has spread, I have realized that my title for my Splinter Group presentation, "Beyond Competitive Bidding," was not quite right. What we builders really wanted then and still want now is to move beyond *bidding for free*. We want to be paid at a professional rate for our skilled estimating and other preconstruction contributions. We want to improve our chances of building a client's project by earning their confidence during the preconstruction phase.

But I now understand that even with all that accomplished, we still don't move beyond competing. As a builder, no matter what path you take to winning clients, you will always be competing one way or the other. With every service you render, every interaction with a potential client, every nail driven and bolt tightened and pipe soldered, you are working to win the continuing trust and support of your clients. In fact, if you adopt cost planning as your approach to getting jobs, you may well have to compete harder. For these days, you will find yourself going up against many other builders who offer preconstruction services and who have won the support of the people whom they have served. Those builders are likely the most highly regarded and most capable builders in your town. Go up against them for cost planning opportunities, succeed, and you have won yourself a place in an elite group.

As you learn about cost planning, you may decide that it is not for you. You may find that you prefer to compete in the old way, by giving free estimates, maneuvering to land an acceptable portion of the projects you bid and to complete them without losing money. If that is your preference, you will find plenty of opportunity. An estimate and bid is a valuable product. It's not hard to give away things of value, and you may even find that your generosity generates satisfactory returns. For even though competitive bidding is in my view a flawed system that produces too much waste, heartache, and animosity, it's also true that it is a system that persists.

It is a system that sometimes produces benefits, even for builders.

A bad system with benefits

Many designers and clients still believe, as Boston area builder Paul Eldrenkamp puts it, that "the best way to choose a builder is to dangle a project in front of three or four of them and see who wants it most desperately." That approach gains ground especially during recessions. Then, many potential clients are paralyzed by fear of financial catastrophe. They are afraid to go ahead with construction projects. As a result, even good builders are hungry for work; and bolder clients, recognizing that, realize that builders are holding a fire sale. They go price shopping.

During recessions, even builders who have previously been able to rely on cost planning are driven backward. Some continue to win projects via cost planning. But they may find themselves unable to charge a professional rate for the extensive preconstruction services they provide. They are rewarded only with an inside track to winning a contract for construction. Cost planning is reduced to a loss leader. Others find themselves forced to return to bidding for free against other builders, all likewise desperate to somehow get enough work to keep their operations alive and to eke out an income.

Because the hard times do come, it may be wise to remain prepared for them by keeping your competitive bidding skills in shape, even when other avenues to work are open. Deva Rajan, a legendary West Coast builder who often won projects during his career via his version of cost planning, advises younger builders against moving entirely away from competitive bidding. It keeps your pencil sharp, he says. It keeps you in touch with the marketplace, allowing you to see how your bids compare to those of other builders. Thereby, it discourages you from relaxing discipline and letting your overhead and your prices drift upward until you cannot compete when work is scarce.

Competitive bidding can also be useful as a way of exploring a new market. If, for example, you have been strictly a residential remodeler but would like to expand into custom home building or small-scale commercial work, submitting free bids for a few such projects will help you get the lay of the land. You will learn to allow in your estimates for the much more extensive coordination with utility companies that homebuilding requires. You will learn that commercial work can involve inspection and site maintenance costs that are well beyond what you've encountered before. You will have a chance to evaluate the pros and cons of subcontracting work that you may have previously done in-house.

Competitive bidding can also provide substantial benefits to builders who rely on it exclusively and learn to work the system successfully. Among them are builders who lock down on a narrow range of projects—for example, second-story additions to traditional houses on flat lots in the same town—and do one after the other, subcontracting almost all the work. They can become so efficient at obtaining prices from their subs that giving free bids burdens them very little. One such builder claims he can crank out a preliminary estimate for his typical project in a few hours. He does so by using (1) square-foot prices from his subs for most of the work (framing, roofing, drywall, etc), (2) knowledgeable guesstimates for the remaining trades (such as electrical), and (3) allowances for materials, such as windows, that have a wide price range.

If clients accept his preliminary bid, he takes a deposit from them to reserve a place in his schedule, works up a final price (which, he promises the clients, will be within 10% of his prelim if the scope of work does not change), and signs a contract for construction. He claims to win half his bids and to enjoy a substantial six-figure income from his projects. Even allowing for some excess self-celebration on his part, I think his approach may exemplify one way in which builders can prosper with competitive bids. I have heard similar claims from other builders who have similarly structured operations—that is, who do the same kind of project over and over and sub out most of the work.

A second way to prosper via competitive bidding is to possess (or cultivate) charismatic sales ability. Charismatic people can exercise their compelling charm to win competitive bids when their price is not the lowest tendered, and even when it is the highest. One charismatic builder says he is not afraid to enter the fray as the high bidder. Indeed, if he is the low bidder, he knows that he "has made a mistake," and walks away from the project (with a charming apology no doubt). If he is a higher bidder, he follows up his bid multiple times, touring the clients through his previous projects, pointing out refined details, and subtly flattering the clients for having the sophistication to appreciate them. Often enough, clients come away from his sales calls convinced that working with him will be a special experience, and he wins the contract for their project. (Or at least that is the case during good markets; during one recession, he did submit 40 bids without winning a single job.)

Still other builders find success via competitive bidding by focusing on specialized work that few companies pursue. William Mitchell, in his *Contractor's Survival Manual*, offers the example of a concrete contractor. The contractor was doing the usual residential work—slabs, sidewalks, retaining walls—and "finding that he had to compete with every joker in town who owned a pickup truck and trowel and thought he could pour and finish mud." The contractor shifted to bidding on work such as installing small dams, communication tower bases, and pumping stations. There he found himself facing far more paperwork but only a very few competitors and was able to push his bids up to an acceptable level. Mitchell's advice: "If you're going to live on competitive bids, find yourself a solid, profitable niche. Otherwise, your best bid will come in second 90% of the time."

Another way of succeeding with competitive bidding is to evolve your company to the point where it is bidding only on projects that require performance bonds. The financial firms that provide bonds winnow out the flakier competition, agreeing to guarantee performance (i.e., actual completion of projects) only by builders with strong track records. On projects with bond requirements, you won't be bidding against those befuddled contractors operating in the wide-open world of residential remodeling and small-scale commercial work, who submit bids barely above your cost for materials.

Bidding on bonded work can be especially productive if you carefully qualify the projects in

Competitive Bidding's Possible Pluses

- A way of scraping up work during a recession
- Encourages disciplined control of costs
- A way of exploring new markets
- Estimating can be done very efficiently for repetitive projects built largely with subs
- Short interval from bidding to construction and arms length relationship with owner
- Presents opportunities to exercise charismatic salesmanship
- Can work well for specialized niches where few builders are competing for the work
- Opens door to extra charges via change orders

which you invest your bidding resources. The chief estimator for one bonded company emphasizes that they do not try to "beat everybody." They choose projects that offer them the opportunity to size up the competition, are in a niche where they experience an acceptable winning percentage, and can be bid at prices that include an adequate potential for profit. With that approach, the builder wins (during good years at least) one out of three of the private projects he bids.

Finally, there are aspects of competitive bidding that, while they may be off-putting for many builders (including myself), are appealing to others. First of all, compared to cost planning, the time between bidding and building is generally much shorter with competitive bidding and requires far less intense involvement with the client. With cost planning, you are involved with a project and its owners all through design, often an emotional period as clients try to give their dreams flesh and bones on paper or a computer screen while not exceeding their financial resources. With competitive bidding, you price out completed designs. If you win, you sign a contract. Then you are on to construction. For some builders that brisker pace and more arms-length relationship is appealing.

Additionally, competitive bidding opens the door to change order charges to a greater degree than does cost planning. Here's why: When clients are paying you for cost planning services, they are trusting you to protect them from cost surprises. You have an obligation to do so, to inform them as fully as possible of costs that may arise during construction beyond those shown on the plans and covered in the initial contract. When you are bidding for free, as one does in competitive bidding, the ethical requirement to prepare the clients for extra costs due to hidden conditions or other unknowns diminishes considerably.

After all, you've been asked to provide, gratis, a valuable product—an estimate and bid that, on a sizeable project, can require a hundred hours or more of skilled work. The client is exploiting you for free labor. It's not on you to do even more uncompensated work by further looking after their interests during preconstruction. You are freer (at least to the extent your conscience and the law allow) to note oversights in the plans, anticipate the consequent charges for extra work, keep the possibilities to yourself, and look forward to profiting from them during construction. In fact, you are almost forced to do so, because if you provide for the additional work in your bid, it's more likely to be too high to win the job.

For some builders, playing the adversarial change order game is appealing, or at least feels like a normal aspect of doing business. They experience it as fair play between grownups and view change orders as financial opportunity. In fact, one builder in the world of heavy construction explains that his company systematically marks up plans for change order opportunities and projects possible profits from those change orders as part of their normal procedure for creating estimates. They do that work in a space they call their "war room."

War, of course, has its costs. For all its possible pluses, competitive bidding imposes a heavy burden on all parties to a construction project.

Free bids and their costs

For clients and designers, the "free" competitive bids turn out not to produce the good deal they were aiming for and not to be free at all. Barry LePatner, the Manhattan construction attorney who makes his living protecting owners from cost catastrophes, describes the hazards for owners in his book *Broken Buildings, Busted Budgets.* Owners, he says, opt for competitive bidding in hope of getting a low, fixed price for their project. However, the supposedly "fixed price" contract they sign turns out to be a "disguised, mutable cost contract" with only a "nominally fixed" price. Once construction is underway, the fixed price is washed away by a flood of change orders. Charges mutate upward so severely that, in LePatner's view, what builders are really doing when they seek work via competitive bidding is competing "for the future right" to increase the cost quoted to the owner in their bid and stated in the contract.

Owners expect that their architects will protect them from cost catastrophe, for historically architects have sold their services to owners with the claim that they could provide protection against those nefarious building contractors. But the owners' assumptions are naïve. Because of fear of being sued, LePatner explains, architects have retreated from the role of protecting owners against builders. In any case, he says, they are just not up to the task. They are not sufficiently expert about construction costs because they are primarily "stylists," products of an "educational system that values abstract theoretical thinking" over practical experience in the field.

Because of the gaps in their knowledge, architects, as well as other designers relying on the competitive bid system, may have little sense of where a project's costs are heading as they dream and draw in their studios. They don't see dollar figures until they have finished the plans and the bids come in. When they do, they often discover, says one architect, that "there is a huge disconnect between a client's budget and the design." Another reports that "again and again" he saw clients of the prestigious firm where he was employed paying large design fees only to "end up with an unbuildable project" when none of the bids they received turned out to be remotely close to their budgets.

When that happens, designers catch hell. Finding themselves facing outraged clients, they may dig around for a cheap builder and take a bid from one who is not qualified to handle the job. The upshot: They see their design realized with mediocre craftsmanship or even end up embroiled in litigation over failing construction. Moreover, designers who turn to or who are required to use competitive bidding can find themselves pummeled nearly as severely as owners in a change order war. An example: A design firm wins a commission for a new elementary school. As is usual for public work, the project is awarded to the lowest bidder. As soon as construction starts, the builder begins imposing additional charges via change orders. By the time the project is completed, the architects have been hit with thousands of RFIs (requests for information) leading to over 600 change orders.

The design firm takes a severe financial hit. It is handling the project for a fixed fee. The fee does not allow for sufficient hours to respond to the torrent of RFIs or to process all the change orders. Reflecting on the experience once the project was finally completed, one of the firm's architects saw it as a sad waste of resources, with the change order struggle diverting attention from the real needs of the project and compromising its quality. The whole miserable process, he believes, was a result of the competitive bid system and its built-in assumption that the lowest bid is the best bid.

As bad as it can be for owners and their designers, the free competitive bid system can be harsher yet for builders. Even LePatner, for all the colorful potshots he takes at alleged builder banditry, recognizes that builders participating in the system are at risk for the "winner's curse." The "winning" builder in many competitive bids has won only because he has bid below the cost of the work shown on the plans. He faces an unpleasant choice. Either he can risk going broke on the project. Or he can attempt to make up the shortfall in his bid and work his way up to acceptable earnings by getting into a change order war. Yes indeed, during bidding the builder was, in fact, competing "for the future right" to increase the cost quoted to the owner in his bid. Once construction is under way he has to exercise that right or take a loss. That's the process the owner, consciously or not, has invited the builder into, in the hope of getting his building put up for a bargain price, perhaps lower than its actual cost of construction.

While there are some who seem to make a go of it, as I have described in the previous section, builders often despise the competitive bid system. I have explored their (our) animosity toward it in the long sidebar on the following pages. In brief, the problem is: If you base your bid on a comprehensive estimate, you don't get the job. Therefore, you

Why builders hate competitive bidding

"Construction is the only business," exclaimed a speaker at a conference hosted by the *Journal of Light Construction*, "where you are upset if you don't get the job and upset if you do." Why is that? Because if you do prevail in a competitive bid, you don't enter a winner's circle to the sound of applause. Instead, you fear you have fallen prey to the "winner's curse," prevailing with your bid only because you have left out something big. Under the competitive bid system, first you bid for free, then, too often, you build for free.

"I hate competitive bidding," says Boston-based builder Paul Eldrenkamp, explaining how clients evaluate the bids they receive: The low bid is assumed to be "fair and accurate." A middle bid, if it is lower than the client's budget and much lower than the high bid, might also be deemed honorable. High bids are always assumed to be unfair and inaccurate.

By corollary, of course, low bidders are assumed to be persons of high character. High bidders are seen as crooks. The "grand irony," says Eldrenkamp, is that incomplete estimates get the prize while "thoughtful, carefully researched" estimates are scorned. In other words, as another builder explains, competitive bidding requires that you "play dumb." You see but don't speak. You see all the oversights in the plans and specs, but you say nothing about them. You bid exactly as instructed by the design documents, raising no warning about other costs that will arise.

You fear you have fallen prey to the "winner's curse," prevailing with your bid only because you have left out something big. Under the competitive bid system, first you bid for free, then, too often, you build for free.

Many builders are disturbed by the moral dilemma inherent in such "malicious obedience." They understand that if they want a reasonable chance of getting a job that's been "put out for bids," they must keep mum about the extra costs they foresee the client encountering during construction. But at the same time, they feel badly about leaving the client in the dark.

Equally, they resent being expected to work without pay for all the labor needed to produce an estimate and bid—easily 40 hours for a custom kitchen, 60 for an addition, 80 to 120 for a custom home, and 150 hours or more for larger light-frame projects. Even the owners' attorney Barry LePatner recognizes the cost: "Working up a real bid is serious work that can cost from hundreds to tens of thousands of dollars depending on the size and complexity of the job."

Too often as a builder, you put in all the work of submitting a bid for a job only to learn that it has ended up "in the drawer" because the design calls for construction far beyond the owner's financial capacity. Your time, along with that of your subs and of all the other bidders, goes down the drain. Even worse, you submit a bid only to learn that you were never really in the running for the project. The owner and designer are interested in your bid only as a control on another builder they have been intent on using all along. They plan to use your bid as a check against that favored bidder's numbers or as a tool for levering him down.

Builders feel not only exploited but also demeaned by the process and the disrespect that underlies it. The American Institute of Architects contract, builder Sam Clark angrily notes, "assumes builders are natural antagonists"; and beneath that assumption is a further assumption: that in the adversarial relationship, it is the builders who are the bad guys. One AIA publication Clark came upon characterized builders as astute manipulators of men, interested only in making money, while architects, on the other hand, were described as "refined persons who are highly trained and put service to society above personal gain."

Worse, you submit a bid only to learn that you were never really in the running for the project. The owner and designer are interested in your bid only as a control on another builder they have been intent on using all along.

"I hate that," remodeler Iris Harrell exclaims. "It drives you nuts," another veteran builder adds. "We in this industry are such chumps that we will spend 60 hours bringing to bear our hard-won experience for free. For free! Guys will keep doing it and doing it."

Not everyone. "Competitive bidding?" muses builder Michael McCutcheon. "Oh, that's when an architect instructs you and a bunch of other contractors to show up for a walk-through, then demands a deposit for a set of plans... You have to pay him for the right to beg for his project." And of course, he expects you to scour his documents, catching all his omissions and discrepancies, then deliver a "tightly sealed bid" that he can "ferret off to his secret chamber" to choose a "winner" so that he and the owner can begin squeezing the poor guy's price down. "What the hell's the point?" McCutcheon asks. Long ago he decided there was none; refused to participate in competitive bidding; and began developing his particular version of cost planning.

must instead practice "malicious obedience," estimating what you are told to estimate by the (inevitably incomplete) construction documents and deliberately leaving out costs for additional work that you know will arise during construction.

Meanwhile, you are kept at a distance from the client all through design and even during bidding. You are treated as an adversary, to be played off against other adversaries, namely the competing bidders, rather than respected as a collaborator. You are denied the possibility of doing your best work because you are forced to submit the low price to win the job and, therefore, cannot bring in your best subs. You are even under pressure to rush your crew and cut corners.

As Paul Cook wrote, with competitive bidding "contracts usually go to the lowest qualified bidder, a practice that fosters indifference to the relative abilities of the bidders." And he's talking about large-scale commercial work. In residential and smaller-scale commercial, the job often goes to an unqualified bidder. You would think that with competitive bidding producing so many "busted budgets" and lousy, "broken buildings" the participants would try to create other approaches to contracting for construction. They have.

Alternatives

Owners, designers, and builders have all attempted to create processes better than competitive bidding. Some of the alternatives amount to little more than historical curios and are not much in use. But at least they suggest recognition that the system of demanding free bids, with the low guy "winning," is a loser. One alternative: switching to middle-bid-wins. That is, three builders are asked to bid a job, with the guy in the middle rather than the low bidder prevailing. But that adjustment does not actually solve the problems. With middle-bid-wins, builders are still left out of the project till the design is completed. They remain incentivized to practice "malicious obedience" to avoid coming in with the high bid. Meanwhile, designers remain at risk of designing a project that is unbuildable for a client's budget. Owners are still exposed to mutable bids and a deluge of change order charges.

Another alternative: Awarding the project not to the lowest or even the middle bid, but to the best bid, with "best" defined as a combination of price and likelihood of reliable performance as determined from the bidder's track record. While attractive on the surface, and maybe an improvement over lowest-bid-wins, the "best" bid approach also fails to solve core problems that arise from competitive bidding—builders being left out of the loop till design is complete, projects getting designed wildly beyond owners' ability to pay for them, change order artistry, and more.

Barry LePatner proposes a more comprehensive alternative solution, which overlaps in some degree with my own cost planning program, (not surprising, because LePatner's book draws heavily on *Running a Successful Construction Company*). His program features.

- Elimination of fast-track construction—i.e., commencing construction before plans and specs have been thoroughly worked out—and replacing it with a requirement that designers provide plans that are 100% complete and that they at least share in financial responsibility for any omissions
- Owners paying their builder for estimating during design with the understanding that they are working together toward a contract for construction
- Genuinely fixed cost rather than mutable cost contracts, which require builders to waive their right to levy extra charges for any items of work that could be reasonably anticipated during estimating
- Employment of an owner's representative—or "intermediary," to use Lepatner's term—who is expert about construction costs and can provide the owner with balancing expertise in case of conflict

"The best thing" about his approach, says LePatner, is that "if a change {exclusive of increase in scope of work} becomes necessary…the owner, is not responsible for it. The contractor and the architect cannot play the blame game… By paying for the bid, the owner has essentially purchased a guarantee that the work will be completed as budgeted."

The worst thing about LePatner's approach is that it overburdens designers and builders while concentrating on benefits to the owner. (That's not surprising either. LePatner, after all, makes his living advocating for owners.) Requiring perfect plans, 100% free of errors and omissions and completely disallowing change orders, is unrealistic. It is all but impossible to anticipate every item in a project in advance. (Are there any endeavors in which we require perfect performance from humans?)

A second problem: the actual availability of competent intermediaries. Perhaps they are plentiful in the world of larger-scale commercial and public construction where Barry LePatner operates. Not so in the world of smaller-scale, light-frame work. Even if they were available, using intermediaries would add to the pile of "soft costs," from permits through special inspections, that already weigh so heavily on projects, soaking up funds needed to pay for good-quality construction. With that said, however, LePatner's approach, with its insistence on collaboration between owners, builders, and designers, is a huge improvement over competitive bidding.

When we turn from alternatives proposed by owners' advocates such as LePatner to those suggested by architects, we find that they range from wishful to comprehensive and valuable. Some architects hope to avert cost surprises by simply hiring a builder to price their plans, just as they would hire "any other consultant," as one architect puts it. Their intent here appears to be to remain fully in control of the project and to stay clear of actual collaboration with the builder. A designer who embraces this approach emphasizes that he does not want builders offering price information directly to owners early in the design process lest it waylay the special, sparkling ideas that are his passion. Given enough time, he hopes, the owners might fall in love with the ideas and be willing to foot the bill.

Builders, however, are not "like any other consultant." Other consultants, from structural engineers to lighting designers, provide information, get paid for it, and then largely go away, leaving their information for use by the designer, the builder, and his subs. By contrast, if builders provide estimates and then go away, their information may prove useless. Why? Because estimates and the underlying assumptions vary so greatly from one builder to another. Unlike other consultants' information, the estimating data and accompanying

Alternatives to Traditional Competitive Bidding

- Middle bid wins
- Best bid wins
- Owner protected by intermediary (LePatner approach)
- Builder as cost consultant to designer
- Integrated project delivery (IPD)
- Slow dancing
- Project proposing for a fee
- Negotiated bidding
- Construction consulting
- Cost planning
- Design/build

information that a builder provides is reliable only for his own work, not for that of others.

From the designers' side of the table, the most promising initiatives are those that call for *full collaboration,* from initial conceptualization to completion of a project, between owner, designer, and builder. Despite the American Institute of Architects' (AIA's) historical disdain for builders, there has been a developing interest in such collaboration in the design profession from one end to the other. Frank Gehry, the "starchitect" known for his design of the Guggenheim Museum in Bilbao, Spain, and a number of other cool contortionist confections, has spoken of his wish "to partner with a construction company rather than being an adversary." A New England designer of more conventional work says, "I love it when I can unite with a builder… Everyone is in the loop from start to finish, and the builder keeps the design from getting overly complicated and expensive."

It's a long way from wishfulness to realization, of course. But systematic efforts are being made in the world of designers to get to a more collaborative way of building. Notably, The AIA California Council proposes a program they call "Integrated Project Delivery" (IPD).

IPD calls for designers to move far beyond not only competitive bidding but even beyond using builders as mere cost consultants held at arm's length. IPD, declares the Council in a publication laying out the idea, "includes tight collaboration between the owner, architect/engineers, and builders ultimately responsible for the construction of the project, *from early design through project handover*… Builders' input is not left until the construction phase when it is typically too late to benefit the design."

Importantly, the Council's manual emphasizes that the builders on the IPD team are to be "ultimately responsible for construction." In other words, they collaborate in the design of the project, then build it. Traditional competitive bidding, as the illustration on the next page of the IPD chart underscores, is thrown in the garbage.

The Council has high hopes for IPD. As they see it, with the traditional approach "construction is often treated as the final stage of design where issues that were not addressed during [preconstruction] are worked out." The result, they note, is to impair speed, raise cost, and reduce quality. During construction the RFI and change order war commences, gobbling up time, piling on costs, and diverting attention from the best possible delivery of the project.

The Council hopes that IPD can simultaneously maximize quality, speed, and cost. That's a huge turnabout from traditional thinking, which holds that if you gain at one point of the quality-speed-cost triangle you will inevitably lose at the other two. But the Council feels that with IPD and the collaboration and transparency it calls for, it is possible to break the old iron triangle and improve on all three factors at once.

As in any such ambitious scheme, the devil surely resides in the details. With IPD, the challenging details appear to be around costs. The Council speaks of "incentivizing" all parties—owners, builders, and designers—to willingly collaborate to improve the project rather than protecting only their own interests. But just how that can be achieved is not really spelled out. At one point, the Council even slips backward toward adversarialism by requiring that builders cap change orders while neglecting to call for balancing obligations on the part of owners and designers.

However, as the Council emphasizes by subtitling its publication "A Working Definition," Integrated Project Delivery is a work in progress. The Council recognizes that making the shift from competitive bidding to working relationships based on "respect and transparency" is not easy. It even suggests that builders, owners, and architects who want to get across the bridge should engage in "collaboration training."

For all the challenges, however, IPD seems to be spreading. The Council claims it is. And it is being promoted not only by designers, but also by builders. I was introduced to the Council's ideas by a builder. He makes copies of the Council's IPD publication and hands them out to architects with whom he is discussing possible projects. "Look," he says as he

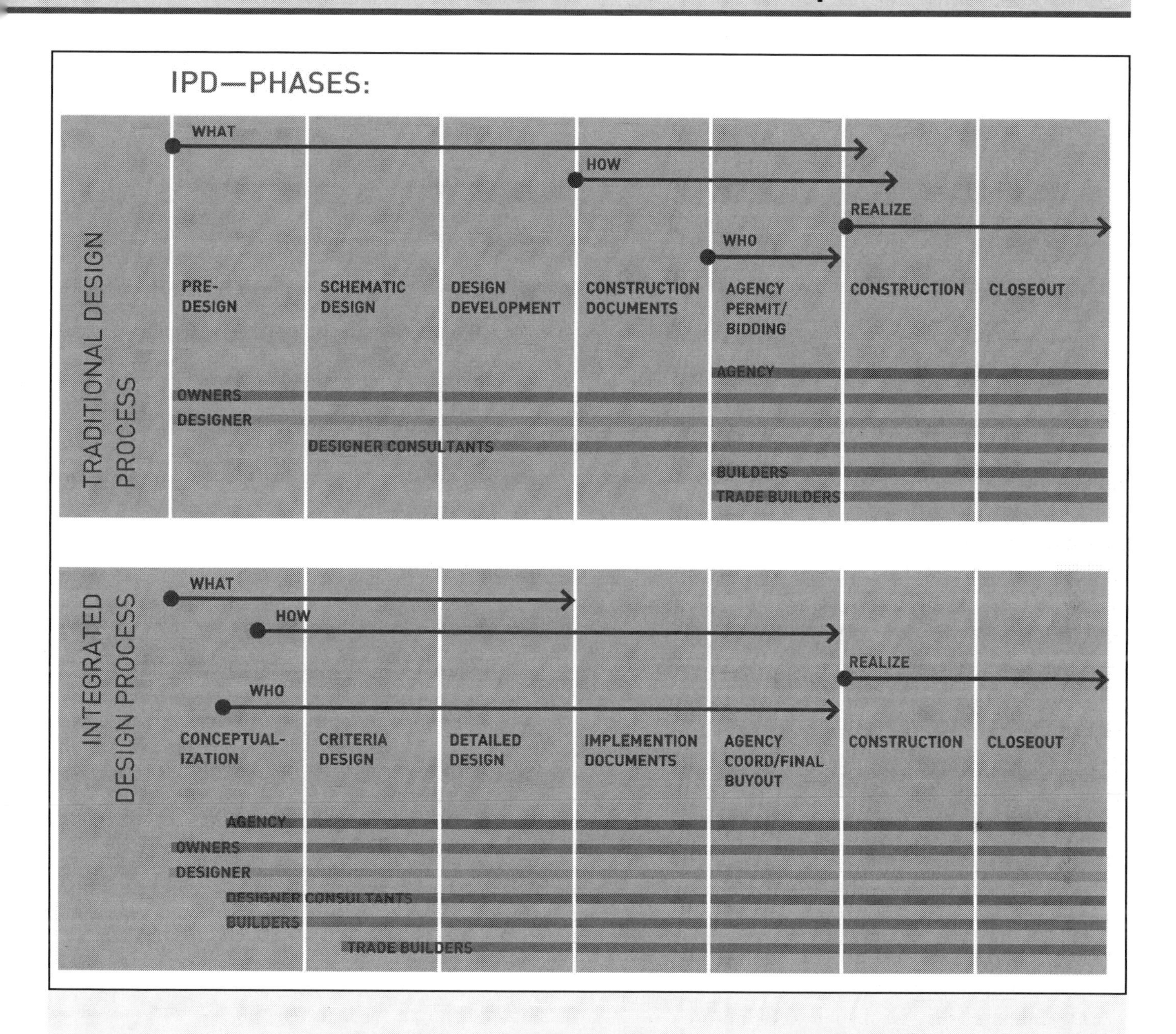

Take a look at and contrast the bars for "builders" and for "trade builders" in the displays of the traditional and integrated processes. You will see how the IPD alternative makes a critical advance beyond traditional competitive bidding. With IPD, builders are involved far earlier during the preconstruction phase of a project. In the traditional process, we are brought in only as permits and bids are being sought. With IPD, we are involved from conceptual design on—and competitive bidding is eliminated entirely.

The chart is reproduced with permission from the AIA California Council's publication, *Integrated Project Delivery: An Updated Working Definition* shows them the chart illustrated here, "Even your own organization advocates we work together from the beginning of the project."

Builders who hate bidding for free have come up with alternatives of their own. While they overlap, and substantially so, they can be broken down into a few varieties, the first of which I think of as "Slow Dancing." The slow dancers move to a tune I first heard from construction industry consultant Theresa de Valence. "Competitive bidding? Don't do it," she warned. "You can't survive." When you are going after a project, "for everything you give, get something back." In other words, you take a step, the clients follow. If they don't, the dance ain't happening.

To designers, he emphasizes the thoroughness of his estimating services and stresses his willingness to work with the designer to arrive at a project the clients can afford.

One master of this approach emphasizes that estimating is so expensive it is the last step you should take in forming a relationship with a potential client. He begins by paying potential clients a visit. He listens attentively as they tell him about their project. Having taken a step toward them, he waits to see if they move with him. Do they express interest in seeing his portfolio? Great! He takes them through it. Do they then follow again, responding enthusiastically that the high quality of service and workmanship he offers—and not price—is their priority?

If they make that move, he might make a dramatic one of his own and ask for their budget. If they continue along with him, discussing budget candidly, he will respond by providing the clients with some broad cost parameters. And so the minuet continues, until he has reached the point where the clients are ready to move to the discussion of preconstruction services. He tells them what he can provide and what his charges will be. They let him know they would like to work with him and assent to the charges. He expresses his gratitude and draws up a contract for the preconstruction services. The clients sign it. And now, finally, he is ready to begin estimating.

A second approach, one I like to call "Project Proposing for a Fee," takes a builder to the same result, but faster, without subtle maneuvering. It's more like a brisk march than a waltz. One practitioner lets clients know politely but promptly, sometimes during a first phone conversation, that even if they already have an architect he's going to have to rework and refine the plans to make their project affordable and buildable. If clients ask him for free price info, a "ballpark" figure, he declines to give it. "I don't want to be wrong," he tells them, pointing out that a casually figured and inaccurate cost projection could lead them to make bad decisions as they go forward. "The only way to know what your project is going to cost is to do a detailed proposal, and it's going to cost some money. Do you want that?" he asks. If they say yes, he begins preconstruction work, emphasizing to the clients that he is not open to negotiating quality or schedule and that cost can go down only in proportion to reduction of scope of work. He is providing a proposal for a fee, he explains, not "negotiating" a price.

In contrast, many other builders embrace a third, more flexible, approach and often refer to it as "Negotiated Bidding." One who sees himself as "negotiating" the path from design to construction leaves all aspects of a job up for discussion—even quality, to a limited extent. "Does anyone ever want a crappy job?" he asks. "No." But he is willing to step down from his preferred subcontractors or install economical finish materials to nudge a project within budget.

The negotiator promotes his services to both clients and designers. To clients who commit to working with him, he offers: A place in his schedule. The opportunity to work with a project lead of their choice, perhaps somebody who did a great job for a friend. Help with understanding costs in order to create a viable budget. Creation of a realistic calendar for the project from schematic design all the way through completion of construction.

To designers, he emphasizes the thoroughness of his estimating services—the careful documentation of all the assumptions and specifications underlying his cost numbers. And he stresses his willingness to work with the designer to arrive at a project the clients can afford.

Interestingly, as he has evolved his "negotiated bidding" approach, he has moved away from architects and toward interior designers. He has been burned by architects too often, he says, echoing a lament I have heard from many builders. He brings an architect a great job. The architect encourages the owner to switch to competitive bidding. He never hears from either the owner or architect again, or at best gets a message telling him goodbye.

With interior designers, the builder finds he does not have that experience. The interior designers do not have the capability to produce drawings sufficient for competitive bidding. They don't know enough about structural or plumbing, mechanical, and electrical work. Their focus is on the finishes, and they are glad to be working with a builder who can take the structural and PME details and specifications off their hands.

In his work with interior designers, the negotiator is moving toward another of the builder-generated alternatives to competitive bidding, one that a builder who has honed his version calls "construction consulting." While this builder does work largely with architects, he observes with amusement that the architects tend to think what they need from him is a preliminary budget. "But honestly," he says, "they need my help with the design...making sure it works, making sure there's a place for the damned water heater, that the ducts will fit into the structural work... Their plans are "really bad, with no HVAC, no plumbing, sketchy specs at best. They are just awful."

That's why he calls his service "construction consulting." For each of his projects, whether a bathroom remodel or a large home, he:

1. Creates a comprehensive statement of the scope of work for each trade.
2. Obtains multiple bids for each trade.
3. Evaluates the bids, making sure that no work has been excluded so as to reduce the potential for change orders and cost surprises to a minimum.
4. Selects one sub for each trade.
5. Prepares a detailed budget for presentation to the client, including in the budget his charges for project management and utility work by his own forces, a contingency allowance, and his fee for overhead and profit.
6. Incorporates the budget into a contract for planning and construction of the project.

This builder works with the ultrawealthy clients often feared and avoided by other builders. He has been so successful at establishing mutual trust with his clients that they have all given him keys to their homes so that he can send over his maintenance specialist to fix any problem detected by the electronic monitors he installs in his projects. He views his clients as friends and regularly hosts them at fundraisers for his favorite charity, giving out bottles of wine made with the grapes from his own vineyard. It is hard to imagine relationships further removed from the adversarial world of free competitive bids, malicious obedience, the winner's curse, mutable contracts, change order war, busted budgets, and broken buildings.

Presentation of an alternative to clients

As I began developing my own alternative to competitive bidding, I realized that I was onto something when I received an enthusiastic letter from another builder. After getting stiffed by two clients for a total of $160,000 and going broke, he had decided to retool his business. As part of the effort, he had read my book on running a construction company and discovered my alternative to competitive bidding. He developed a similar program and laid it out to potential new clients. The result: On two of the next three jobs for which he produced estimates he was awarded the contract for construction. For the third, he wrote, the clients "hired someone else, but at least I was paid to bid it."

Encouraged by that news, I focused on evolving my own program—both the nuts and bolts of the service, which we will go over in the next section, and the presentation of the service to clients, which I will cover here.

I have found it best, when paying clients a first visit, to refrain from aggressively pushing cost planning. I avoid a hard sell. Instead, I listen, ask questions, and then explain that there are a variety of paths to contracting with a builder for construction.

As I listen to clients during a first meeting, I occasionally ask for a clarification or additional explanation. I want to understand the clients' "program"—just what it is they wish to achieve with their project. And I also hope that my attentive listening and careful questioning will convey to the clients that I am there to collaborate with them, not to manipulate them or engage in the sort of cagey poker game that competitive bidding entails.

I listen for as much time as the clients want to spend talking with me about their project. Then I ask if they would like me to lay out the optional paths to contracting for construction. Generally, they do. I give them a concise overview of competitive bidding: plans drawn, bids solicited, and one bidder—low, middle, or best value—chosen to negotiate for construction of the project. I also emphasize that construction projects are successfully completed using the competitive bid approach, and I stress that the chances of success are improved by adherence to a few guidelines: No rushing; avoid fast-tracking the project. Pay for a thorough set of plans. Seek three competent builders to bid on the project. Choose the bid that offers the superior combination of price and likelihood of reliable performance. Have contingency funding in place for change orders. Lastly, I summarize the hazards of competitive bidding, from costly designs ending up in a drawer to change order wars.

Then I move on to cost planning. I keep my presentation of it short and vivid, simply summarizing key aspects of the service. I emphasize that for cost planning to work, it must benefit everyone involved—the clients, the designers, and the builder. For the clients, I explain, there are these key benefits:

- Estimates at three stages of design—conceptual, preliminary, and final—to safeguard against the danger of investing in a fully detailed set of plans for a project that will be too costly to build
- Value engineering—collaboration between the owner, designer, and builder to devise the most cost-effective methods and means to achieve design intent and fulfill the owner's program
- Limitation of cost surprises and change orders
- Transparent estimates and bids backed up with documentation, so that clients can see that no markups for overhead or profit are being concealed in cost lines for materials, subs, and labor

As for benefits to designers, I point out that cost planning goes a long way toward eliminating the danger of unaffordable designs going "in the drawer," with all the disappointment and ill will that brings.

And for the builder, I explain that cost planning offers: The pleasure of being treated as a respected collaborator rather than as a source of free estimates. Involvement in the advance thinking about a project. The likelihood that the multiple estimates required for cost planning will reduce estimating oversights and increase the chances of a decent financial result. And finally, the satisfaction of building a structure that you have helped midwife from conceptual design forward.

When clients sign up for cost planning services, I always feel good. Their acceptance implies trust and respect—and also a sense of fair play. Clients are often surprised to learn how many hours of work thorough estimating, value engineering, and the other steps in

The Cost Planning Process

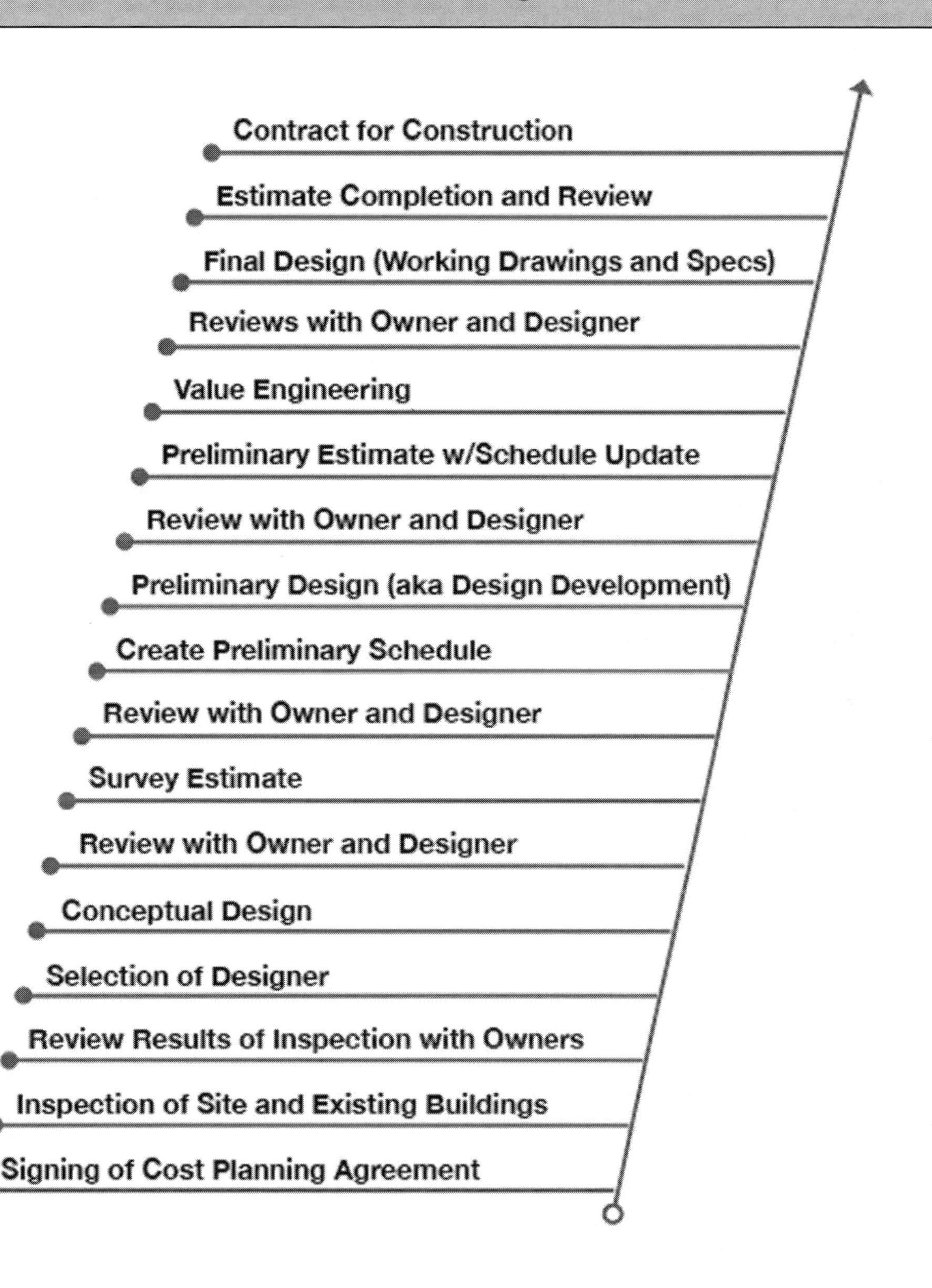
Contract for Construction
Estimate Completion and Review
Final Design (Working Drawings and Specs)
Reviews with Owner and Designer
Value Engineering
Preliminary Estimate w/Schedule Update
Review with Owner and Designer
Preliminary Design (aka Design Development)
Create Preliminary Schedule
Review with Owner and Designer
Survey Estimate
Review with Owner and Designer
Conceptual Design
Selection of Designer
Review Results of Inspection with Owners
Inspection of Site and Existing Buildings
Signing of Cost Planning Agreement

cost planning demand. Once they do know the facts, fair-minded clients find it entirely reasonable to pay for the service. Those who prefer to extract free estimates and bids from you rather than pay you for your work are exactly the people you do not want to do business with. Contracting for jobs via cost planning helps you sift out those folks.

A very high percentage of the clients who ask me to explain cost planning have decided to work with me. If they decline cost planning and invite me to be one of several bidders, I express appreciation for the invitation, but tell them my company is usually fully booked by clients who favor the cost planning approach. That's my polite way of saying no. I have not submitted a free bid since the early years of my career, and hope I will never again have to.

If you are interested in developing a cost planning approach to bidding for work, I think it will likely be well worth your while to invest time in developing a presentation of our own. When a young builder asked recently for advice on how to get out of the competitive bid rat race, I told him about cost planning. He thought he would like to give it a try. In that case, I urged him, develop a clear presentation of your program for potential clients. Make it clear, make it candid, keep it concise. Practice it in front of any friend who's willing to listen. Practice it a few dozen more times in front of a mirror. Then take your show on the road.

If you are also just beginning to venture beyond bidding for free, the same advice may apply. Whatever preparation you make, once you have clients who want to avoid the potential quagmires of competitive bidding and are attracted to your alternative approach, provide them with an agreement for cost planning. It can include an appropriate selection of the services laid out in the next section.

Up the cost planning ladder

The process for cost planning different projects does vary. It will not be identical for a small job, such as a bathroom remodel, and a large one, such as a new house. Even so, for any project, cost planning is likely to involve most of the steps laid out in the illustration on the previous page.

As you can see in that illustration, cost planning can begin with a targeted site inspection. The goal of the inspection is not to take the sort of detailed look (described at the end of Chapter Three) necessary to closely nail down direct costs. That comes later. Rather, it is to zero in on conditions that have the potential to generate significant costs beyond those for the work the owners had in mind when they first called you—or, in a better case, to suggest opportunities for cost savings.

If the project is a remodel, you will check all the major existing systems—foundation and frame; plumbing, mechanical, electrical (PME); exterior skin from roofing through windows, siding and trim; interior finishes; and so on. You will be looking for signs of failure, such as plumbing so decayed it may give out under the impact of construction and should be replaced—especially since replacement can be less costly when done as part of the main project rather than as a stand-alone job.

Depending on your own knowledge level, you may or may not find it necessary to arrange for other inspections to complement your own. You may, for example, need a structural engineer to evaluate the foundation. The extra cost of such inspections can be money well spent. Owners need to know if their building systems are nearing the end of their life or failing. You do not want them to exhaust their financial resources for a new addition or other improvement only to be hit with huge bills in the following years for overdue replacement and repair.

There's risk here, of course. The clients may conclude that they must fix their buildings rather than doing the nice improvements they originally called you about. But if you have signed on to provide cost planning services, you have given up the competitive bidder's prerogative of playing dumb. The clients are paying you to help them plan the use of their money. You are obliged to give them the best information you can produce.

One step up the ladder from site inspection is a crucially important phase of cost planning—the selection of a designer. That's a task with which owners often need help. Frequently they do not realize how greatly designers vary, not only in their capabilities but also in their priorities.

Roughly speaking, designers fall into two basic camps. In one we find those who think of themselves as service providers. They are primarily interested in meeting their clients' practical needs—around functionality, around financial limits—with aesthetics an important but not overriding consideration. In the other are designers whose primary commitment is to what one architect calls "Big A" architecture—work that is "striking, innovative, and fresh." For those guys, art comes first.

Clients are often not aware of the differences between designers and assume they are all eager, first and foremost, to provide clients with the services they want. They do not realize that some feel, on the contrary, that clients should be encouraged to support Big A endeavors—or even tend to see themselves, rather than their clients, as proprietors of the project. One architect, who worked in prestigious design firms before quitting to become a builder, tells me that he often heard his colleagues referring to a project as "our house" rather than as their clients' home. Some clients are happy, even eager, to serve as patrons of architectural innovation. But others are not or just cannot afford to do so. All need to be informed of the difference between the Big A and the service-oriented

Early in my career, I remodeled these two homes. For one, the architect produced a design thoughtfully developing the traditional style of the original construction. For the other, the architect, an avowed modernist, designed a minimalist box and stuck it on the backside of the mock-Tudor-style home. I helped the owners of the traditional home choose their architect. The modernist had been hired for the other project before the owners contracted with me for cost planning services. If he had not been, as a first step in my cost planning work, I would have made sure the owners understood the modernist's aesthetic commitment. That information might have helped them make a better choice of designer and spared the neighborhood the obnoxious pink box.

guys, and given choices among several designers whose practices suit their preferences.

Likewise, as a builder who aims to take on projects on a cost planned basis, you will want to find the right designers to work with. There are designers who are just not capable of working with builders in a collaborative fashion. They are unable or unwilling to make the shift that the AIA's Integrated Project Delivery proposal calls for and to work in partnership with builders. I have tried to do cost planning with such designers. One avoided getting input from me even as he raced toward a design pleasing to himself but far beyond the owner's intended budget.

Such designers are not good choices as cost planning partners. You want instead someone like the architect Mike Thomas, who I had the pleasure of working with for a year and a half on a large residential project. All the way through cost planning he always gave thoughtful consideration to my input even if it impacted his vision of the project. During construction, when we needed to solve a problem, he came to the job site, pulled out his pencil, sketched up an idea on the back of a 2x12 scrap, and asked for my crew's thoughts as well as my own. In the end, we produced a new home that was both affordable for the owner and among the most striking projects I have ever built. Mike's willingness to collaborate with my cost planning efforts helped to free up the funds needed to produce his imaginative design.

Before introducing a designer to a client, go over your cost planning program with him or her. Make sure he or she understands it and welcomes it in its entirety. Make sure the designer is willing to take each step in the process with you to deliver optimal results to the client.

Once the designer has been chosen and the conceptual phase of the project is underway, it's important to emphasize to the designer that all that is needed at this point are minimal drawings. Some designers have a hard time holding themselves back once they begin drawing. They may resist your insistence that they do hold back. However, the responsibility of the builder acting as cost planner is to protect the clients from investing excessively in a design—often charged by the hour—before they have some assurance they will be able to afford to build it. They should have a preliminary estimate based on the conceptual design before design goes further.

At the conceptual level, all that is typically necessary in the way of design work is: A floor plan. A section that shows any involved structural work such as a vaulted ceiling. A couple of schematic exterior elevations that indicate the volume and roof structure of the proposed building. Interior views *only* of those surfaces that will involve substantial cost not visible in the floor plans. For example, if the floor plan indicates kitchen cabinet layout, elevations of the cabinets are not necessary; if an elaborate ceiling is a possibility, a reflected ceiling plan will be necessary.

In other words, the conceptual sketches need show only minimal detail. Any additional information necessary for estimating at this stage can be gathered in brief conversations with the designer, at a lower cost than would have been incurred if the designer had put that information into the drawings and specs. With targeted questions, you can quickly learn all the specs likely to have high impact on costs: Exterior cladding materials. Roofing. Door and window style and quality. Interior floor, wall, and ceiling finishes. Specialty items such as such as cabinets, appliances, and elevators. And unique details requiring high-cost materials and labor.

Engineering essentials can be gathered up with even fewer questions. For light-frame structures in a given locale, engineering requirements do not vary that much. Primarily you only need to know: Type and depth of foundation. Drainage requirements. Any major variations from standard stick framing, sheathing, and hardware. And, very important, any requirements for structural steel—as I learned from one of the two cost planning breakdowns I have experienced.

The engineer had made no mention of steel during conceptual or even preliminary design. I did not press the issue, but too readily assumed there would be none, for projects such as the house we were designing never did, in my experience, require steel. In his final drawings, however, the engineer, to my astonishment, required that the complex roof frame—ridge, hips, valleys, common and jack rafters, even fascia—be constructed entirely of steel beams welded together on site. He would not budge from the requirement or even consider my suggestion that we look for a way to stick frame the roof. The steel busted the owners' budget so severely they decided to shop around for someone who would build their new home for their budget. With the help of their architect, they did. My cost planning effort was a failure. It had driven an owner back to competitive bidding. And the failure led to a sad result. The owners did turn up a builder who submitted a bid within their budget. It was far too low. She had to mortgage her own home to fund completion of the project.

Survey Estimate for: Mode/Jacobs Date: 9/16/2014

Item Cost Level: $ -Low $$ - Moderate $$$ - High

Impact on Project Cost: ! – Low !! – Moderate !!! - High

General Requirements	$$		!!
Access	$$$	Long + Steep Drive	!!!
Foundation	$$$	Sloping Site	!!!
Frame	$$		!!
Windows/Xt. Drs	$$$	New and replacement of existing	!!!
Siding	$$		!!
Roofing	$$		!!
HVAC	$	} Tie into (E)	!
Plumbing	$		!
Electrical	$		!
Thermal	$		!
Drywall	$$$	Level 5	!!
Flooring	$$		!!
Finish Carpentry	$$		!!
Paint	$$		!

Notes: Other costs - Design, Permits and Inspections

A survey estimate, such as the abbreviated sample illustrated here, can be especially helpful to clients if you fill it out together with them, explaining the cost levels and impacts as you go.

The survey estimate and the preliminary schedule

After the owners have approved a conceptual design, it is time for a first round of estimating. I call it a "survey estimate." Its purpose is to determine whether or not the conceptual design is within reasonable range of the clients' budget.

You can create a form for the survey estimate, complete it in your office, then bring it to the clients for review. But there can be some advantage to filling out a form together with the clients, even by hand as suggested in the illustration on the previous page. You can work your way slowly and patiently down the form, explaining the cost and impact of each step in the project, filling in the cells and columns. That may help them grasp more firmly what is involved in the project and the impacts of their choices and decisions, while also vividly demonstrating your grasp of construction and its costs.

In the illustrated survey estimate, the dollar signs in the second column indicate cost/quality levels for divisions of work. The third and fourth columns provide space for comments and a notation of the impact of a division's cost on overall project cost. Please note that cost/quality and impacts are not necessarily the same: Exterior doors and windows will be both high cost ($$$) and high impact (!!!), since good windows are expensive, and the project requires a lot of them because existing windows are to be replaced. Drywall will receive a high-quality level 5 smooth finish ($$$), yet its impact on the overall cost of the project will reach only midlevel (!!) because drywall is relatively low-cost work. The mechanical, plumbing, and electrical work is both lower cost ($) and low impact (!) because it will only extend the straightforward PME work already installed in the existing structure.

With the survey estimate sketched out, you can talk about a project budget with your client. But do so carefully. There is a danger that if you give too low a figure, clients will latch on to it and then be disappointed when final costs come in substantially higher. Thus, for the illustrated survey, you might note for the clients that their cost/quality and impact levels are largely in the mid to high range, so their overall project will likely run in the same range. If your costs for comparable additions are running around $300 per square foot for midlevel work and around $400 for high-end work, then by multiplying $350 by the square footage of the project you can produce a prudent survey-level figure for your client's project.

When you project a budget figure, emphasize to the clients that tempting ideas that add cost will likely come up during design, and that the clients may face a choice of expanding their budget or resisting the temptations. In my own work, I underscore the point by explaining that the survey figure is intended to be within roughly 15% of the final cost. I emphasize that the lower end of that range will be achievable only with cost discipline. Meanwhile, I add, even the upper figure may not be adequate if the clients and designer take their eye off the cost ball, for budget realization requires budget discipline.

Make sure, as well, that the clients understand that your survey projection is for building costs only—that design fees, permits, and other soft costs are separate considerations. In short, do the best you can to discourage your clients from unrealistic optimism. If their budget allows only for costs to the low side of the survey range, make sure they understand that keeping costs down to that level may be possible, but only with restraint on their part and, quite likely, some reduction in scope of work.

After the survey estimate is completed, if the clients continue to feel their project is feasible and wish to continue with cost planning (they almost always do, in my experience), you can then create a preliminary schedule. Together with the project designers, set target dates for each stage of design, making sure to be conservative and steering clear of any temptation to fast-track the project. Ask the building department how long they are taking to process permits. With the designer's target dates and an approximation of permitting time in hand, and on the basis of your experience with estimating similar past projects, you can produce a preliminary schedule for design through construction.

The preliminary estimate

With the next step up the cost planning ladder, you come to a crucial phase of the process: building a preliminary estimate from a preliminary design. Now you are moving toward nailing down your numbers and collaborating with the designers to optimize value for the clients. As in the conceptual phase, you should discourage the designer from providing more detail than necessary. Again, the designer may not want to hold back. But if she or he does not, the owner may be burdened with fees for design work that will never be utilized for construction.

Typically, a preliminary design will incorporate enough detail to be roughly halfway to full working drawings. You want to see more developed floor plans, elevations, and sections than you received at the conceptual level and, along with those drawings, details of any unusual and high-cost items, such as an elaborate custom stove hood or an undulating soffit with concealed lighting. From the engineer, you want a section showing the scope of the work necessary for the primary structural components and details for anything that is much out of the ordinary—including, most especially, large custom-made structural components such as steel beams.

But you should not yet need all the info that typically shows up in full working drawings and specs—door, window, and fastener schedules; every item of structural hardware; exact specs for trim materials, paint colors, and other finish work. From the preliminary drawings and an efficient run-through with the designers of your assumptions regarding specifications, you should be able to anticipate the costs of construction with sufficient accuracy for a preliminary estimate.

You will know what the building needs. You will see the needs in the plans even if they are not shown. Clients, or even designers, may find that difficult to understand. It's as if you are claiming supernatural powers, an ability to see the invisible

Pay your subs for cost planning?

One reward of cost planning is that you are paid for your estimating and other preconstruction contributions to a project. A question naturally arises: Should you extend the financial reward to your subs? Should they also get a check? Generally my own answer has been, sometimes to the consternation of my subs, "No."

Why "No"? Because if you obtain work on a cost planning basis and work with the same subs repeatedly (as I do), you are already paying them. You are bringing in work for which they have a very high win ratio and for which they spend, compared to competitive bidding, little estimating time. In other words, they are being rewarded with greatly reduced marketing and sales costs.

Perhaps most importantly, the pay for cost planning does not come because it is deserved for some abstract moral reason. If builders want to be paid for cost planning they must offer a program that adds value much greater than does competitive bidding. The same holds true for subs. To date, however, I have never heard from a sub who has developed such a program.

Details for unusual items, such as this mantelpiece created by Peter Witti, will likely be only hinted at in preliminary drawings. But you can come up with reasonable allowances for your preliminary estimate on the basis of brief discussions with the designer and fabricator.

future. But you know light-frame construction, and the components of light-frame buildings simply do not vary that much, even for buildings with fanciful architectural flourishes.

There's an 80% to 90% overlap in the types of items (though not necessarily the quantity) from project to project, as if details and specifications came out of a pattern book—because in essence they do, for there are many graphic guides to light-frame construction. Thus, if you are shown only the floor plan of an addition, you will instantly know all the items (though not the exact quantity of those items) needed for the foundation from lumber through rebar and concrete; for the frame from mudsill through sheathing; for cabinetry from boxes to tops and hardware; and so on.

To create a reliable preliminary estimate from the preliminary drawings:

1. Do a thorough takeoff and fill out a takeoff sheet as discussed in Chapter Four.

2. Estimate material costs using current quotes from suppliers.

3. Estimate costs for labor by your own crew using your own historical labor productivity records.

4. Do a site walkthrough and a plan review with all the major subcontractors, pointing out obscure items that might elude them. Provide them with a *complete* set of the preliminary plans—hopefully in PDF form to save printing costs and waste.

5. Include generous allowances for correction of hidden conditions and for elaborate details that have been discussed but not yet drawn or specified.

6. Create a summary of your estimate and mark up for overhead and profit on two separate lines to arrive at a preliminary selling price.

As you build your preliminary estimate, create a detailed list of the assumptions you are making. Examples: Concrete mix to be standard PSI. Roof frame to be manufactured trusses. Holdowns at exterior corners only. Attic insulation to be 22" of blown-in cellulose.

Especially important, spell out the assumptions underlying allowances for the hidden conditions and other items whose costs cannot be exactly known in advance of construction. For a thorough preliminary estimate, the assumption list will be long. It will be detailed. Even so, as you gain experience with cost planning, you will be able to produce assumptions and efficiently verify them with the designers. In time, you will find that you can develop an assumptions template and, at least for conventional work, will need merely to edit it for each new project.

In my experience, a thorough preliminary estimate almost always comes in within the range of costs established in a thoughtful survey estimate. It also typically comes within 10% of the final estimate or even closer—assuming the scope of work is not significantly changed.

Value engineering

Once preliminary estimating has verified that project costs are within the survey estimate and within budget, one of the most beneficial phases of cost planning can get underway. And that is "value engineering." In the world of large-scale construction, the value engineering process has been developed to the point of intimidating complexity.

Sitting on my office shelves is a book by Alphonse Dell'Isola entitled *Value Engineering: Practical Applications,* which describes goals and methodologies for big commercial projects. It is 464 pages long. But its basic proposition can be boiled down to a few sentences. Value engineering should aim to "optimize" design decisions. Its purpose is not simply to rein in cost, but to *maximize* project quality for the dollars available. In other words, the purpose is to get more bang for the buck—to deliver better functionality, greater pleasure of use, more desirable aesthetics, and enhanced operating efficiency, all the while staying within budget.

In Dell'Isola's opinion, we fail to notice opportunities to maximize value because we:

- Are too lazy. We just take the first workable solutions that come along and do not make the purposeful effort necessary to come up with better solutions.
- Fall into old habits, dismissing better approaches. "That's how I have always done it; I am comfortable with it," we say.
- Become defensive and react to ideas differing from our own as if they were accusations of incompetence.

Dell'Isola describes several projects where the preconstruction team—architects, engineers, builders, and consultants—break out of the grip of laziness, old habits, and defensiveness to generate dramatic savings and improvements in the project. Value measures developed for one office tower included: Eliminating an entire basement level by moving the heating system that was to be housed there to another level. Changing the floor structure from cast-in-place concrete to less costly hollow-core planks. Changing the configuration of the power distribution. All told, value engineering generated some fifty ideas that reduced cost by $12 million.

At a glance, Dell'Isola's example may seem irrelevant to light-frame work. But if you look closely, you will see that that the measures mentioned above—eliminating unnecessary spaces, searching out alternative structural systems, efficient layout of PME—all have close parallels in light-frame construction. I made use of just such value engineering strategies when building the house described in the sidebar on the next pages. Light-frame work offers us endless opportunities to optimize design/cost decisions—and not only in design, but also in planning of construction procedures. Here's one that caught my eye: The construction of a steeply pitched roof (10 in 12) for a two-story house. The builder decided to frame and even shingle the roof on the ground and then lift it into place with a crane, rather than hauling all the heavy material up to the roof and maneuvering it into place from there. He reports that he not only saved money, he eliminated the danger of working on a steeply pitched roof twenty to thirty feet above the ground.

Smaller and less dramatic optimizations can also be made, especially in remodeling, and they add up. For example, in reconstruction of older

Value engineering a home

When I set out to design and build the home that I have written about in *Crafting the Considerate House*, I had several goals in mind. I wanted to build a house that was environmentally considerate—that would use a minimum of material for construction, be so durable it would need no significant maintenance for half a century, and consume minimal energy for its day-to-day operation. I wanted the house to be affordable for working people, yet comfortable and visually appealing.

To accomplish all those goals, I had to relentlessly value engineer the project, optimizing every design decision to restrain cost while pushing up quality. Here is a sampler of the decisions chronicled in the book:

To begin with, I chose a form that would be attractive yet minimize the use of material and labor consumed for initial construction. That form was a two-story structure, fairly close to cubic in its dimensions, but with pop-outs for the garage and stairway.

Developing the floor plan, I strictly minimized space that would be used only for circulation. In many homes, a large percentage of the floor area provides little more than a pathway from one part of the house to another. In the Considerate House, 2% of the floor area is devoted to circulation-only space.

Structural work offered particularly striking opportunities for value engineering. Collaborating with the soils engineer, I specified a pier and grade beam foundation of the kind usually used on steep hillsides. Relative to a conventional spread footing foundation, it reduced excavation by 90% and concrete work by roughly 50%, even while providing superior strength and stability.

With framing, I incorporated all the elements of what I call "frugal framing" (also termed "advanced energy efficient framing") from two-foot on-center layout through drywall backing made from plywood waste. Relative to standard framing, material and labor was reduced by roughly a third even while the frugal frame, because it eliminates useless lumber in favor of cavities for insulation, supports a superior thermal boundary.

Infrastructure was likewise designed to minimize consumption of material and labor and of operating costs. The bathrooms were placed in a two-story cluster together with the water heater.

That made for short drain, waste, and vent (DWV) and supply runs. Those short runs in turn reduced the time between the turning on of a faucet and the arrival of hot water at a fixture. And that held down water and energy usage while adding a bit of convenience and comfort.

HVAC installation and operational costs were optimized by carefully installing an uninterrupted thermal boundary and by locating most windows on the south wall to provide passive heating. Those measures combined to keep the heating requirements very low. And the low requirements enabled the use of a compact combi-system—i.e., a hot water heater that provides for bathing and washing but also feeds a coil in an air handler to provide space heating. The result is a heating system that was inexpensive to install, is only occasionally needed, and is therefore very inexpensive to operate. Even as it saves money, it adds to comfort. It gushes warm air so quietly into the living spaces it is almost inaudible.

Value engineering is not about going cheap. It is about cost optimization. I went a little too cheap with that flooring choice.

Finishes also yielded to optimization. For example, only the kitchen and bathrooms were finished to level five. An orange peel, so light it was indistinguishable from level 5 from a few feet way, was applied elsewhere. That change saved nearly a thousand dollars in construction cost. Over the life of the house, it will save more because orange peel is much easier to touch up than a level 5 smooth wall. At the countertops, we installed beautiful two-foot-square Italian tile diagonally with minimal grout lines (to make for easy cleaning). It cost a fraction of even the sort of low-grade and often garish granite that is now ubiquitous.

After I completed the house, I realized that I had missed other opportunities for value engineering. One example: upgrading for a modest cost from bamboo flooring made of strips to the more durable and more elegant stranded version. Value engineering is not about going cheap. It is about cost optimization. I went a little too cheap with that flooring choice.

Even so, the overall results have been satisfying. Visitors find it hard to believe that the four-bedroom, two-and-a-half bath house is less than 1,400 square feet. Because so little of it is gobbled up by circulation-only areas, and because the extra inches of ceiling height—achieved by using full eight foot rather than standard studs—add a sense of spaciousness to a surprising degree, the house feels much larger than it is. Even more startling to visitors: the heating bill. It has run around 80 dollars. That's annually, not monthly.

structures in my area, builders often leave patches of stucco in place, though removing all of it is probably the better solution. Yes, that entails additional demo and off-hauling costs. But they can be more than made up for by the benefits. You get an all-new building skin with a longer life expectancy. And you save construction costs, because installation of the additional new material requires so much less labor than creating tie-ins from old to new stucco. In short, you can get better value. Generating the numbers that allow you to weigh the costs and benefits with some precision is the actual value engineering.

On the other hand, one cost-reduction move that may seem appealing but in fact offers little benefit is substituting cut-rate for first-rate subs. Owners often ask for "sub shopping." My response is, sure, we can look for less expensive subs for a few of the trades if necessary. Going with cheap subs, I emphasize, is likely, however, to deliver less than the client hopes for. Why? Because by taking on an inferior sub, I increase my supervision burden as well as my short- and long-term liability. I have to charge for that. So finally, with relatively little savings, the clients can end up with noticeably less quality. That's not penny-wise. That's reverse value engineering.

With preliminary design, estimating, and value engineering completed, you can move up to the last steps of the cost planning ladder, namely to final design and estimating. In my experience, those last steps are often the easiest for the cost planner. If there are no substantial changes in scope of work, and if you have produced a really thorough preliminary estimate—one that anticipates almost all the items that appear in the job—you will be able to produce a final estimate with minor adjustments of your preliminary estimate. Here and there the quantity or quality of an item may have changed between preliminary and final design. For example, a cabinet may have been added while the quality of the cabinet doors was pushed up somewhat. All you need do is make appropriate adjustments to the costs in your spreadsheet and to your written assumptions regarding quality.

You will need to get a complete copy of the final plans and specs to the subs for their final estimate. And if much time has passed between the preliminary and final designs, you will have to update your supplier quotes and labor rates and adjust material and labor costs in your estimate accordingly. But if your preliminary estimate is thorough, with value engineering integrated, even those adjustments can be made efficiently. You'll quickly move to completing your estimate, reviewing it with the owner, and signing a contract for construction.

Your path beyond bidding for free

If you are still in the early stages of your career, a cost planning program as fully evolved as the one described in this chapter might seem overwhelming. It would have been for me when I was starting up. In fact, when I first stumbled onto the approach, it did not even have a name. My "company," meaning me, an ancient pickup truck, a wooden box of hand tools, and a few power tools, had been asked to renovate the top two stories of the Victorian shown in the photo on the next page.

I jumped at the offer. Then I was shown the plans—rumpled sketches produced by the owners' roommate who had studied architecture a bit between sessions with his bong at a hippie college south of San Francisco. I realized it would take me forever to figure out the costs for building everything that was vaguely implied in those "plans." I hesitated. I couldn't give away the time, I told Rob and June, the owners. For every hour I spent on their plans, I would be out the $13 an hour I could have been earning at a job site, and I needed to earn that money to pay my share of my family's expenses.

It was looking like I would have to bow out of the project. Then I had a brainstorm. "How about this?" I said. "I'll charge you guys $12 an hour to do the estimate. I can get by on that." Rob and June, who were as naïve as I was, and did not realize that paying a builder to create an estimate was unheard of, thought over my proposal and agreed that it was fair.

As I proceeded with the estimate, talking with subs and specialty suppliers, discussing options with the owners, I realized I was providing service beyond mere number crunching. Some of it was simply design and specification, but beyond that, I was trying to come up with ideas for the most cost-effective ways of building the project—value engineering, in other words, though I had never heard the term. I liked the work. Rob and June liked the results. I liked getting paid to produce them. And so—propelled by a mixture of naïveté, economic necessity, and pleasure at delivering benefit to my clients, I had started up the cost planning path. As opportunities for other projects arose, I honed my presentation of what I came to call my "cost planning" alternative, and was soon acquiring all the work I could handle without having to resort to competitive bidding.

Of course, to acquire projects via cost planning, you need more than a sales pitch. You need the knowledge and skills to back it up. You need the thorough understanding of estimating and the historical cost data that enables you to produce reliable estimates and useful value engineering. (And, it goes without saying, you need to provide reliable project management and good-quality work once construction is underway.) Hopefully, the information available in this book will help you compress the time needed to get your skills and knowledge up to speed. But you do have to put in the hard miles. You don't get cost planning opportunities because it is unfair that owners and designers exploit builders for free bids (even though it is unfair). You get the opportunities because you have worked to offer clients something better than competitive bidding and can present your alternative in a compelling way.

The owners of this Queen Anne Victorian in San Francisco's Haight Ashbury district had purchased it for $10K (around 1975). It was a worn-out mess in need of complete renovation. I asked them to pay me $12 an hour for developing an estimate of costs from their sketchy plans. They agreed. I had taken my first, almost accidental, step away from bidding for free and up to cost planning.

As you gear up for cost planning, bear in mind that you can use it for any size or type of project. My first venture into cost planning happened to be for a job, Rob and June's Victorian renovation, that was the largest I had ever done. In retrospect, that does not seem ideal. I might better have started with smaller projects. It's possible. Even a deck job or interior remodel benefits from thoughtful specification, value engineering, and estimating. You can reasonably ask for a fee for delivering those benefits.

Here's an example: A builder who was a couple of years into developing his own business gave me a

Cost planning for smaller projects

After I suggested to a fellow builder that he ramp up a cost planning program by starting with smaller projects, he wrote this letter to a client with whom he had been discussing a deck renovation:

> Hi Walter-
>
> Now that I've had a chance to review the project to some degree and understand the general scope, I wanted to propose how to proceed.
>
> To do an adequate job in providing you with an estimate that you can feel comfortable with—covering all reasonably assumed costs—I am estimating that I'm going to need to commit to about 15 hours of work, including: Assessing the existing structure. Arranging and meeting with a structural engineer. Navigating the county requirements. Bringing in potential subcontractors, and building a number. I charge $60/hour for this work and can cap it at $1,000.
>
> Please also know that I am a small organization—just myself, but can "labor up" when necessary—and that I am on the job 8 hours a day focusing on the small details to get it done right, the first time. My time away from the job working on estimates is limited. On projects requiring this kind of work, I therefore need to be adequately compensated. Please let me know your thoughts on this.
>
> Thanks,
> Gordon

call about a bid he was developing for a tricky deck rebuild. As we chatted, I realized he was spending a lot of time with the client simply in hopes of contracting for the project. Had he considered charging for all that preconstruction service? No, but he liked the idea. He sent the client the proposal reproduced here and eventually collected his fee.

It's a damned good proposal, short and sweet. If you're hesitant, it might provide just the encouragement you need to take the first steps into charging for your preconstruction work. Many builders, even highly experienced guys, are hesitant. One veteran builder says, "To this day, I have to get my courage up every time I am on the phone and asked for an estimate. But I know I need to say, 'I will be glad to work up a budget for this project, and our rate for that is such and such.'"

Paul Eldrenkamp, the Boston-area builder whose thinking I admire, abandoned bidding for free when he realized he was winning only one out of fifteen of the jobs that were sent his way by architects. He says that he often meets builders who are having a similarly horrible bidding experience but who are scared to try another way. Don't be, he urges them. "Clients are mostly good, fair, and reasonable people. If you explain to them that you play by different rules, and that you do it in the interests of providing better service, you will have their attention. They will not run away or laugh. If they do, they are probably not qualified prospects anyway."

Why are builders hesitant to require pay for producing a valuable product—a reliable, fact-based estimate? Some industry educators who are more willing than I to venture into pop psychology think it's a "self-esteem problem," that builders have a hard time asking for fair pay for their work because "they don't think much of themselves." Maybe so. I have no idea, really. All I can say is that I balk at the very thought of doing a thorough estimate without compensation. I'm with Paul Winans on this: When you do cost planning, says Paul, "you are bringing practical solutions that help people realize their dreams. Why do that for free? I want to be paid to provide people with solutions."

Sample Cost Planning Agreement

YOUR CONSTRUCTION COMPANY (YCC)

License #045678 Ph: xxx-xxx-xxxx Email: You@YourCompany.com

Date:____ Project:____

COST PLANNING AGREEMENT

Working in collaboration with the owner and the owner's designer, YCC will provide the following services in order to prepare the above-named project for construction by YCC beginning approximately _____________.

- Assist owner in selection of designer by suggesting three possible designers and interviewing them together with the owner
- Inspect existing building(s) for signs of substantial failure
- Create survey estimate based on conceptual design
- Create preliminary schedule for design, permitting, and construction
- Create preliminary estimate based on preliminary design
- Revise schedule as necessary
- Together with owner and designer, value engineer project to optimize quality while keeping project within owner's budget
- Create final estimate and bid based on final design
- Upon completion of each of above estimates, review it with owners and designer(s)
- Prepare contract for construction

YCC will maintain transparency in the estimates and bid by providing the client with: 1) Copies of quotes from the suppliers, subcontractors and service providers that will be involved in the project. 2) A labor rate summary sheet showing a detailed breakdown of wages, labor burden, and margin of safety for all levels of workers to be employed on the project. 3) A bid which shows all markups for overhead and profit separately from direct costs (No overhead and profit will be folded into or concealed in direct costs).

Charges: $__________ per hour. Charged against a retainer of ____________.
Charge for above named services not to exceed $____________.

Dispute Resolution: Any disputes that cannot be settled via nonbinding mediation will be settled in a small claims court. YCC will not be required to pay damages exceeding payments collected for cost planning up to the time of the dispute.

Signatures: Builder: ____________ Owners: ______________________________

Note to the reader: For your own cost planning agreement, make sure you are in conformity with legal requirements for the area in which you work. .

Ethical obligations, legal boundaries

Competitive bidding, I sometimes caution clients, is like a poker game. The guys on the other side of the table are pros whereas you are amateurs. Those pros want to rake in as many chips as possible. They have no obligation to protect your chips. They have been asked to provide a valuable service—an estimate and bid—at no charge; and they are free, therefore, to provide it with an eye toward their own interests.

That's not the case with cost planning. The builder is not at liberty to play his cards close to his chest to maximize his own winnings. He is obliged to provide cost information that is not only complete and detailed but also transparent.

In particular, builders who practice cost planning (or like services) are not free to conceal their overhead and profit in their lines for direct costs. The clients are paying for transparent numbers that will enable them to make the best decisions possible. Builders who charge for their estimates and then manipulate the numbers in those estimates to conceal their markups are not only muddying the numbers and doing their clients a disservice but are committing fraud—a wrongful deception intended to result in personal financial gain.

> ***Builders who charge for their estimates and then manipulate the numbers in those estimates to conceal their markups are not only muddying the numbers and doing their clients a disservice but are committing fraud—a wrongful deception intended to result in personal financial gain.***

Certain builders, when confronted with the demand for transparency, howl in protest. When a builder posted his opinion at an online construction forum that "the days of hiding costs and protecting information are coming to an end… No more games, everything must be disclosed," readers reacted hotly: "Insane," said one, while another reader called the builder an "idiot."

One builder who charges substantial fees for estimates urges giving only a single lump-sum figure to the clients who pay for the estimates. If they ask for a breakdown, she instructs, provide no more than a breakdown by areas of the project and never give them more than that. "Why, it's like giving them a stick to beat you with," the builder insists. "They will say, 'I can get that 2x4 cheaper somewhere else.'" If they continue to press, just say no and walk away from the job, she advises. "It's not that we are trying to hoodwink people," she says. "We are not trying to deceive anyone. We are not trying to gouge our clients. But our figures are proprietary."

The fact is, however, that this builder so adamantly refuses to offer transparency because she is worried that if she did, her clients would see her real margins for overhead and profit and rebel. The margins are close to 45%. In other words, 45 cents of every dollar she collects from clients goes for overhead and profit. She is concerned that if that became widely known, the profitability of her business would be jeopardized.

Other builders who capture similar markups from the jobs they win via cost planning or parallel approaches to contracting for projects are more candid in describing their motives. One told me that he does provide clients with detailed breakdowns of cost, but that the breakdowns "are bullshit," deliberately distorted numbers designed to mislead. He offers them to clients, he says, because they "build trust," and in an unguarded moment he laughingly described his practices as "fraud, lying, and cheating." In his breakdowns, he explains, he shows "whatever people want to see," say 10% for overhead and 10% for profit. The rest of his markup (which in his case can bring the total to 40 cents or more of every dollar he charges) he hides in his lines for direct costs.

What's sad about all this is that the lying and cheating is not necessary, even if you do want to make a lot of money.

Why? He wants to make a lot of money, he says, and if you do want to make a lot of money in the construction business, there is "no way" you can honestly express your full markups. In short, he chooses riches over integrity.

What's sad about all this is that the lying and cheating is not necessary, even if you do want to make a lot of money. Paul Winans, the highly successful remodeler who became the national president of NARI, also collected close to 40 cents for overhead and profit out of each dollar he charged his clients. But, he says, he did not hide the charges. He "expressed" them openly in his estimates. If clients were taken aback, he explained "that's what it takes to run a real company that's going to be in business to take your phone calls 15 years from now." His clients accepted his explanation. Paul built a respected company and prospered.

Another builder, Kevin Ireton, who returned to working as a general contractor after editing *Fine Homebuilding* for two decades, expressed disgust when I asked him about the endemic hoodwinking in our industry. "That's untenable. Being dishonest with people is unacceptable." I agree. People who have seized upon the cost planning approach as a new way to fleece clients are worse than the change order artists they have replaced. They are cheats. The change order artists are just hardball poker players.

A saving grace: Their dishonest practices put the cheats in a vulnerable position should they end up in a courthouse. I asked a veteran construction attorney what he thought of my harsh view of builders who phony up estimates in order to deceive clients who are paying for those estimates. "I could not agree with you more," he answered, and told me that the hoodwinkers were on shaky ground legally. "Guys who use those very high markups and have not discussed profit margin are setting themselves up" for trouble in the event they get into a lawsuit.

"You were playing hide the ball," the attorney tells such builders when they come to him for help, "You knew what you were doing and if you get into litigation it's going to come out in discovery." In other words, the numbers in an estimate for which a client has paid a fee to a builder are not proprietary, slick rationalizations notwithstanding. The client has a right to them and to transparency.

Reasonable expectations, ample rewards

Some guys are happy to contract for preconstruction services with a handshake. My preference: A written agreement, such as the one shown on page 327 that describes the services you provide, places a limitation on liability, and declares a commitment to transparency. While you will want to get input from an attorney regarding liability (as well as other aspects of the agreement), the goal is to avoid being held responsible for the cost of specific items of the construction, especially correction of existing conditions, because you did not spot them during cost planning. Planning does not equal perfection. Protect yourself against being required to provide it. As for the transparency clause, it will distinguish you from builders who offer cost planning–like programs but then manipulate the numbers that they give clients in order to conceal overhead and profit markups.

Finally, you want to set charges for your services. So we arrive at the question: How much should you charge? At the outset, when you are just working out your process, you should probably ask for only modest payment, just enough to represent acknowledgement from the owner that there's value to preconstruction services. What might that be? Maybe the equivalent of wages for a couple of hours of project management work? As you gain experience, you can gradually move your charges up to the level charged by established pros. And how much do they charge?

With striking consistency, I have found the charges to be in the neighborhood of 1% of construction cost for all projects, from small through very large. For a $20,000 bathroom in Alaska, one builder charges around $200 for estimating. A Silicon Valley company charges approximately $2,800 for a $300K remodel. A builder of large projects requiring a year or more to construct sets his preconstruction fee right at one percent of his survey estimate of construction cost, or $15K for a $1.5 million project.

My own cost planning fees have run around 1% for large projects such as whole house reconstructions. But for small remodels I have charged more, because they require cost planning hours disproportionate to their construction cost. If I were to take on a custom kitchen these days, I would set my fee a 2%–3%.

Some builders provide that a portion of their fee, or even all of it, will be applied to construction cost if they do sign a contract for construction with the client. Their aim here is to incentivize the owner to contract with them. I balk at that. In my view, cost planning is a valuable service that should be honored with professional-level compensation rather than being reduced to a loss leader and a marketing ploy.

Some builders include in their agreement a clause that links cost planning to construction: It makes clear that the cost planning is provided as a prelude to construction, that the owner is intending to work with the builder who does cost planning, and that the builder is holding a place in his schedule for the project. I have always used such a clause. A similar understanding is built into the IPD program and the LePatner approach discussed earlier. My clients have been okay with a contractual agreement that links my preconstruction work to construction. But another builder who has used the cost planning approach for decades reports that he has not been able to include such linkage in his preconstruction agreements. His very wealthy clients won't go along with language that binds them to hire him as their builder once cost planning is complete. To them, it apparently implies too much surrender of control.

He, too, has built nearly all projects he has cost planned (about $250 million worth). A few times, however, he not only did not get the project, but the clients refused to pay his cost planning fee. They stiffed him. He's philosophical about that; he shrugs his shoulders and moves on. He did not expect cost planning to move him from the hell of competitive bidding all the way to heaven. It won't. It may make for a much better experience than bidding for free, but you've got to keep your expectations reasonable:

- You may find that cost planning is of limited value in your locale. If you work in a small town where there are, say, only three reputable builders, and you attempt to begin charging for your preconstruction work while the other two continue along the old path, you may find that your opportunities for work dry up. Your fellow townspeople may not be willing to let one of their builders stop working for free. (Solution? Take the other two out for a lunch and work out an agreement that you are all going to slowly begin charging for preconstruction services. That's how cost planning took root in my area. Builders who regularly got together to discuss business development decided they were done with being exploited for free bids and educated clients about the better options they offered.)

- Even in larger metro areas, as word gets around that you are no longer operating as a charity and doling out bids to anyone who asks you for one, some of the people in your client network may stop calling you. Almost certainly, some designers will.
- In certain industry niches, you may run into greater resistance than in others. A builder specializing in foundation, drainage, and structural repair work says that he has asked for a planning fee on large and complex jobs. The result? The clients never call back.
- Regardless of what niche you are in, when recession hits, you may meet strong resistance to preconstruction charges. Most of the relatively few clients in the market for construction are bargain hunting. They think that they will find one via soliciting multiple free bids. As one builder laments, "Nobody knows you when you're down and out. All your reputation gets you is an opportunity to bid against six other guys. A few bucks high, and you lose." Another builder reports that even during the Great Recession he continued to get projects via cost planning. However, he had to offer cost planning as a loss leader. He was unable to collect a fee for his preconstruction services.

Even at its best, cost planning does not offer a magic carpet ride. It is not for everybody. It's not a good option for people who prefer a more arm's-length relationship with clients—who like, as one such contractor says of himself, "to get the job, get in, get it done, get the money, and get out." Cost planning requires much longer and more intense involvement as you proceed from initial inspections through the multiple estimates and reviews with the clients and designers. Or as Paul Eldrenkamp puts it, "The sales cycle is a lot longer and requires much more work at each step."

Eldrenkamp also reminds us that cost planning demands much greater transparency than is required in competitive bidding. "You cannot charge whatever you want," he says. "You have to be a lot more open about where the money goes." I agree. With cost planning, you assume ethical and legal obligations beyond those required by competitive bidding—an issue I have explored in the commentary on ethical boundaries and legal obligations on pages 328 and 329.

Hardest to weather of all the challenges in cost planning are the failures, disappointments and breakdowns that inevitably will occur. I once had clients terminate cost planning because of what amounted, in their opinion, to a mismatch of our personalities; that is, I liked them but, as they candidly told me, they found me a bit intimidating. On another project, I did complete the cost planning, but as the clients and I sat down to sign the contract for construction, the wife hesitated, turned to her husband and said, "I can't do this. We can't afford this work; I have decided to leave you. We need to save our money for the divorce proceedings."

My worst setback ever: I was well into a cost planning on an exceptional project when the architect, whom I had brought into the job, persuaded the client to switch to competitive bidding. It was, he informed me, his "fiduciary" duty to shop for the best deal he could get the client. I have spoken with many other builders who have had a similar experience, repeatedly, in some cases. It enrages them. "I felt the architect was throwing me under the bus," said one.

Sometimes cost planning works well but ends in disappointment. The owners appreciate and pay for the cost planning but cancel their project. The cost planning has shown them that the project is going to impose more stress, financial or otherwise, than they are willing to take on. But at least in such an event, regretful though you may be, you don't come away empty-handed. You've rendered valuable service and helped the owners make an important decision. You've earned a fee. And possibly you have built a strong enough relationship with the clients that they recommend you to their friends. I've had good jobs fall through only to lead to better ones.

In a nutshell, stuff happens with cost planning as with everything else in life. But for all the challenges, if you do move away from bidding for free and travel the path to cost planning, you've moved beyond the off-putting roles that competitive bidding drives builders into: Sly practitioner of malicious obedience. Bullying monopolist. Change order artist.

With cost planning, *if you practice it conscientiously and ethically*, you move to a role more like that of a caregiver. You are like the doctor who prescribes what he sincerely believes a patient needs, not the procedures that will bring him the biggest payday. Or, as Fernando Pagés says, you are like the trusted neighborhood mechanic, the guy his customers recommend to their friends to pay him back for his honest and able service.

That's money in the bank—actual financial capital along with the emotional satisfaction. Because with cost planning, once you are good enough at it to collect a professional fee, estimating and bidding moves from being an overhead burden to an earnings center. It also increases the chances that you will build the projects you do bid and that you will enjoy decent financial results, because with the repetitive estimates produced during cost planning you increase the likelihood that you will nail your numbers.

A Note to Readers:
We have now completed our discussion of the principles and practices of estimating and bidding. In fact, if you have got your arms around all that we have covered so far, it's not absolutely essential that you go any further. You understand what you need to do as an estimator. You know why you need to do it. In fact your estimating knowledge is on par with that of the top flight estimators who were practicing the craft toward the end of the twentieth century.

However, there's been a major development since that time as you have undoubtedly noticed. Desk top computers have been invented. You can buy a powerful one for a $1000 or less. Now any pro estimator deserving of the title does his work on a computer.

You will want to do the same. We will discuss why and we'll explore just how best to make the move to electronic estimating in the fifth and last part of this book, beginning with the next chapter.

PART FIVE

AUTOMATE YOUR ESTIMATING

CHAPTER 17

Power Up

Short and sweet

You are nearing the end of your path to skilled construction estimating and bidding. There's one more step to take. It is a step that can make a substantial difference in your efficiency as an estimator and your enjoyment of the work. It is the step to automating your estimating, to moving it from a pencil-and-paper procedure to one that is executed entirely with computer applications.

If you have acted on suggestions made in earlier chapters, you have already moved in the direction of automating your estimating. You are receiving plans for projects in PDF format and sending them to your subcontractors as e-mail attachments. You are using your word processing application to record those invaluable historical labor productivity records and to create the statements of assumptions that accompany bids and contracts. You have even used your word processing software to build a detailed estimating checklist/spreadsheet — one that would have been considered the pinnacle of professionalism just a few decades ago.

COMING UP

- A last big step on your path to skilled estimating and bidding
- The power of electronic spreadsheets
- An escape from tedium
- Painless "what-ifs" and revisions
- The electronic spreadsheet as a marketing tool
- General principles for automating your estimating
- Electronic spreadsheet options

All that work will prove invaluable as you move forward. But the word-processed checklist/spreadsheet is now seriously outdated. When you are ready, you want to move your checklist onto an electronic spreadsheet. Here in Part V, I will look at alternative ways of doing that.

The move to an electronic spreadsheet will take work. But the payoff will be sweet indeed. On the other hand, not taking the step to electronic spreadsheets would be like becoming a skilled carpenter yet never acquiring facility with power tools other than a Skilsaw, drill motor, and table saw—ignoring the compound sliding miter, laser level, nail gun, and impact driver. You might continue to like building. You might even do good work, and make a living from it. But you would forgo great gains in productivity and competitive position.

Electronic spreadsheets

We have already taken a look at illustrations of electronic spreadsheets in Chapter Five. However, those illustrations are static, just still pictures of abbreviated versions of spreadsheets that have been formatted and filled out. They do not show the dynamic power of the electronic spreadsheet. That power can be summarized in a single sentence: An electronic spreadsheet takes the tedious work off your shoulders and does it instantaneously while leaving the challenging and pleasurable part of estimating to you.

An electronic spreadsheet takes the tedious work off your shoulders and does it instantaneously while leaving the challenging and pleasurable part of estimating to you.

When you estimate with an electronic spreadsheet, you still enjoy figuring out how you will build a project and constructing it in your imagination. You continue to meet the challenge of making sure you have accounted, one way or another, for every item in the project. You must still apply your knowledge of labor productivity rates and gather reliable subcontractor and supplier prices. But you are spared the repetitive math—the endless multiplying of quantities times costs, the interminable adding up of columns of figures, and the boring task of transferring figures from your main spreadsheet to your recap (i.e., summary) sheet.

You are done with the hours upon hours of punching numbers into a calculator. As your skill at using your electronic spreadsheet increases, you can even develop it to the point where, if you type in the square feet of an exterior wall, the spreadsheet will then display your material and labor costs for framing the wall, sheathing it, and installing the water-resistant barrier and siding. And, of course, the computer will automatically tally up the total costs for divisions of work and for the project as a whole. When you first pump out an estimate using an electronic spreadsheet, leaving it to take care of all that number crunching, you feel as if you have been released from prison into paradise. As Wayne DelPico emphasizes, "Nowhere in the construction industry has the efficiency of computers simplified and improved the process as much as it has for estimating."

Not only do electronic spreadsheets speed the work by automating the math, they greatly ease the creation of revised estimates, the "what-ifs" that display changes in cost resulting from changes in scope of work. If you want to know what the cost impact will be if you increase or reduce scope, you simply type in the new quantities of work. The computer instantaneously updates your estimate, recalculating all your subtotals and totals so that you can see the impact on cost of the adjustment in scope, whether it's a reduction or an addition.

Say, for example, you are cost planning a job. The architect gets carried away during design development and the project is coming in over budget. You generate a series of what-ifs: What would the cost change be if we made this change to the design? What if we made that reduction? What if we substituted pavers for the wooden deck? How about if we reduce excavation and concrete work for retaining walls by shifting the site of the building? You create a takeoff for the revised scope of work and make an electronic copy of the original estimate spreadsheet. Then you substitute the new quantities for the original ones in the copy. Instantly, the spreadsheet will generate new totals for each item, each division, and the project as a whole. You can see whether you have reached budget or whether more value engineering is necessary.

Similarly, you can adjust a spreadsheet for a late-arriving subcontractor bid. All you need do is substitute the sub's figure for a placeholder figure you had been using while you waited on the sub. The spreadsheet will recrunch the numbers instantly and produce new totals.

Construction business consultant Michael Stone maintains that these days you really have no choice but to use electronic spreadsheets to speed up your estimating. He believes that in the twenty-first century, clients have been conditioned to expect instant turnaround on estimates, just as they want instantaneous service with every other consumer product. That may often be the case, though I have found that good clients are generally willing to wait for a good builder.

What is definitely the case is that if you move beyond bidding for free to cost planning, the use of electronic spreadsheets is hugely helpful. You will be showing your clients, in detail, the costs for their project. A crisply printed spreadsheet will make the rows upon rows of numbers much more accessible to your clients than would a spreadsheet with numbers scrawled out by hand. In fact, the spreadsheet can be rendered not only clear but also graphically friendly with the use of varying print sizes and styles, boldfacing, and color. Thus, the electronic spreadsheet can be not only a hyperefficient estimating tool, but also an effective marketing aid. It conveys your commitment to being considerate of your clients' needs while providing them with professionally done work.

General Principles

- Unless you are experienced at using heavy-duty software applications, optimize your manual estimating system before automating it.
- Automate only estimating of direct construction costs.
- Do not automate marking up for overhead and profit, which involves judgment calls and leadership decisions that you should reserve to yourself and not hand off to your computer.
- Control automation costs. Managed inattentively, technology costs can drain a company's cash reserves and seriously impair long-term financial results.

Making the move

Exactly how you go about learning the new skills required to build and use an electronic estimating spreadsheet will depend on your personal inclinations and capabilities. But there are a few general principles that should govern your transition to automated estimating:

- As a stepping stone to automation, you should first put into place a sound manual system—one that includes not only a comprehensive estimating checklist/spreadsheet but also procedures from project and client qualification through the production of historical labor productivity records. In other words, as the AGC (Associated General Contractors) advises, "systems should be optimized before being automated." That's especially the case if your use of computers has been limited to such everyday tasks as texting, e-mailing, web browsing, and social networking, but you have little experience with heavy-duty work applications such as a word processing program. You may be that exceptional builder who has extensive prior experience with heavy-duty applications. *If not, avoid learning to use your electronic spreadsheet while you are still assembling your estimating system and procedures.*

- Electronic spreadsheets are an invaluable tool for estimating the direct costs of construction. However, setting of markups for overhead and for profit should not be handed over to the spreadsheet. As stressed earlier, cost estimating is a *management* task. It is about establishing facts, the hard costs of work at the job site. It can be automated. Marking up for overhead and for profit is, on the other hand, a *leadership* task. It involves judgment calls based on consideration of many factors, including the complexity of a project, the economic environment, and your sense of where you want to take your company. You can tell your spreadsheet to give you costs for 100' of exterior wall framing. Provided with current labor rates and material quotes, it will reliably churn out the costs of framing the wall. But you do not want it to robotically add a percentage for overhead and another for profit. You may let the computer do the math once you have decided on your markups. But you want to determine those markups yourself, job by job, with forethought.
- Manage the costs of automation vigilantly. Automation of estimating can be costly. Over the years, it can inflict undue financial burdens.

As discussed in Chapter Fourteen, construction can too easily become a not-for-profit business. Your chances of decent financial performance will be enhanced if you attentively manage automation costs just as you do all other overhead. Say, on the one hand, that your out-of-pocket costs for electronic estimating average $200 a year—roughly what you might spend for the electronic spreadsheet included with a software suite such as Microsoft Office and for education in the use of that spreadsheet. Say, on the other hand, that you spend $1,500 a year, roughly what it might cost you to own and operate a construction estimating package sold by a construction software vendor. The lower-cost option results in a net savings of $1,300 annually, $13,000 per decade, and $52,000 over a 40-year career. Socked away in a retirement account, and enjoying average historical rates of growth, that savings can grow to $400K over the course of your career. The $400K can make the difference between a comfortable versus a threadbare retirement for a decade or more. That is doubly a shame because, as you will see in the next chapters, the best choice for automating your estimating costs very little and is no harder to master than the more costly approaches.

Manage the costs of automation vigilantly. Automation of estimating can be costly. Over the years, it can inflict undue financial burdens.

The best choice of software for the automation of estimating, to my way of thinking, is Excel, the electronic spreadsheet included with the Microsoft Office package. Quite likely, you already have it installed on your office computer as it is the go-to software for office work. Yes, I know, there are other spreadsheets such as Google Sheets and iWork Numbers. However, Excel is generally the choice of builders. And, importantly, the best education in the use of spreadsheets that I have come across is widely available and inexpensive for Excel, while I have found few good instructional materials for the other options.

Other alternatives to Excel include construction estimating packages—spreadsheets combined with electronic catalogues of construction costs—sold by a variety of vendors. There may be some out there worth investing in. But I have not found one that I would willingly use. The fully capable construction companies that I have encountered do not use them. They use Excel. The alternative packages appeal to people who are, for one reason or another, hesitant to take full responsibility for the process—estimating their costs of construction—that lies at the very heart of their business. They want someone to hand them a fill-in-the-blank-program that requires only that they provide quantities of work and that will then tell them how much to charge for that work.

To be sure, there are experienced construction industry consultants—often, it should be noted, guys who make their living selling the estimating packages—who disagree with my preference for Excel. They claim the fill-in-the-blank programs are the way to go. Certainly, many thousands of builders have elected to make use of one of these packages. For that reason, before introducing you to Excel and a low-cost way of learning it, I will briefly offer an overview of the other options, noting their potential advantages as well as their drawbacks.

CHAPTER 18

Programs from Free to Not So Cheap

Poor Choices

There are three overlapping groups of prepackaged programs created specifically for construction estimating. I call the groups "freebies," "loss leaders," and "insurance adjuster spin-offs" (IASs). None that I have reviewed compare favorably, in my opinion, to the sort of Excel-based estimating system that organized and disciplined builders can create for themselves. A sales rep for one of the programs characterizes a prime competitor's program as "lame." He's right about his competitor's program; it is weak in important respects. But the trouble is, so is his offering, and, to varying degrees, so are all the others I have looked over. In fact, I view these programs as such poor choices that I am going to largely avoid even mentioning them by name. My intention in this chapter is not to provide a buyer's guide to specific programs, but to suggest a means of evaluating the prepackaged programs, should you elect to search them out for yourself.

COMING UP

- Learning from the freebie estimating programs
- Hazards of free programs
- The program that went poof
- Benefits and hazards of low-cost estimating programs
- Insurance industry estimating software
- Coming around to Excel

Freebies

If you run a web search for "free construction estimating programs," a surprising number of offers will turn up. Some are only for free trials, but other offers are for indefinite free use of a program, or for programs that are so inexpensive they are virtually free. If you are eager to jump into the world of automated estimating, one of these freebies may tempt you. You might just want to grab hold of it and get going.

Resist the temptation. You do not want to adopt your estimating software on impulse. One builder told me that when it came time to automate, he "went at it without a vision of where we wanted to be and just said, 'Let's try some of this and some of that.'" He was still untangling the resulting mess a couple of years later. He would have been better off had he chanced upon and heeded this advice from Fernando Pagés: "The biggest mistake you can make when evaluating software is a hasty decision. New software takes months to learn and fully integrate." You don't want to impulsively jump into a program, invest time getting up to speed with it, then find out you've chosen the wrong program.

That said, it might be worth your while to explore freebies. You probably will not find one you can reasonably adopt for the long, or even short, term. But you will get a look at alternate ways in which estimating programs are designed and how they appear on your screen. You will have a chance to observe spreadsheets in action. As you take a freebie for a spin, you will experience the core power of electronic spreadsheets—their ability to take over those tedious, repetitive math calculations that estimating demands. All in all, the freebies can give you a sense of how electronic spreadsheets are set up and how they function.

A freebie can also introduce you to the use of "templates." After creating a spreadsheet/checklist for tenant improvements, a kitchen remodel, or any other type of project, you can then copy and store that estimate. It becomes a template, a guide or pattern, for initiating any similar estimate in the future.

Some freebies come equipped with links to YouTube videos that show you the spreadsheets in action and introduce you to their operation. All that learning is to the good. But I have to caution you: Even learning from the freebies can be time-consuming and frustrating. I often found them to be counterintuitive, running against the grain of how builders actually think about construction. And their online manuals and Help functions are sometimes so clumsily organized that they are more confusing than helpful.

If you feel tempted to adopt one for use in your company, be on the lookout for these serious weaknesses:

- *Opaque math.* All the math takes place behind the scenes. You don't see the formulas that are being used to generate the prices you are going to offer to clients.
- *No visible proofs.* Because the math is opaque, you can't determine if it is accurate. You can't run checks, as you can in Excel, to make sure the right formulas are being used and that the costs for all items in a project are included in the total for the project. Personally, I find that downright scary. If an estimating program has a bug in it, if it fails to get the math right or leaves something out of its totals, it could do some serious financial damage, repeating the error in estimate after estimate.
- *Complex procedures for building your list of items.* Rather than simply being able to insert a blank row into your spreadsheet and then type in the item, you have to open a drop-down menu, ferret out the item you need, and then insert it into your spreadsheet. That's an irksome chore, especially when you have to do it over and over.
- *Inadequate lists of items.* In a free program offered by a national building materials supplier, the General Requirements drop-down menu of items offered only a single item, "miscellaneous general conditions," instead of a comprehensive selection of general requirements. Thereby it reinforced the already costly tendency among builders to overlook general requirements costs in a project.
- *Missing items.* In one program I reviewed, the items menu for concrete installations did not include reinforcing bar—just forming material and concrete!
- *Wasteful material lists.* One program I reviewed fills in the blanks for quantities of studs and plate stock when you type in the length of a wall. At a glance, that looks like a helpful feature. However, the program provided for 12' and not 20' sticks for plates, which would result in waste of material and labor as well as walls weakened by unnecessary joints.

Another troubling feature you should be on the lookout for if you sample freebie programs is a tendency they can have to push you toward violating two fundamental principles of skilled estimating and bidding: Putting someone else's costs into your estimates. And using the same uniform percentage markups for overhead and profit rather than figuring them one project at a time.

When you type in quantities and the programs fill in the costs, they are getting those costs from

Gone!

During research for this book, I came across a free estimating program offered by a national building supplier. It was cloud-based. You could not download it to your computer but could access it only via the web. It was a decent enough program, maybe useful as a start-up package for readers who needed a stepping-stone to Excel. Problem is, when I went back to look at it a year later, as I was writing this chapter, it had disappeared. I could not find it at the supplier's site or track it down on the web.

I called the supplier to find out what had happened to the program. They were no longer making it available, a customer rep apologetically told me.

So, I asked, what happened to the estimates I had created using the program? Where were the templates I saved? Where could I find them?

Perhaps, responded the sales rep, I could get them from the vendor who had provided the software to the supplier.

And how could I find those folks? I inquired.

The rep told me he would get back to me with an answer. He never did. I searched the web myself but could not find a way to access my work.

Since I was just exploring the program as research for this book, not for use in my business, its disappearance cost me only a few hours of my time. But for builders who had actually integrated the program into their business operations, the disappearance of their program, estimates, and templates would have been a serious setback.

Their situation illustrates a danger of using one of the numerous free or very cheap estimating programs, especially if they are cloud-based and cannot be downloaded to your computer and backup systems. You could put in a great deal of effort into mastering the program and building up your file of estimating templates, only to have your work evaporate into cyberspace. Almost certainly, that will not happen with Excel. It will not go poof.

somewhere. The companies that produce the programs may be gathering cost data for themselves, or they may be leasing it from another publisher of cost data. But either way, when you rely on their data, when you let them stick costs into your estimate, you are letting someone else set your labor and material charges. And that is something you should avoid, for all the reasons discussed in Chapter Ten, if you are out to nail your numbers.

Equally concerning is a feature that steers users into figuring overhead and profit markups with a uniform percentage applied across all projects. For example: You type in 100' of wall framing. The freebie program fills in $1,000 for labor and another $1,000 for material in the cost columns. Then the program adds markup for overhead and profit to arrive at a selling price for framing of $2,500. The markups may be based on a default percentage set by the program or on percentages you have placed in the program yourself.

Either way, by using fixed percentages, the freebies introduce a serious weakness into your estimating. They treat overhead and profit as "closed costs" like material and labor when in fact overhead and profit are "open costs." As you recall from Chapter Fourteen, open costs are best figured with respect to the time a job takes and the portion of your capacity it utilizes. They should not be robotically figured using a fixed percentage—appealing though the seeming simplicity of that approach might be. By doing so, the freebies eliminate your opportunity to make considered judgments about the likely duration of a project, market conditions, client and designer reliability, and the other factors that should go into the calculation of markups.

Finally, if you do decide to adopt a freebie and then use it for a couple of years, you will have a lot of labor invested in it. You will have customized it to suit your needs and created valuable templates. However, if it is cloud-based, not downloadable to your computer but stored instead on a remote server, and if the company that sponsored it decides to stop doing so, then it can go poof, just disappear. Suddenly you find yourself with no access to the program you had placed at the center of your business.

The possibility that a freebie may go poof is real. For the freebies, you see, are actually offered as enticements. The folks who offer them, quite naturally, want something in return for giving you their program. They want to draw you into using their products—whether building materials, cost data catalogues, or a variety of consulting and other add-on services. For example, for a time a major building supply chain offered a free estimating program. It was set up to funnel you into ordering materials for your projects from their stores. When you completed an estimate, the program automatically generated an order for materials and offered you the chance to submit the order to the supplier with a few clicks of your mouse.

Who knows how many builders integrated that freebie into their operations? All I know is that builders who did build their businesses around it went to work one morning and found they could no longer access the program. The supplier had decided against continuing to make it available. Presumably, it was not proving to be effective enough at creating sales of materials to justify the costs of maintaining it. As described in the sidebar on the previous page, the program just disappeared—along with estimates and templates that had been built with the program.

Loss leaders

A step over (though not necessarily up) from the freebie estimating programs and similar to them in many respects are what I call the "loss leaders." They are inexpensive, costing less than a thousand dollars or even just a few hundred dollars. Their low cost reminds me of printers, relatively cheap because what their manufacturers are really interested in is capturing you as a long-term customer for their ink cartridges.

The vendors of loss-leader programs want to recruit you via the software as long-term customers for their other products—books, consulting services, and especially their construction cost data. Like vendors who pass out freebies, they are interested in selling you services you will need in perpetuity once you have placed their estimating program at the heart of your operation.

That's not to say the loss leaders are worthless. Those I have looked over are somewhat more flexible than the freebies I've seen, allowing you to modify your lists of items, and to bundle items into assemblies. With them, you can create templates or even draw on a large library of ready-made templates. At least some of the programs are downloadable or even available as CDs. You can install them on your computer rather than running the risk of their suddenly becoming unavailable because they are cloud-based.

However, to my way of thinking, the downsides of the loss leaders outweigh their strengths. As with freebies, you run into the robotic marking up for overhead and profit that treats them as closed costs, ignoring the crucial time and capacity factors and judgment calls. There is also the risk that support for your program may suddenly become unavailable because the vendor goes out of business.

To builders who rely on these programs, the suggestion that their vendor might abruptly close its doors and shutter its web site may seem a stretch. They have been relying on the programs for decades. But here is sobering fact: During the research for this book, I talked to a salesperson for one of the most venerable of the loss-leader vendors. She told me that the company was barely surviving when its founder sold out to new owners. At this point, whether the vendor stays in business and remains available to support the thousands of programs it has sold is anyone's guess.

If you do decide to consider loss leaders, each vendor will tell you how easy it is to generate estimates using their program. One vendor claims that you can be up and running, creating reliable estimates in as little as six hours. Would that not be wonderful? You would not have to build up records of your historical labor productivity costs or obtain up-to-date quotes from your material suppliers. All you have to do is spend a Saturday morning, coffee drink of your choice on the desk beside you, getting acquainted with the program, then punch up a takeoff for a prospective job, plug quantities into the program, and—viola!—it will crank out labor, material, sub, service, and overhead costs and provide you with a profit as well.

Sound too good to be true? It is.

In my attempts to test-drive loss-leader programs, I found them not at all easy to learn—harder than Word, which I use to write books; InDesign, which I use to design books; QuickBooks Pro, which I use for accounting; and Excel, my beloved estimating program. I not only found the programs difficult to learn (as have other experienced builders) but help from the vendors very hard to come by.

For one loss-leader program, getting help was like squeezing sap from a kiln-dried 2x6. I called repeatedly. Got only voice mail answers. Sent e-mails. Back they came tagged undeliverable. Decided to call one last time. Reached a delightful woman who seemed to know the program inside out and who did clarify things for me. Called back with another problem, but could not reach her again. Got only her assistant, whom I will charitably describe as a grumpy character who seemed even more befuddled by the program than I was. So, I looked for videos. I found none. I did locate a manual, but it was so badly written that I had trouble locating the instructions I needed and then more trouble following them when I did.

My skepticism about them notwithstanding, some loss-leader vendors have been in business for decades and claim to have tens of thousands of customers. Even though that amounts to only a miniscule percentage of construction firms in the country, the figures suggest several possibilities: Some guys are finding solid value in the programs. Or, there are a lot of suckers out there. Or, most likely, is my guess, there are a lot of builders who bought the sales pitch for a loss-leader program as they made their way into automating their estimating, managed to squeeze acceptable results out of the program, and have stuck with it out of inertia. After all, once you have sunk time and money into something, it's hard to toss out the investment and start over.

You will find plenty of testimonials from users of loss leaders on the web. Believe them if you like. You may decide that one of the (initially) low-cost programs is for you. Before you make a purchase, however, you should carefully evaluate the cost data that the vendor will be feeding into your computer. Obtain the program for a free trial period, a lengthy one. Compare the costs it uses to those that you produce using your labor productivity records and to the charges you experience at your building materials supplier. If you see that the program's costs do not match up with your costs and you are, therefore, not going to use them, then buying the program rather than using Excel really does not make sense. Essentially, you would be forking over cash for a spreadsheet that is probably inferior to and costlier than the one already included in your Microsoft Office suite.

It is not likely, given the means by which it is collected, that the cost data provided by a loss leader program will align closely enough with your own costs to serve for accurate estimates. Although the data providers I spoke with were secretive about their collection methods (one calls it his "secret sauce"), now and again I was able to get a good look. A rep for one vendor explained to me that the company employed a dozen staffers ("cost engineers," they are grandly called) whose job is to achieve the following feat of research: Track cost data for material and labor costs for over 15,000 items in each of 250 different areas of the U.S. and Canada and update all costs monthly. Do the math and you will see that the dozen staffers are tracking 4 million costs per month with each "engineer" updating about 2,100 per hour, or more than one every two seconds.

Who knows what sorts of algorithms and web crawling software the staffers use to generate those 4 million costs per month? It is all but certain, however, that they are not obtaining the kinds of dialed-in costs you get when you figure labor with the help of your historical labor productivity records. They are not getting the reliable figures that you obtain by sending a materials list to your building supply yard for a quote. And it surely does not seem likely that data churned out at the rate of a cost every two seconds will take you down the path to nailing your numbers.

Even the founder of one of the loss-leader vendors allowed that the data he fed into the programs that he sold to his customers was good for only 80%, or at the outside 90%, of the costs of a job. I suspect those figures are highly optimistic at best, or outright baloney sales promotion at worst. But whatever might be the correct percentage, how do you know where the numbers churned out by the program are off? Or way off? How do you correct for the errors if you don't know where they are? Even if you do know where they are, how do you correct for them other than with WAGs (wild ass guesses) since, as a result of relying on the program's data, you do not have your own reliable records? And, finally, if you don't correct the errors, how badly will they skew your estimate and how can you protect yourself against the

Factors to Consider When Selecting an Estimating Program

Your Costs

- What are the annual costs of ownership and maintenance?
- How do the costs compare to the costs of other options?

Evaluation Opportunity

- Does the program's vendor offer a sufficiently long free trial period for you to thoroughly test-drive the program before committing to purchase?
- Will the vendor give you contact information for other users in your area so that you can get their assessments of the program?

Features

- Is the program easy to read and manipulate on your computer screen?
- Is the program flexible? Can the program be tailored to fit your own operation?
- Does the program allow you to create and save templates for estimating your repetitive projects?

Cost data

- If the vendor leases cost data for use with the program, how frequently is it updated?
- How closely does the cost data sold with the program resemble your actual labor and material costs?

skew other than by heaping on arbitrary "contingencies" and "fudge" in the manner of the struggling estimators described in Chapter One?

If you rely on the data provided by the vendors of loss-leader estimating programs, you are back to using a cost catalogue for your estimating. For that is all the data bundles from the vendors are, digitized versions of paper cost catalogues. Other experienced builders have told me the costs in these digitized catalogues don't match up with their actual cost experience any more than do the costs in the paper catalogues. One builder did, however, come up with what he calls a "solution" to the problem. He bought a loss-leader program, used it and the accompanying data for a while, then ditched the data and continued to use the program's spreadsheet, but substituted in his own costs for the vendor's.

In response, I again come to the question: Why fork out good money for an estimating program when you already have Excel available at no extra cost, packaged in the Microsoft Office suite that is already on (or that should be on) your computer? Why pay a loss-leader vendor for an inferior program when a fabulous, well-supported spreadsheet program is already at hand?

Insurance adjuster spin-offs

Beyond the freebies and loss leaders, we come to a third group of prepackaged estimating programs—those initially created for insurance adjusters (the guys employed by insurance companies to appraise flood and fire damage to buildings) but adapted for builders. These insurance adjuster spin-offs (IASs) differ from the loss leaders and freebies in one striking way. While they allow you to estimate in the conventional fashion, by putting quantities into a spreadsheet that will then fill in the costs for you, they also provide you with a very different way of inputting the scope of work. Rather than doing a takeoff and entering quantities in the spreadsheet, you can create a sketch of the project right on your computer screen, incorporating dimensions and adding specifications. When you have completed the sketch and specs, the program will read them and pump out a detailed estimate of labor and material costs along with markups.

IASs are impressive (and seductive; one industry consultant, after seeing an IAS in operation for the first time, giddily exclaimed, "It does 90% of the work for you!"). A builder who uses (and swears by) his IAS values it not only because, in his opinion, it makes estimating easy. He has also found it to be a valuable sales tool. He can take along a laptop loaded with the program on a sales call to a potential client and on the spot—at least for a small remodeling project—sketch up the project, agree on specifications with the client, and generate a price.

But, of course, IASs can also have drawbacks that are similar to those that come into play around freebies and loss leaders: Overhead and profit bundled up with direct costs and thereby treated as closed costs. Cost catalogue data that may be substantially out of line with your own costs of construction. Moderate initial purchase price but an additional steady and substantial overhead burden over time. A steep learning curve. Some risk of the vendor going under after you invest in their product and become dependent on it.

The promotional material for an IAS will, of course, tell you that it is easy to learn. One video claims that within days of being hired, new employees of a remodeling company were cranking out estimates for bathrooms in five minutes. However, in YouTube videos of training classes for the program, you get a very different picture. You see not only trainees, but also their instructor, struggling with rudimentary aspects of estimating-by-sketching, such as placing a door in a wall.

An experienced builder, who went bankrupt during the Great Recession and earned a living as an insurance adjuster until the economy revived, expresses admiration for the program he used. It allowed him, he reports, to churn out five appraisals a day. But, of course, cranking out an appraisal entails much less personal risk than committing to build a project for a given price. And even that builder-turned-appraiser says that the program is the most challenging to learn of the half dozen construction estimating software packages that he has tried.

However, the greatest concern I have about insurance adjuster spin-offs arises not from the perspective offered by a builder who was using one during an interlude in his career. It results from an extensive interview I conducted with a builder who first discovered the program 20 years ago, recalls that when he first came upon the program he felt he had found "the last piece of the puzzle," and remains an avidly enthusiastic user of the program to this day.

I happen to admire this guy. He loves building and runs a classy, compact company that produces good work. But I have wondered if he would have built an even stronger company had he chosen a different estimating program.

I happen to admire this guy. He loves building and runs a classy, compact company that produces good work. But I have wondered if he would have built an even stronger company had he chosen a different estimating program. For his IAS program, he says, regularly degrades his selling prices. The prices for material that the data package feeds into his estimates, he complains, are for the low-quality goods available from the cheapest suppliers. Labor costs, he feels, are too low for the complex work turned out by his highly skilled and well-paid crew.

The builder tries to compensate for the shortfalls. If the costs the program churns out for 100' of wall framing look low to him, he throws in a line named "additional framing" with additional costs for labor and material. Even so, when he tallies up the total direct costs for a completed project, he regularly finds they run substantially above his estimate and that he has earned little if anything in the way of profit from the project.

There are a number of possible reasons for the shortfall: The sketches generated with the program may be too sketchy and obscure or leave out small items whose costs add up to significant amounts. Or the numbers coming out of the data package are just too low, and the builder's intuitive, WAGish adjustments via fudge-like "additional framing" entries don't make up for the inadequate figures produced by his program.

Whatever the exact dynamics, the root problem would seem to be that the builder has relied on a cost catalogue rather than on current supplier quotes and on labor productivity records for his own crew. He's building estimates for his jobs using someone else's numbers, not his own, and getting the all-too-typical results.

But perhaps there's another side to the story. Because my disinclination toward the prepackaged, fill-in-the-blanks estimating software packages and preference for Excel is so strong, I have tried to find people who could speak convincingly for the other side. I called up a veteran industry consultant, a guy I like and respect, who of late has been making his living as a vendor of IAS training. He argues that any spread between the IAS data and actual costs experienced by the user is not of consequence. You can, he says, compensate for such a spread via thorough and relentless job costing, substituting your recorded job costs for the IAS data.

But I'm afraid that just does not make any sense to me. If you are going to have to create reliable item and assembly cost records of your own to use the IAS effectively, why buy it in the first place? Possibly, if you run a business that does only small simple remodel jobs that need no more design than a quicky sketch, the estimate-by-sketch feature may make purchase of an IAS program worthwhile. However, if you are doing or plan to do more complex jobs that require full blown professionally produced plans, then all the program provides you with is a spreadsheet inferior to Excel for a good deal more than Excel will cost you. Why choose it? Why spend all that extra money on it?

My consultant friend, however, has a response to my befuddlement. Builders, at least those who come to him, he declares, are "artists," restless souls who don't have the patience and determination

Good enough for you?

Xactimate is a well-known vendor of a construction estimating program that has been adapted from a program for insurance adjusters. When you use Xactimate, all you need do is input quantities of work via entries in a spreadsheet or by creating a sketch of the project. Drawing on its cost database, Xactimate will pump out a detailed estimate.

That's so appealing. Estimating made easy! But at what price? A cautionary note, posted on the web, comes from United Policy Holders, a nonprofit, consumer advocacy organization: "Xactimate's estimates often look impressive because they are well organized, professional looking, and lengthy, but they are often inaccurate… Pricing is often too low and generic… [It] is best suited to tract homes. For property that is custom-built, historic, or located in a high-value area, Xactimate's numbers will likely not be sufficient. Underestimating on material costs and lowballing resulting from the improper use of Xactimate has factored into many claim disputes and lawsuits."

To be fair, I need to emphasize that the lawsuits referred to here go back some years. However, there is still, in my opinion, cause for caution in the use of Xactimate's cost data. One builder in my area—and he is not a critic of Xactimate, but a long time and enthusiastic user of its program—told me that for his projects, typically higher-end remodels, the numbers Xactimate pumps out are just too low and that he must jack them up.

A former estimator who has extensive knowledge of Xactimate, and who is also a staunch advocate for the program, vigorously defended it in an e-mail exchange with me. But, in fact, his argument confirmed the possibility of the very problems that United Policy Holders describes. Xactimate, he told me, gathers cost reports from a wide range of builders across a variety of markets in a particular location. Then it averages their reported costs. The result for, say, drywall installation in Cincinnati, will be a square foot cost roughly halfway between the highest end and lowest end costs encountered in that city. Thus, a custom builder in Cincinnati—like the high-end remodeler in my own area—may find the costs for drywall pulled out of Xactimate's database to be seriously low. On the other hand, a builder doing production work might find them off to the high side.

So, what are we to make of this? My take: Xactimate may well be producing costs that are reliably average—just as the costs found in a variety of cost catalogues may be reliably average. But how many builders consistently experience costs that are exactly average rather than costs that are uniquely their own and lower or higher than average?

Maybe your costs are exactly average. Maybe they consistently come in very close to the costs provided in the database from Xactimate or some other software company. In that case, you might come reasonably close to nailing your numbers by relying on Xactimate's or one of its competitors' cost numbers.

If not, if you persistently need to adjust from the averages, there does not seem to be much point in paying for that averaged data in the first place. You are going to have to compile your own data to accurately modify the data provided by the outside vendors. So, would you not then be better off just going directly to that data to produce your cost estimates? Would you not save yourself the time spent relentlessly comparing your actual costs to the costs from the database? Would you not also save yourself the annual subscription fees for the database—fees that add up to substantial amounts over time? To my way of thinking, the answers are all "Yes."

necessary to run their company in the way "that a David Gerstel runs his company." His guys suffer from Attention Deficit Hyperactivity Disorder, he told me. And he meant it literally. He was not being ironic. When I first heard it, my friend's opinion about his clients startled me. Then I heard the very same opinion, including the dead-serious allusion to ADHD, voiced by the lead sales rep for another well-established vendor of a prepackaged estimating program. Both men insist that their clients do not have what it takes to build their own estimating systems on Excel, that they need someone else to create an estimating system for them.

As I thought about their perspective, I realized that though it seemed contemptuous at first glance, it is descriptive of some builders I know. They are restless souls; they want to get on with construction. They grow uneasy at even the thought of sitting down and building an estimating system. If that description does fit you, maybe you do need an off-the-shelf, prepackaged, fill-in-the-blanks estimating program. My guess, however, is that if you have come all the way to this page of *Nail Your Numbers,* you've got the right stuff to build an Excel-based system. And in my experience, that is exactly what women and men who have what it takes to build enduring construction companies usually do.

So let's get to it.

CHAPTER 

Excelling with Excel

The indispensable tool

You've already had a sneak preview of Excel in earlier chapters where Excel worksheets (i.e., electronic spreadsheets contained in a "workbook") have been used to illustrate estimating fundamentals. Now we'll take a closer look at the use of Excel for estimating. Here in this chapter, I will give you an overview of setting up Excel and putting it to work as an estimating program. Additionally, I will suggest a high-quality but brief course of study, largely of videos, that will get you up to speed in the hands-on operation of Excel.

In brief, to use Excel effectively you will need to learn to:

- Design a comprehensive direct costs worksheet for construction estimating and bidding.
- Use the fundamental commands and tools that come packaged with Excel.
- Create and enter formulas into your worksheet in order to automate your estimating math.
- Maintain the security of your worksheets and estimates.
- Run audits (checks and proofs) to catch and correct errors in your estimates.

Once you have a good grasp of Excel basics you may want to develop other worksheets within your workbook. First and foremost, you may want to create a stand-alone worksheet to track and update overhead costs. You may also find it well worth your while to set up additional worksheets for labor rates, for direct cost details (for example, for the per-foot cost of installing a one-story foundation), and for estimate summaries.

COMING UP

- Why Excel?
- A high-quality, low-cost approach to learning Excel
- Recommended videos and references
- Designing Excel worksheets (i.e., spreadsheets)
- The fundamentals of using Excel
- Crunching numbers with Excel
- Maintaining the security of your estimates
- Auditing (i.e., checking and proofing) your Excel estimates
- Creating a worksheet to track overhead
- Creating additional worksheets for labor rates, direct cost details, and estimate summaries
- An advanced topic: Linking the additional worksheets to your direct costs worksheet

Such worksheets can be linked to your comprehensive direct costs worksheet, which covers all items in a project. Thereby, as you will see toward the end of this chapter, the linked worksheets can enable you to create an estimate and the bid that flows from it much more efficiently.

Each of those broad categories of necessary learning will include many discrete steps, because Excel is powerful. It provides a wide variety of ways to format, enter, and manage numbers and text within worksheets. Gaining command of all that power takes effort. You won't become skilled at operating an Excel estimating workbook in a day, any more than you became skilled with all the tools of your trade in a day.

On the way to mastering Excel, you will probably, at times, become frustrated. My advice (or plea, really) is this: Bull through the frustration. Fight through to the rewards as have countless other builders. (They did it. You can do it. Right?) Once you have learned to create and operate an Excel estimating workbook, you will produce your estimates far faster, with greater precision, and with more enjoyment.

How long will it take you to get to the rewards, to become fluent enough with Excel to build a basic but powerful estimating program? The answer depends in good part on the depth of your prior experience with computers. (If you have worked with other heavy-duty work applications, you'll move more easily along the path to Excel fluency.) But I am certain of this: If you are a capable builder, you can master Excel. Doing so may not be easy, but it will not be as challenging as learning to construct even a small remodel project. Not even close! The effort you put into learning Excel will be well spent. One builder who mastered Excel rather than buy a prepackaged program explained, "I wanted to create, own, and control my estimating system." That is the position you get to by learning to use Excel for estimating and bidding.

You will find that Excel can also be used for many tasks besides estimating—such as working out job schedules and creating task lists. Excel is a program you should not be without. "Knowing Excel," declares one experienced builder, "is a matter of basic business literacy." He's right. It's the indispensable tool.

A lousy way to learn Excel, and a good one

In my experience, books are not good venues for learning the nitty-gritty of software operation. I have tried to learn about Excel from books. Even the best and most popular of them give me a headache. So you won't bump into sentences in this chapter that take you through all the points and clicks necessary for composing Excel formulas like this: =SUMPRODUCT(B8*B3:D3). To the extent that it is practicable, I am going to refrain from giving you step-by-step instructions. Instead, I will leave it to you to master the details of using the Excel commands for formatting your worksheets, to enter numbers and text, and for performing calculations. In my experience, the best way to do that will be (1) playing around with and test-driving your own Excel worksheets and (2) watching a series of videos. (Please note, if you own Microsoft Office, a suite of software you do not want to be without if you are serious about achieving long-term success as a builder, then you already own Excel. It is part of Office. If you do not own it, well, then it is time to buy Office.)

The videos, with their voice-over explanations, clearly demonstrate the use of the Excel commands for tasks from formatting text to the installation of mathematical formulas. Unlike a book, which can only *tell you* how to use Excel, the videos can simultaneously *show you.* As you watch the videos and test-drive your worksheets, you will be learning in much the way that you learned building—working behind people who know what they are doing and imitating them.

My recommended sequence for learning Excel is summarized in the sidebar titled "A high-quality, low-cost course for learning Excel," which you will find on the next page. As a first step in the course, and even if you have never used the program before,

A high-quality and low cost course for learning Excel in 10 manageable steps

Part I

1. Open an Excel workbook on your screen and play around with a worksheet.
2. Watch *Microsoft Excel Tutorials for Beginners* #1, #2, #3, and #4 from Motion Training on YouTube.
3. Establish an account at Lynda.com.
4. At Lynda, watch *Excel for Mac 2016* by David Rivers. (Rivers's superb instruction will prove invaluable whether you use a Mac or a PC and regardless of what version of Excel you use.)

Part II

1. Read this chapter, with its focus on the use of Excel specifically for estimating, to the end.
2. Continue to play around with Excel on your computer as you read.
3. Watch selected sections of *Excel 2016 Essentials* by Dennis Taylor at Lynda.com.

Part III

1. Watch Estimating with *Excel for the Small Contractor* on YouTube.
2. Watch Excel Basics #22: IF Function Formula Made Easy! on YouTube.
3. Watch Lynda.com videos on use of IF function

References

- *Excel for Dummies* (Preferably the edition that matches up with your version of Excel)
- *Construction Estimating* by Adam Ding

See Resources for reviews of these two books. While not nearly as effective in teaching the use of Excel as the videos suggested above, they are handy to have around when you have very specific questions about the operation of the program.

you will find it helpful to take Excel for a spin. Open the program. Open a workbook. If they don't come up automatically, figure out how to get the Excel commands onto your screen. Mess around with a worksheet. Scroll back and forth.

Click in a cell, one of the little boxes spread across your worksheet, to make it active. Type in a few numbers. Put in a bit of text. Try out a few of the commands such as the curved arrow pointing left that you click on to undo a previous command, the "B" for making text bold, the scissors for deleting text or numbers, the straight arrows for repositioning decimal points. Try out as many commands as your patience will allow.

Try, in fact, to go over the entire array of commands that Excel, one way or the other, will present to you. I say "one way or another," because Microsoft issues a new version of Excel about every three years. The commands necessary for estimating don't change much, if at all, from version to version. Sometimes the changes in the on-screen presentation of the commands from one version to another are minor. But sometimes they are big.

When I first began using Excel, the commands were arrayed in a compact set of bands called "toolbars." (You will see them in a number of the illustrations included in this chapter.) In a later version, commands were arrayed into somewhat more complex bands Microsoft then called "ribbons." And while this book was being written, Microsoft came out with a newer version of Excel with yet another presentation of the commands. Here the command ribbons can be accessed only by clicking on one or another of a series of drop-down tabs at the top of the screen.

These changes (and no doubt more to come) create no obstacles for learning Excel. Whatever version of the program you have on board your computer, you will be able to learn how to use it by learning about other versions. Essentially, in all versions you will be making use of the same set of commands for estimating and bidding. You may have to search them out as you move from one version to another, for they are displayed differently in the different versions. But you will spot them because you will have seen them in other versions, including the one on board your computer and those in the videos you have watched.

In fact, if you have used Word, the word processing program included with Excel in the Microsoft suite, you will recognize many of the commands in Excel even if you have never seen Excel before. Other commands are Excel-specific and will be new to you. Let me urge you again to try them out. Use them to format or manipulate the numbers and text. Tinker with the commands you do not recognize. If you run into a dead end, as you likely will, don't worry about it. Try something else or close the workbook, open another, and start afresh. Playing around, as you will see, is a good way to learn about Excel, even as you systematically absorb more formal instruction.

To go further with my suggested course of instruction, move to step 2 of Part I. Watch the Motion Training videos, *Microsoft Excel Tutorials for Beginners*. The four videos I recommend are only 45 minutes long in total and they are free on YouTube. They will provide you with an introduction to the fundamentals of Excel and a good deal of the operational know-how you must absorb to set up a workbook.

Once you have the Motion Training videos digested, take steps 3 and 4 in Part I. Go to Lynda.com, a website that offers excellent Excel instruction (along with instruction on numerous other subjects). As of this writing, you can sign up at Lynda.com for a free 10-day trial. Ten days is likely all you will need to review the recommended videos. If not, you can sign up for additional time at a modest cost.

Lynda's Excel 2016 video by David Rivers, which is the Lynda video I strongly recommend you view first, is only an hour long and is broken into easily digestible three-to-four-minute segments. It goes over Excel's primary features systematically, and it is awesomely good. Watch it twice, all the while test-driving your own Excel workbook, and you will have a handle on nearly all the commands you need to build a basic estimating worksheet.

After having spent a few hours with the Motion Training and Rivers videos and acquiring some facility with Excel, you will be prepared to move on to Part II of the course, reading the rest of this chapter and focusing on the use of Excel for estimating. With the chapter completed, you will likely find it worth your while to continue with Dennis Taylor's *Excel 2016 Essentials* videos from Lynda.com and to watch selected segments. Taylor's video is six hours long, although, like David Rivers's, it is also broken into very short segments. While it is not quite as tight as Rivers's video, it is very good. It will deepen and extend your knowledge of Excel.

If you prefer, you can depart from the sequence I have suggested here in the text and in the sidebar. Learning Excel is not like following a recipe for baking

Electronic spreadsheet vocabulary

Absolute reference—A reference within a formula to a particular cell that does not change when the formula itself is moved from one cell to another (in contrast to a relative reference, defined below)

Active cell—The cell you have clicked or just placed your cursor on in order to enter text, numbers, or a formula, or to view a formula that has already been placed

Auditing—Checking over your work for errors, i.e., proofing

Cells—The small rectangles (i.e., boxes) that populate a spreadsheet grid

Cell address—A column letter and row number, such as D20, that indicates the location of a cell within a spreadsheet grid

Cloud—Industrial facilities that store and deliver data and programs used by computers

Columns—The vertical divisions of an Excel grid

Command button—An icon, such as the B for boldfacing or the Σ for summing a column of numbers, that you click on in order to instruct Excel to perform a certain action

Extending—Multiplying two numbers located in different cells and placing the result in another cell. For example, multiplying 100' of wall framing by $50 per foot labor cost and extending the result, $5,000, into an adjacent cell

Fill handle—A tiny square at the lower right corner of each cell that becomes visible (barely) when you click on the cell to make it active, and that can be "grabbed" by placing the cursor over it, so that information in the cell can be dragged to and placed in an adjacent cell or group of cells

Format—The display options—such as boldface, font (letter style), dollar signs, or decimal place—for text or for numbers

Formula—Information that instructs Excel to perform a calculation and display the result in a cell or cells. Example: =SUM(F4+F8) means sum, i.e., add, the values in cells F4 and F8. (Note: All formula entries start with an = sign.)

Freeze Pane—An Excel command that keeps a column or row displayed during scrolling

Function—A formula, such as the formula for autosumming, that is already set up in Excel and that you can place in a cell simply by making the cell active and then clicking on a command icon rather than having to type out the formula

an apple pie. It's not that linear a process. It's more like learning to drive a new kind of vehicle through a maze with many interconnections and many paths that will take you to the same destination, some more rapidly than others. You drive the vehicle around, i.e., test-drive your worksheet and issue commands, until you know how to make your way through the maze and get where you want to go. You can learn how to navigate the maze with whatever sequence of learning you prefer. You won't end up in a wreck.

Whatever sequence you choose to learn Excel, the one I have suggested or another you prefer, you will, as you watch the videos and read my text, find yourself going over the same topics multiple times. That will be worthwhile. With each repetition, you will experience the use of Excel a bit differently.

Hiding—Instructing a spreadsheet to not display selected columns and/or rows

Merged cells—Cells that have been joined to accommodate lengthy text, symbols, or numbers

Navigation—Moving around in Excel via tabs, by scrolling, with the cursor, or with keystrokes

Operator—A symbol that calls for a particular math operation, such as *, which in Excel serves as the multiplication sign

Painting—Clicking on a formula in the top cell of a column, then dragging it down the column to place it in subsequent cells

Program—Software used for a particular task such as estimating

Relative reference—A reference to a cell made inside a formula that changes when the formula is moved from one cell to another

Ribbon—A band displayed above a worksheet that displays command icons used to format and manipulate data in a worksheet (The ribbon has replaced the toolbars that appeared in earlier versions of Excel.)

Rows—The horizontal divisions of an Excel grid

Software—Information and instructions used by your computer to perform tasks such as word processing or estimating

Spreadsheet—A piece of paper or a computer screen divided into columns and rows of rectangles, i.e. boxes, called "cells"

Status bar—A narrow band at the bottom of a worksheet that displays command icons and information about the worksheet

Templates—Estimates that have been saved so that they can be used as starting points for estimating the cost of similar future projects

Toolbars—Collections of icons, typically displayed above a worksheet, that you click on to initiate various operations from boldfacing text to inserting formulas into cells (Note: The toolbars that accompanied earlier versions of Excel have been supplanted by what is now called the "ribbon.")

Workbook—An Excel file that contains at least one worksheet (and may contain many)

Worksheet—An Excel spreadsheet (that can be one page or many pages in length)

You will become acquainted with different pathways. Uncertainty that may have developed on a first take will clear up by the second or third, and you will improve your ability to navigate the program and to get where you want to go.

If there is a drawback to the course I am recommending, it is that the videos may discourage you a bit even as they educate you. The presenters are so skilled with Excel you may feel like a dunce by comparison. That's why watching another video, the one titled *Estimating with Excel for the Small Contractor*, may be helpful.

In that video, available over YouTube, a builder who goes by the handle "rcargin1" attempts to demonstrate his homespun use of Excel. He offers ideas about the design and formatting of an estimating worksheet, acquaints you with basic commands used for estimating, and takes you through the calculation of labor costs for a construction job. Just as you will do in the beginning, rcargin1 makes plenty of errors, and he unapologetically and efficiently corrects them. Watching him work, a little clumsily at times but nevertheless successfully, can be encouraging. You realize you don't have to reach the level of fluency with Excel demonstrated by the maestros at Lynda.com. You can use the program effectively even if you sometimes take wrong turns or select less than optimal paths through the maze.

Worksheet design

Excel has vast capabilities. It's used, among other things, for engineering calculations and complex statistical analyses. However, you need utilize only a small, select portion of its capabilities to design and create a workbook for estimating and bidding—and the presence of the other stuff won't get in your way at all.

To begin with, understand that when you open an Excel workbook you are opening a computer file just as you do when you open a Word document. Each Excel workbook is saved and stored as a separate file, just like Word documents. Each *workbook* file contains a minimum of one *worksheet*—that is, an electronic version of a spreadsheet—and it can contain multiple worksheets.

As you can see in the illustration on the next page, the columns in an Excel worksheet are lettered A, B, C, D, and so on. The rows are numbered 1, 2, 3, 4, 5, and so on. Each cell in a worksheet grid has an "address" made up of a column letter and a row number. For example, C6 is the address of the cell that lies at the intersection of column C and row 6. You make use of these addresses when you are setting up your worksheet to do your math for you. For example, you could instruct your worksheet to multiply the number in cell C6 by the number in E12.

As you begin setting up Excel for estimating, your first big step will be to create a master workbook with a comprehensive direct costs worksheet. Just like a spreadsheet created by other means (handwritten, with Word, etc.) your Excel worksheet should list, row by row, all the divisions, assemblies, and items of work that you encounter at your projects. Likewise, it should include columns in which to record the quantities and costs of that work. If you are a bit rusty on the design of spreadsheets, you might, at this point, want to go back to Chapter Five and review the subject. By spending a few moments there, you will be able to visualize the various row and column labels estimating worksheets typically include, for they parallel the division/assembly/item breakdown used for a spreadsheet created by other means.

When you want to create an estimate for a specific project, you simply make a copy of your master workbook with its comprehensive direct costs worksheet, name it for the project, "Jones Street Renovation" for example, then save it using the Save As command. Once you have saved your Jones Street workbook you can fill in all the lines for quantities and costs as appropriate for that project. Further, if you are satisfied with the design of your Jones Street workbook, after you have completed your estimate for the Jones project you can save it as a "template." As explained in the sidebar on page 358, you can use the template as a starting point for similar projects in the future, thereby saving yourself a good deal of work.

You'll notice, once you have built your own worksheet, that it is not initially divided into pages. Excel worksheets can go on close to forever. If you were to scroll all the way down a blank Excel 2016 worksheet, you would discover it contains over a million rows, and if you scroll left to right, that it contains over 16,000 columns.

If you print a worksheet, however, it emerges from your printer not as an endless scroll but broken up into individual pages. You can place page breaks in the information shown on those printed pages wherever you like. For my own estimates, I place a page break at the end of every division from General Requirements onward. As a result, in a printout of the estimate, the last page of a division is usually only partly filled.

In my version of Excel (yours may present somewhat differently) you can use icons included in the so called "status bar" at the bottom of the worksheet to preview page breaks before printing out or sending off an estimate. Just above the status bar you will see another feature of Excel that you will regularly make use of. That feature is a series of tabs (labeled Sheet1, Sheet2, Sheet3 in the illustration below). By clicking those tabs, you can move from one worksheet to another within a workbook. You will make use of the tabs as you advance your skill with Excel and create the overhead sheet and other aforementioned additional worksheets.

When you design your Excel worksheets, you can choose to view and print them in one

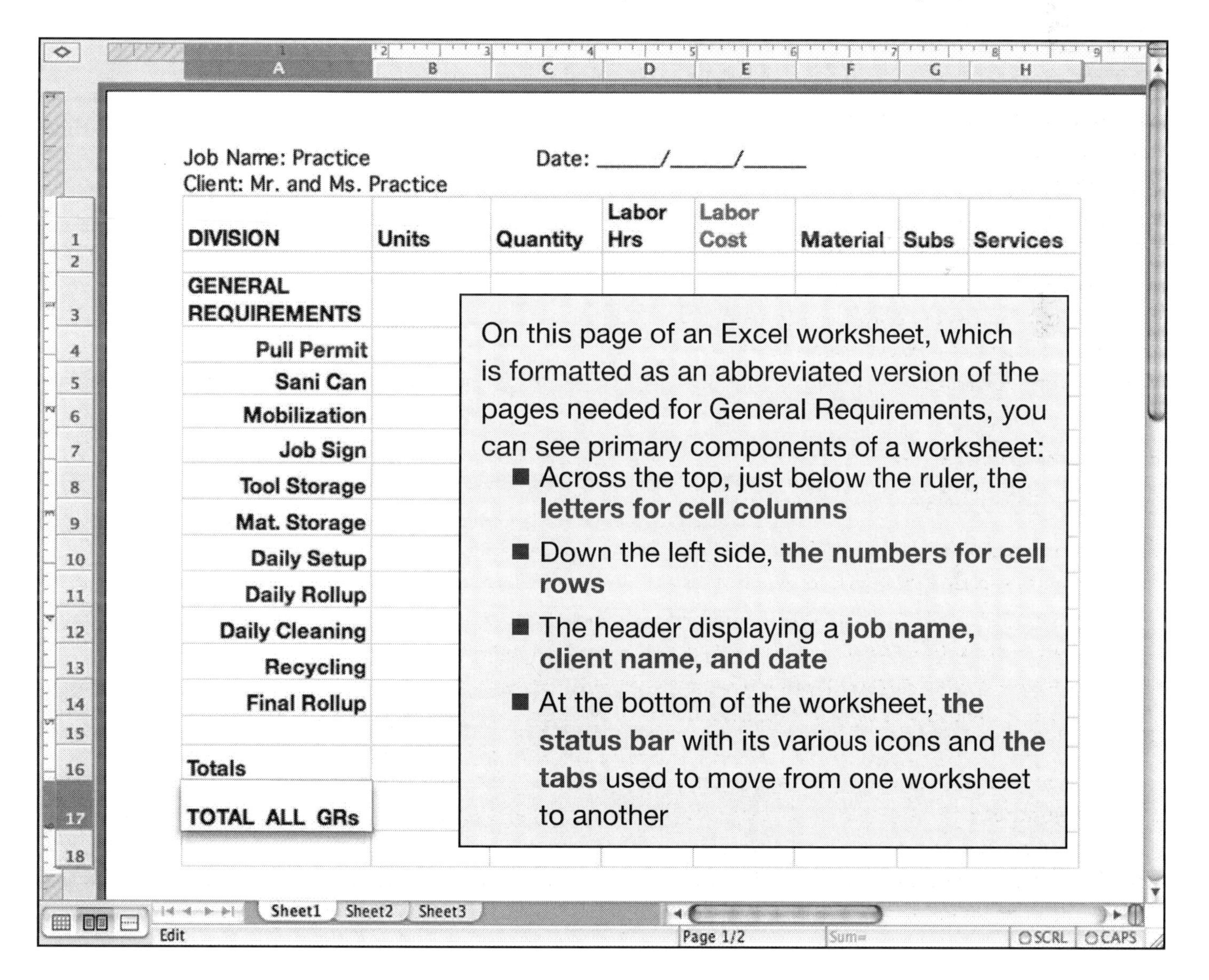

On this page of an Excel worksheet, which is formatted as an abbreviated version of the pages needed for General Requirements, you can see primary components of a worksheet:

- Across the top, just below the ruler, the **letters for cell columns**
- Down the left side, **the numbers for cell rows**
- The header displaying a **job name, client name, and date**
- At the bottom of the worksheet, **the status bar** with its various icons and **the tabs** used to move from one worksheet to another

Creating Excel Templates

Excel enables you to easily build up a library of estimating templates—workbooks for particular types of projects that, once created, can be used repeatedly.

Say, for example, that you are estimating the costs of constructing a high-end kitchen for Luis and Maria Alvarez. To create the estimate, you begin with your master workbook for direct costs and save a copy with a new filename such as "Alvarez Kitchen Remodel." Next you customize it for the job—that is, you hide items like foundation work that are not relevant, and add items that are unique to kitchens or the particular kitchen you are estimating. Then you create your estimate row-by-row, item-by-item and assembly-by-assembly—with Excel automatically producing your division totals for you. When the estimate is complete you save two copies, one for the Alvarez project, the other as a template, which you could name "High-end kitchen template." At a later time, when a new high-end kitchen comes your way, you can start with your template rather than with your more generic master workbook, making a copy of the template and naming it for the new job.

Of course, if much time has passed since you created the template, you will have to update material and labor costs. Finally, because no two kitchen jobs will be identical, you will need to adjust quantities of work. But you will find that a good many items, particularly general requirements such as project management, setup, and cleanup, and even various rough and finish items, will be virtually identical in quantity and perhaps even in cost. Therefore, using the template, rather than starting from scratch with your main workbook, will enhance estimating productivity. (That will especially be the case if you have advanced to using the linked worksheets discussed toward the end of this chapter, for then labor and material updates will likely already be in place.)

Over time, you can build up an impressive range of templates. You'll create an Excel templates master folder on your computer desktop. Within the master folder, you can place subfolders for all the types of construction projects you take on. And within each subfolder will be a series of templates. For example, in your folder for kitchens you might have templates for economy, mid-level, and high-end kitchens. Likewise, you could create subfolders for any category of projects you bid for, from new homes or commercial structures to remodels, structural repairs, and window replacement projects. Your collection of electronic templates, each available with a couple of clicks of your mouse or taps on your computer screen, will prove a treasure, substantially increasing the efficiency of your estimating.

of two ways. You can view them in portrait orientation, i.e., with the longer side vertical, as in a typical letter. You can also view them in landscape orientation, with the longer side horizontal. Landscape view is generally better for estimating, as it allows room for more column headings. The room for extra column headings will come in especially handy if you elect to design your worksheets with multiple columns for labor hours and costs as illustrated in the worksheet below. Multiple labor columns enable you to estimate labor cost with greater precision. For it is often the case that different tasks within a project are done by crews of differing size and, therefore, entail differing costs. Multiple labor columns give you a place in your estimate to include costs specific to those different-sized crews. Of course, you could use multiple labor columns with other types of estimating systems, such as word-processed spreadsheets. But I have never seen anyone do that, perhaps because it's just too much additional work without the automated multiplication and addition that Excel offers.

In fact, utilizing a series of labor columns for a range of crew sizes may seem cumbersome to you even with Excel. If so, you can simply use an average hourly rate for your workers and multiply that average rate by your estimated hours for each task to get labor costs. Many builders do just that. They use an average labor rate to figure the cost for all items of work in a project no matter which workers will actually do them.

Project: Practice Addition **Date: April 29, 2018**
Clientd: Mr. and Ms. Practice

	A	B	C	D	E	F	G	H	I	J
1										
2	DIVISION	UNITS	QUANTITY	HOURS/ 1 CREW	LABOR COST/ 1 CREW	HOURS/ 2 CREW	LABOR COST/ 2 CREW	MATERIAL COST	SUBS	SERVICES
3										
4	GENERAL REQUIREMENTS									
5	Pull Permit									
6	Project Manage									
7	Job Site Sign									
8	Tool Storage									
9	Material Storage									
10	Daily Setup									
11	Daily Rollup									
12	Daily Cleaning									
13	Recycling									
14	TOTALS									
15										
16	TOTAL ALL GENERAL REQUIREMENTS									
17										
18										

In other spreadsheets in this chapter, I have shown only a single column for labor hours and a single column for labor cost.

Here I have an hours column and a cost column for a one-person crew and also an hours column and a cost column for a two-person crew.

I have seen worksheets with hours and cost columns for as many as half a dozen crews of differing sizes and composition.

Multiple labor cost columns will help you to nail your numbers.

However, on most projects, some tasks will involve only the crew lead. Others, such as forming and framing, will involve the lead and his entire crew. Still others will involve the lead and only part of his crew or just a couple of crew members, say a carpenter and an apprentice. The hourly cost of these (and still other) different crew formations can vary significantly. For a given task, therefore, you will generate a more accurate estimate of labor cost if you use a labor rate for the people who will actually be doing the task, rather than using the average labor rate for your entire crew.

For tasks like project management and material ordering done by the lead alone, your estimate of cost will be low if you use the average rate because the lead will cost you significantly more per hour than the average. On the other hand, if you use your average rate for cleanup or digging a ditch, work that will largely be done by lower-cost apprentices or laborers, you are likely to run to the high side.

Here's an example of the different estimates of cost you can arrive at by using an average labor rate or the specific rate for the workers actually doing a task: Estimating on the basis of average hourly cost per worker for framing layout, you could figure as follows. Your average per worker is $40 per hour. You figure the layout will take 22 person-hours. Therefore, you estimate that the layout will cost $880 (40 × 22 = 880).

In contrast, estimating on the basis of a right-sized crew, you figure as follows: The layout work will be done by a two-person crew consisting of the lead and an advanced apprentice. Their combined cost per hour is $94. You estimate that working together they will spend 11 hours (or the same 22 total person-hours) on layout. Therefore, the layout cost will be $1,034 (11 × 94 = 1034). Using a crew-specific labor rate, you have arrived at an estimate that is $154, or 17.5%, higher than you got using the average rate. That's a big spread for one item; and the spread can grow greatly when it occurs for many items across an estimate.

In short, by designing your worksheet so that it provides different hour and labor columns for different combinations of workers, you will get significantly different results than you would by using a single hours column and a single cost column. Excel enables you to design your worksheet to efficiently produce that greater precision.

Fundamental commands and tools

If you have elected to use the sequence for learning Excel that I suggested, by now you have spent enough time with the videos to have gained a good grasp of the fundamentals of Excel operation. You have a handle on the basic commands and tools that you will need for designing Excel worksheets and using them to produce estimates and bids. Therefore, you may find this section, which is focused on those very tools and commands and is intended only to make double sure you do have a handle on them, largely redundant.

In order to enter text or numbers into a cell, you must click on the cell to make it active. Once you have made a cell active, the border around it becomes more pronounced. That lets you know the cell is ready to receive text, numbers, or a formula that will do your estimating math for you.

As you move the cursor around your spreadsheet, it takes different forms—a flashing vertical bar, an arrow, a thin cross, and a cross with a double arrow. These different forms of the cursor are used to accomplish different actions, such as inserting a letter into a line of text, widening a column or row, or—very valuable for speeding up entry of formulas into cells—grabbing the corner of a cell and dragging the formula in that cell down a column in order to place it, modified as necessary, in other cells of the column.

To place and adjust information within a cell, you will make use of a variety of commands. They are available to you in several forms: Icons, also called "control buttons;" menu selections; and "dialogue boxes." Often, you can execute a given operation in multiple ways, choosing between an icon, a menu, a dialogue box, or even just a keystroke. (As I mentioned earlier, Excel is a maze with many paths leading to the same place.)

E16 =SUM(E4:E15)

Job Name: Practice Date: _____/_____/_____
Client: Mr. and Ms. Practice

DIVISION	Units	Quantity	Labor Hrs	Labor Cost	Material	Subs	Services
GENERAL REQUIREMENTS							
Pull Permit	Events	1	4	300			
Sani Can Install and Remove	Events	2	2	150			
Mobilization	Events	1	6	450			
Job Sign Install and Remove	Events	2	2	150			
Tool Storage	Days	35	18	1350			
Mat. Storage	Events	3	6	450			
Daily Setup	Days	36	18	1350			
Daily Rollup	Days	36	18	1350			
Daily Cleaning	Days	36	18	1350			
Recycling	Events	3	6	450			
Final Rollup	Events	1	6	450			
Totals			4	7800			

FUNDAMENTAL EXCEL CAPABILITIES

- **Formatting**: The Labor Cost column has been formatted so that the values have been highlighted in red and the decimal point has been removed so that only whole dollars, not cents, are shown.
- **Text Wrap** has been used to fit longer item names such "Job Sign Install and Remove" into a cell.
- **Formula display**: The total of 7800 shown in cell E16 was produced by the formula (=SUM(E4:E15). It is visible in the formula bar—the long white box to the right of the equals sign—which you can see above the worksheet. The formula is instructing Excel to add up all the values in cells E4 through E15.
- **Trace Precedents**: The blue box and arrow indicate the numbers (precedents) that have been summed, i.e., added up, to produce 7800 in E16.
- **An error message** consisting of an exclamation mark inside a yellow diamond has popped up at D16. (If you were to open it, you would see that it is alerting you that cell E15 is empty. While that's potentially an error, it is not one that throws off the calculations made here.)

Also, as stressed earlier in this chapter, the various icons and menus are presented to you somewhat differently in different versions of Excel. However, in all of the versions available at the time this book was being written, the commands needed for estimating can be located with a bit of searching and are virtually unchanged, and they will likely remain so in the future. They include:

- Commands for opening, closing, saving, or printing a file
- Commands called "Text Wrap" and "Merge Cells" that you use to fit text, such as a lengthy division name like "General Requirements" or an item name like "Job Sign Install and Remove," within a cell.
- Commands in the form of curved arrows for Undo and Redo, which are useful for correcting (undoing and redoing) typos and other mistakes
- A command that looks like an "M" standing on its side—Σ, the Greek letter Sigma. It is Excel's command for automating a math process—most importantly in estimating, for automating addition, or "autosumming." It is the single most valuable command that Excel provides to estimators, because we do an awful lot of addition.
- Formatting icons that will be familiar to you from Word or even emailing. They are used for such things as italicizing or boldfacing text, choosing font size and type, and inserting color highlights or lettering in your worksheet.
- Additional formatting icons that you can use to increase or decrease the number of decimal places in a number or to insert dollar signs.

Along with making sure that you can make use of such basic commands, you should acquire working familiarity with that crucially important tool, the formula bar. You will use it to construct some of the most important commands you issue when using Excel—namely the instructions to multiply costs and quantities that you embed in your worksheets. Along with the autosumming accomplished with the Σ icon, it's the use of embedded formulas that makes estimating so much less tedious once you move up to using Excel. In the next section we will go into the use of AutoSum and also of formulas.

If there is a feature of Excel that deserves as much emphasis as autosumming and the creation of formulas, it is Excel's auditing tools—the commands you can issue in order to proof, i.e., check, the math calculations in an estimate. Auditing is so important we will take it up in a section of its own further along in this chapter. For now it will be enough to just take note of one of the most useful auditing commands, namely "Trace Precedents," which is illustrated in the sidebar of key Excel capabilities on the previous page.

Undoubtedly, as you explore the Excel commands and tools you will occasionally find yourself getting jammed up. At such times, the Excel Help feature will generally be a big help. So that you can get further acquainted with Help, I suggest that you use it to further explore a few of the commands we have discussed, such as Save, Save As, Page Breaks, AutoSum, and Trace Precedents, or any others that may especially interest you.

Crunching numbers

Once you have established familiarity with Excel tools and commands, you will want to focus closely on Excel's greatest gift to estimators. That is its ability to take over the endless tedious math calculations estimating requires and to accomplish them all but instantaneously.

Using Excel, you can automate calculation in three useful ways. You can create a math formula and place it in a single cell. You can create a formula in one cell, and then drag it into a series of other cells. Or you can select a function, one of the formulas that come prepackaged with Excel, and place it in a single cell or multiple cells. Let's look first at the creation of formulas. For the purposes of estimating, you can enter them for:

- Multiplication using a star (*) rather than the usual "x" as a multiplication sign

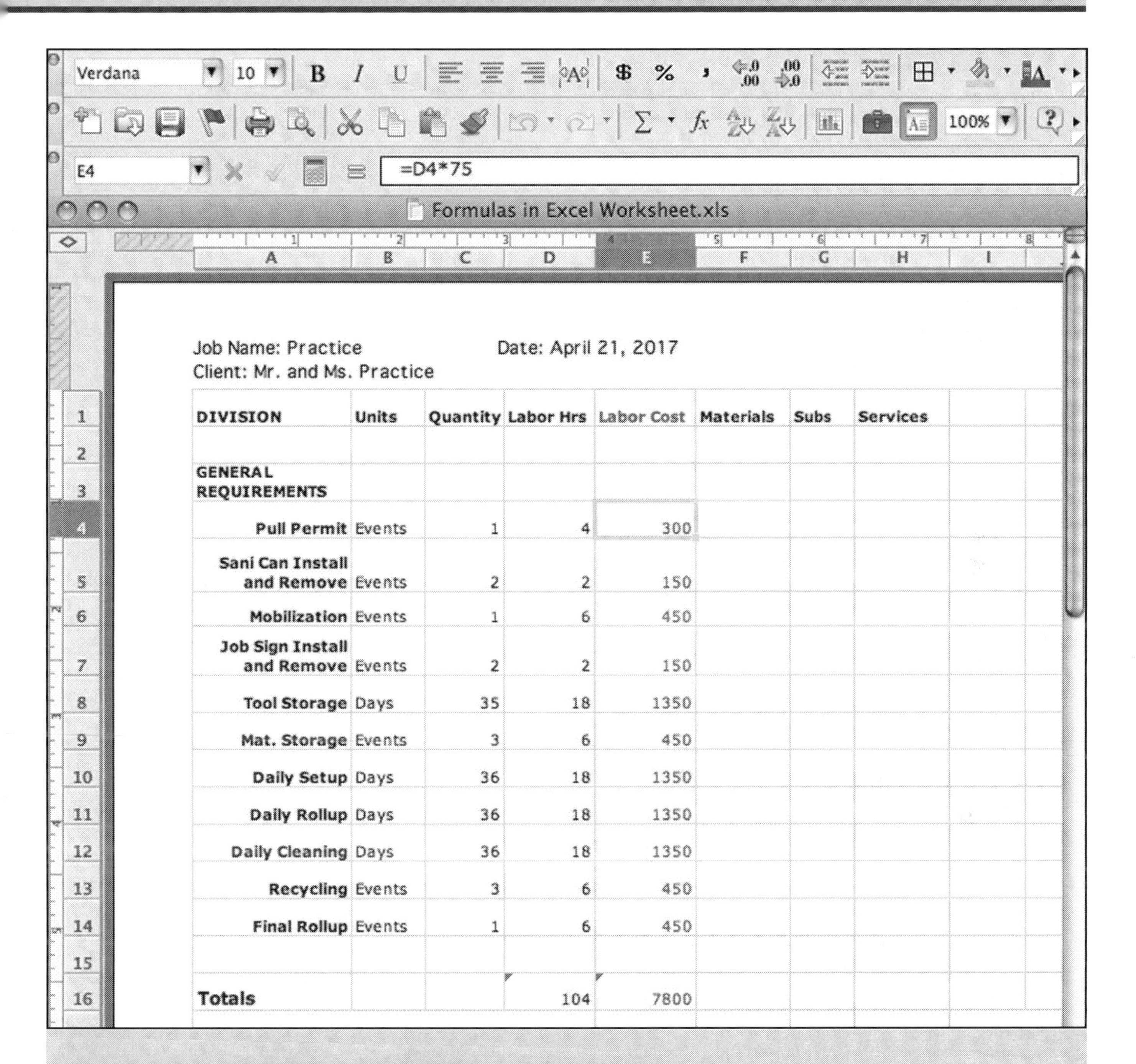

Job Name: Practice　　Date: April 21, 2017
Client: Mr. and Ms. Practice

DIVISION	Units	Quantity	Labor Hrs	Labor Cost	Materials	Subs	Services
GENERAL REQUIREMENTS							
Pull Permit	Events	1	4	300			
Sani Can Install and Remove	Events	2	2	150			
Mobilization	Events	1	6	450			
Job Sign Install and Remove	Events	2	2	150			
Tool Storage	Days	35	18	1350			
Mat. Storage	Events	3	6	450			
Daily Setup	Days	36	18	1350			
Daily Rollup	Days	36	18	1350			
Daily Cleaning	Days	36	18	1350			
Recycling	Events	3	6	450			
Final Rollup	Events	1	6	450			
Totals			104	7800			

PLACING A FORMULA IN EXCEL

When an Excel formula is created in an active cell it shows up in the formula bar (the white band within the toolbar that is located above the worksheet). Once you have created a formula you must accept it in order to embed it in your worksheet. In this version of Excel that's done either by pressing the Enter key or by clicking on the checkmark visible to the left of the formula bar. After acceptance, the formula will appear in the formula bar while the result of the calculation the formula makes will appear in the active cell. Thus, in this worksheet, the E4 cell is active, as evidenced by the light grey box around it; the formula for figuring the labor cost to pull a permit is shown in the formula bar; and the result of 300 is shown in cell E4.

- Division using a backslash (/) as the division sign
- Addition, using a plus (+) sign
- Subtraction, using a hyphen (-) as a minus sign

To enter a formula, you can simply click on a cell to make it active and then type the formula into that cell—*always* beginning with an equal (=) sign. Say you want Excel to figure the labor cost of pulling a permit, as I have done in the Excel worksheet illustrated on the previous page. You have estimated that the permit work will take you four hours and that you will charge $75 per hour. You could, therefore, type out =4*75 in the appropriate cell, E4 in this case, to get the 300 that is visible in the labor cost column. Likewise, working your way down the column, you could enter one equation after another: =2*75 to get $150 for the Sani-Can Install and Remove; =6*75 to get $450 for Mobilization; and so on, all the way through =6*75 to get $450 for Final Rollup. There is, however, a much better option in Excel. It involves writing a formula that refers to cell addresses rather than a formula with specific numbers as described in the preceding paragraph. Very importantly, once a formula involving cell addresses has been created, you can rapidly place that formula in multiple other cells. You don't have to type out a formula separately for each cell. You simply create the formula in the first cell of a column or row, then drag the formula down the column or across the row. The formula *will automatically adjust at each cell in the column or row* so that it is multiplying the appropriate numbers.

Let me illustrate the process with an example. For the worksheet we have just been discussing, you would place the formula =D4*75 in the labor cost cell for Pull Permits (E4). When you drag that formula down the "E" column it changes (as you would see in the formula bar though not in the cells themselves, which display only the results) to =D5*75 in the next cell down, =D6*75 in the next cell, and so on all the way down the column until it becomes =D14*75 for Final Rollup.

When figuring the labor costs to be placed in each cell in column E, Excel will automatically refer to the number of hours in the named cell—2 hours in D5, 6 hours in D6, and so on all the way down to 6 hours in D14. Thanks to that marvelous autofilling, as Microsoft calls the process, you are saved the work of laboriously typing out a formula for each and every item in your estimating list. Autofilling is so powerful, is such a time-saver during estimating, I suggest you try it out on your own worksheet now to make sure you have a handle on it. If not, you would do well to go back to the videos and go over autofilling until you understand how to accomplish it step by step and are ready to make use of it. An hour or two spent on autofilling right now could easily save you 1000 hours of estimating work across the next few decades of your career.

You can, in Excel, create formulas not only for multiplication but also for addition, subtraction, and division. However, the formulas you will be embedding in your estimating workbook will be almost entirely multiplication or addition formulas. Subtraction and division will rarely come into play; for, alas, in construction the quantity and cost numbers almost always mount up rather than diminish as we proceed through our estimates.

When it comes to addition, you won't even need to create formulas. You will rely instead on the AutoSum command (Σ) we mentioned in the previous section. AutoSum is one of hundreds of the functions, i.e., prepackaged formulas, that come with Excel. For estimating it is by far the most important. When you place AutoSum in a cell, it automatically adds up all the numbers in the designated columns or rows. For example, with one application of AutoSum, you can add up the entire column of labor costs for the framing of a new building. With a second application, you total all the labor, material, sub, and service costs for the whole framing division.

On the next page you will find an example of the use of AutoSum along with an explanation of the avoidance of errors in its use. . We'll get to the possible errors in a minute. But first please note in the illustration that the totals in row 15 for labor and material costs for General Requirements were produced using AutoSum. Thus, AutoSum was placed in cell E15 at the bottom of the column for labor cost.

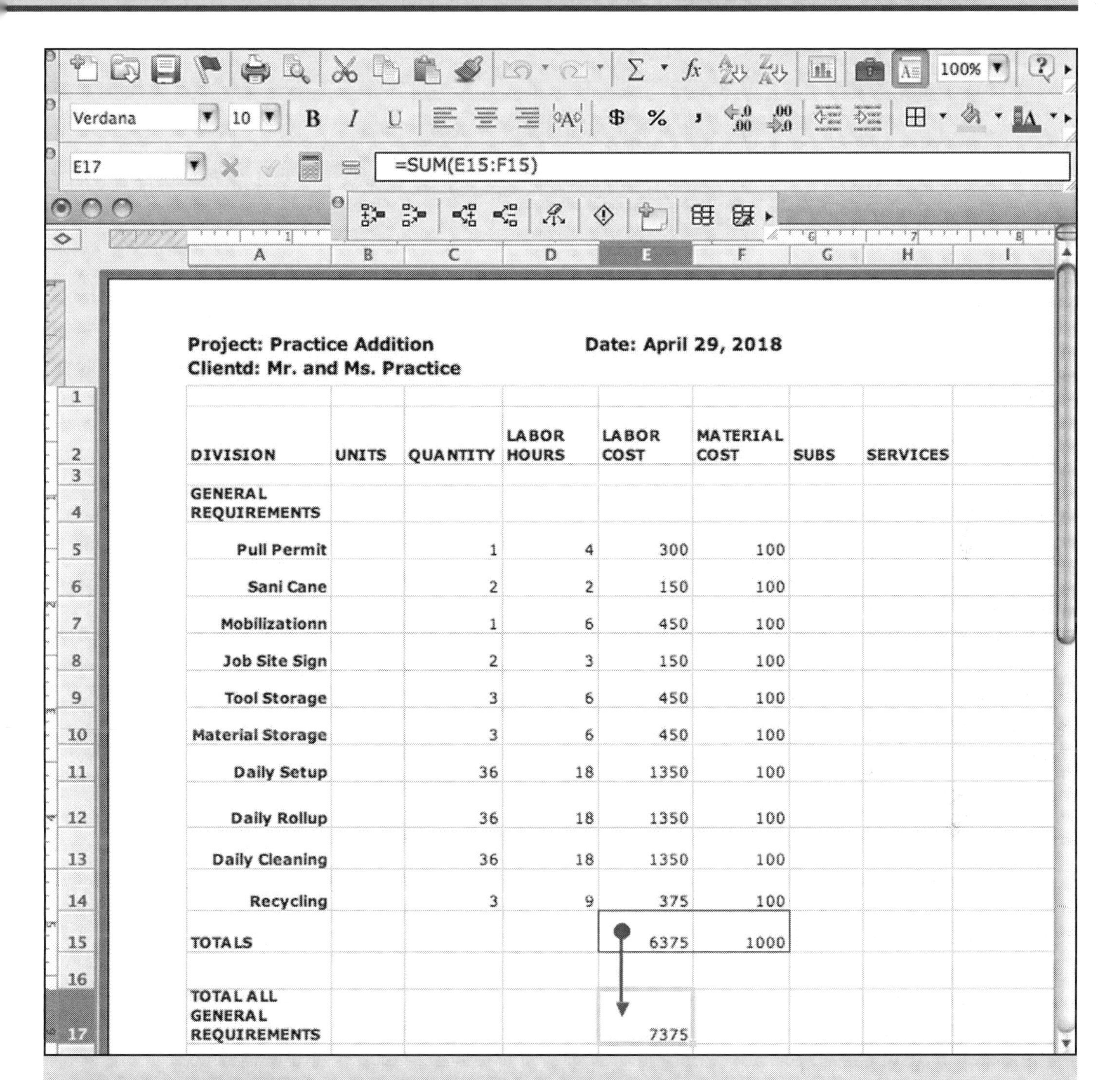
Verdana 10 E17 =SUM(E15:F15)

Project: Practice Addition   Date: April 29, 2018
Clientd: Mr. and Ms. Practice

DIVISION	UNITS	QUANTITY	LABOR HOURS	LABOR COST	MATERIAL COST	SUBS	SERVICES
GENERAL REQUIREMENTS							
Pull Permit		1	4	300	100		
Sani Cane		2	2	150	100		
Mobilizationn		1	6	450	100		
Job Site Sign		2	3	150	100		
Tool Storage		3	6	450	100		
Material Storage		3	6	450	100		
Daily Setup		36	18	1350	100		
Daily Rollup		36	18	1350	100		
Daily Cleaning		36	18	1350	100		
Recycling		3	9	375	100		
TOTALS				6375	1000		
TOTAL ALL GENERAL REQUIREMENTS				7375			

USING AUTOSUM

When using AutoSum, you must make sure it is actually adding up the numbers you want it to add up. In the operation illustrated here, I clicked on ∑ with cell E17 active. If left to itself, AutoSum would have automatically referred to cells in the column above E17—i.e., cells E5 through E15—and added up the numbers in those cells. However, I wanted AutoSum to produce not a total of the labor costs in column E but a total cost for General Requirements. To have AutoSum do that, I had to over-ride its instinct to add up the costs in column E and instead have it add the total costs of GR labor and GR material as shown in E15 and F15. So that AutoSum would total the amounts in those two cells, I clicked on E15 and then dragged my cursor through F15. As the blue precedent box and arrow show, I thereby totaled the correct numbers.

It added up all the costs in that column and produced the total of 6375 that you can see in E15. Similarly, AutoSum was used to add up the costs in the materials column and produce the total of 1000 at the bottom of the F column.

Finally, AutoSum was placed in E17 to produce a total 7375 (6375 + 1000) for all General Requirement costs. (Note: In the illustration, since cell E17 is the active cell, you can see the formula produced by AutoSum in the formula bar).

Like AutoFill, AutoSum is such a valuable tool for estimating that you want to make sure you have a handle on it. Test-drive it in your practice workbook. Try this: 1) Place numbers in a series of cells in any column of a worksheet. 2) Click on the next cell down the column to make it active. 3) Click on the AutoSum symbol to place AutoSum in the active cell. Enjoy the magical results and imagine how you are going to love AutoSum when you are adding up long columns of labor, material, sub, and service costs for actual estimates in the years to come.

Great tool though it is, AutoSum is also a little dangerous. Why? Because while it knows *how* to add, it does not know *what* to add. And while it will always make a guess as to what to add, it can guess wrong. So sometimes you have to tell it what to add. Otherwise it may sum a different set of numbers than those you intended. Specifically, if you put ∑ at the bottom of a column, it will sum the numbers in that column. In some cases, those are *not* the numbers you want to add.

For example, in the illustration on the previous page I wanted to total up not a column of figures but instead just two figures, one a labor total and the other a material total, namely the figures in cells E15 and F15, to arrive at a total for all General Requirement costs. As the figure for the total and the blue precedent arrow in the illustration show, I used AutoSum correctly. However, to do so I had to highlight the relevant two cells in order to direct AutoSum's attention to them. Had I not highlighted those cells, AutoSum would have referred to the column above the active cell. In short, AutoSum does have a mind of its own, but you do not want it thinking for you; so sometimes you must override its decisions.

Maintaining security and reliability

Every benefit, it seems, has a cost. A cost of reaping all the benefits of estimating with Excel is that you must invest some time in maintaining the reliability and security of your estimates.

First of all, you must bear in mind that as you create your estimate you are working with an electronic file. Therefore, you want to save your work frequently using the Save command. Otherwise, you run the risk of having your hard work stolen away by a computer glitch, a power outage, or your own inadvertent error. I managed to lose portions of my estimates a couple of times to such thieves when I was first using Excel. It was a miserable experience, seeing hours of work disappear from my computer drive. These days I save constantly, every few minutes, and not only to my computer hard drive, but to a backup zip drive. One experienced Microsoft Office user tells me that is not enough. She saves after every change she makes.

In addition to Save, Excel offers several methods for enhancing the security and reliability of your workbook that are particularly useful during estimating. In this section I want to acquaint you with three of the most important methods: Freeze panes. Hiding and unhiding. Protection.

Freeze Panes. If you estimate complex construction projects that take several months or longer to build and that involve multiple trades, your direct costs worksheet could easily come to include dozens of pages of items. That results in a problem. As you scroll from the first page of your worksheet to the second and beyond, your headings for the unit, quantity, cost, and any other columns scroll away and out of sight. Consequently, you can easily lose track of which column a number should go in and mistakenly place, for example, a labor cost number in the material cost column, thereby undermining the accuracy of your estimate.

Excel offers a solution to the problem. Using a command called "Freeze Panes," you can freeze the row of headings in place. Once frozen, they remain visible no matter how far down you have scrolled in your worksheet. Search out the Freeze Panes command on your version of Excel and give it a try in your practice worksheet. You will see what I mean.

Hiding and unhiding. At times, you may want to hide columns or rows in your workbook. For example, assume that you are estimating costs for an interior renovation for Henry and Shulien Chang, have made a copy of your workbook, and have named the copy the "Chang Project." The Changs' renovation involves no site or foundation work. So, in order not to repeatedly scroll through the lines for that work as you create the estimate—and run the risk of accidentally putting in a cost for site or foundation work when there is none— you hide them in the copy.

Of course, you may need to unhide the hidden rows. Excel will allow you to easily do that, too. If is the Chang project is changed—say an addition for a breakfast nook is added so that now the project does involve site and foundation work—you can, with the click of a key, unhide your site and foundation divisions.

Protecting. To prevent yourself (or someone else) from changing a workbook, you can "protect" it by using an Excel setting that prevents unwanted changes. You may want to protect it from yourself so that you do not inadvertently strike a key and change, say, a $500 labor charge to $50 when you are opening the workbook merely to take a look at it. You may also want to protect a workbook from clients. For example, if you are working with clients on a cost planning basis, you may have come to the point where you want them to look over your estimate. But you want the clients to only review the numbers, not tinker with them. So you protect your workbook before you send it off to the clients. Likewise, your company may grow to the point where several people may review a given estimate. In that case, you may want workbooks protected so that only the estimator who is responsible for the estimate can make changes in it.

Auditing your numbers

Along with providing Save, Freeze Panes and Hide commands as well as protection settings to help you prevent unwanted changes to your worksheets, Excel offers you several ways of ensuring the accuracy of the numbers you deliberately place in your worksheets. I have been surprised by how little use even skilled estimators fluent with Excel sometimes make of those safeguards. They seem to have so much faith in Excel and their facility with it that they feel immune from making mistakes. They are not immune.

Exhibit A: A certain East Coast builder is so good with Excel that he teaches courses in using it for construction estimating. But he candidly admits to having inadvertently set up his program so that it was, by default, including a price for only *one* window in his estimates even when the projects required multiple windows. He overlooked the error in estimate after estimate, taking a financial hit each time he won a job, until he was estimating a project that required 40 windows. Now the error in his workbook was too large to miss, because it was including the cost of only one window and omitting the cost of 39 others. That got his attention.

Because a potential for serious errors does come with using Excel—as it does with any estimating system or program—it's absolutely necessary that you audit, i.e., monitor, the calculations produced with your workbooks. Here are three methods (again, you may find it beneficial to try them out on your practice worksheet as you read):

- *Auditing with Trace Precedents.* We have already touched on precedent boxes and arrows in the "Crunching numbers" section. (If you need to refresh your memory, turn back to the illustration on page 361 titled "Fundamental Excel Capabilities.") As explained, precedent

Job Name: Practice Date: April 21, 2017
Client: Mr. and Ms. Practice

	A	B	C	D	E
1	DIVISION	Units	Quantity	Labor Hrs	Labor Cost
2					
3	GENERAL REQUIREMENTS				
4	Pull Permit	Events	1	4	=D4*75
5	Sani Can install & Remove	Events	2	2	=D5*75
6	Mobilization	Events	1	6	=D6*75
7	Job Sign install and Remove	Events	2	2	=D7*75
8	Tool Storage	Events	1	4	=D8*75
9	Mat. Storage	Hrs	3	6	=D9*75
10	Daily Setup	days	36	18	=D10*75
11	Daily Rollup	days	36	18	=D11*75
12	Daily Cleaning	days	36	18	=D12*75
13	Recycling	Events	3	6	=D13*75
14	Final Rollup	Events	1	6	=D14*75
15	Final Cleanup	Events	1	6	=D15*75
16	Totals		=SUM(C4:C1	=SUM(D4:D15)	=SUM(E4:E15)
17					

Sheet1 Sheet2 Sheet3

DISPLAYING FORMULAS

To review the formulas you have embedded in a worksheet, click on any cell in the worksheet. Then press the appropriate keys (for my iMac that means simultaneously pressing Control and the key labeled with a tilde (~) and a short downwardly slanting dash. For your computer it may be other keys.) Like my formulas in the labor cost column shown here, your formulas will be displayed so that you can review them for correctness.

arrows and boxes allow you to see that a formula you put in an estimate is actually referring to the cells you want it to refer to. Thereby Trace Precedents enables you to make sure you are actually adding or multiplying the numbers you want to add or multiply. Using Trace Precedents you can, in a few minutes, check all the totals in an estimate to see whether they do include all the numbers and only the numbers you intended them to include. That's time well spent.

- *Displaying formulas*. To display all the formulas you have embedded in a worksheet, make any cell in the worksheet active and then click the appropriate key or keys. For my iMac, that involves simultaneously pressing the Control key and the Tilde key, namely the key labeled with a squiggly dash (~) and a tiny downwardly slanting dash. All formulas used in the worksheet will then be visible, as in the illustration on the this page, so that you can look them over and make sure you have put the right ones in the right cells. After reviewing the formulas you can readily clear them from your worksheet. Just press the same keys you pressed to display the formulas in the first place, and they will go away (in other words, the keys act as a toggle, an on-off switch).

- *Setting up automated addition proofs*. This is just an electronic version of the method of checking—i.e., of proofing—addition that we were taught in elementary school back in the day. Then, to check our addition, we added up our numbers two different ways. If you got the same result both ways, you were good to go. In Excel, you can perform a proof by using the IF function, as laid out in the illustration on the next page. It will enable you to set up a formula that instantaneously adds up your numbers two ways and lets you know if the results are the same. For example, you can set up IF to figure the total costs for an estimate by (1) adding together all the totals for all the divisions from General Requirements through Specialties in the estimate and (2) adding up one by one the cost of each and every item in all of the divisions. If the two numbers match and the IF function delivers the message "verified" you know that your math checks out. But should the message be "false," (or any other message, such as "no way, pal," you choose to have IF deliver), you know that there's an error or errors lurking amid your calculations.

When the IF function does turn up an error or errors in your worksheet, Excel does not just hang you out to dry. Rather, it offers a variety of ways for tracing the errors that have resulted in the failure of the proof. Even so, setting up an IF function is a bit complicated. It's one of those Excel operations where the videos provide much better step-by-step instruction than any written text possibly could.

I have met builders who blow off proofing their estimates with IF. I think that is reckless. I would not consider a worksheet to be adequate if it did not provide for an audit via an IF function. So I urge you to use the IF function, learning how via the video suggested in Part III of the course laid out at the beginning of this chapter.

To make sure that you are not making errors when you are estimating with Excel, you may need to monitor not only your worksheets, but also yourself. Here's why: As you build an estimate in an Excel workbook, it will, via the AutoSum formulas you have placed in your worksheet, keep track of the mounting costs of work in the project. If you allow yourself to constantly glance down at the total, that can adversely influence your estimating.

You are liable to find yourself thinking, "It can't be so much!" Or, "I'll never get the job at that price!" So you arbitrarily and impulsively tighten up your estimate of labor hours and material costs or even sub and service costs. "It won't take that long," or "I can get a lower price," you tell yourself as you trim your numbers. In the end, you produce a competitive estimate and "win" the bid. But your price is so woefully low that you get financially hammered on the project.

So, in summary, use precedent tracing, display of formulas, and the IF function to audit your estimates. And keep a check on any tendency you might have to skew costs, by keeping your eyes off the running totals as you estimate. Or better yet, use the Hide command to conceal the rows for division and project totals until all item costs have been put into your worksheet. Then unhide rows to see the totals, sit back, and decide whether or not to run the risk of cutting your profit markup—*not* your estimate of direct and overhead costs—in order to improve your chances of getting the job.

Additional worksheets

After putting in the study and practice suggested in the preceding sections, you will be able to build a comprehensive and powerful worksheet for estimating the direct costs in your projects. There are, however, additional components you can add to your workbook to increase the convenience, efficiency,

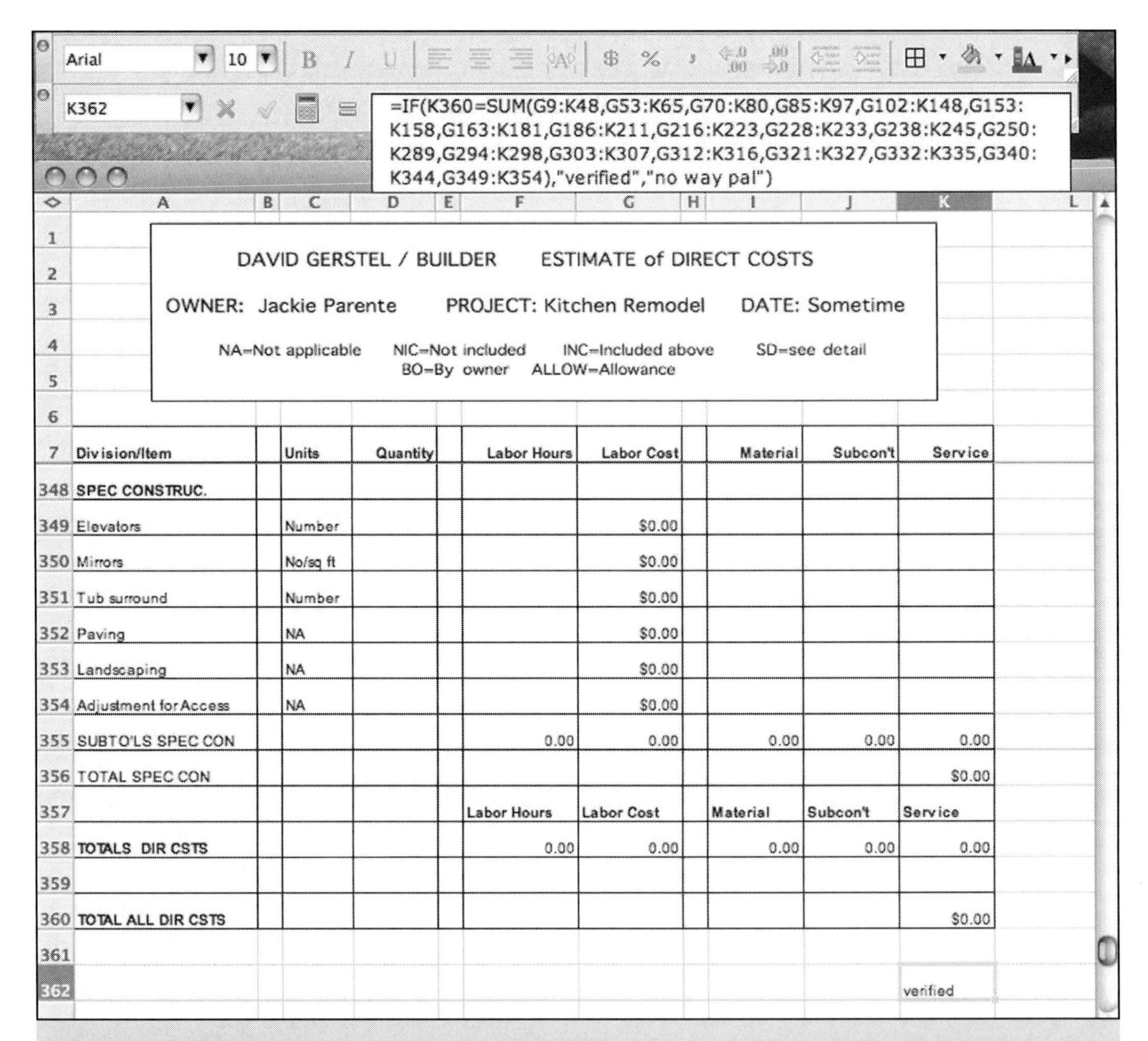

Arial 10

K362 =IF(K360=SUM(G9:K48,G53:K65,G70:K80,G85:K97,G102:K148,G153:K158,G163:K181,G186:K211,G216:K223,G228:K233,G238:K245,G250:K289,G294:K298,G303:K307,G312:K316,G321:K327,G332:K335,G340:K344,G349:K354),"verified","no way pal")

DAVID GERSTEL / BUILDER ESTIMATE of DIRECT COSTS

OWNER: Jackie Parente PROJECT: Kitchen Remodel DATE: Sometime

NA=Not applicable NIC=Not included INC=Included above SD=see detail
BO=By owner ALLOW=Allowance

Row	Division/Item	Units	Quantity	Labor Hours	Labor Cost	Material	Subcon't	Service
348	SPEC CONSTRUC.							
349	Elevators	Number			$0.00			
350	Mirrors	No/sq ft			$0.00			
351	Tub surround	Number			$0.00			
352	Paving	NA			$0.00			
353	Landscaping	NA			$0.00			
354	Adjustment for Access	NA			$0.00			
355	SUBTO'LS SPEC CON			0.00	0.00	0.00	0.00	0.00
356	TOTAL SPEC CON							$0.00
357				Labor Hours	Labor Cost	Material	Subcon't	Service
358	TOTALS DIR CSTS			0.00	0.00	0.00	0.00	0.00
359								
360	TOTAL ALL DIR CSTS							$0.00
361								
362								verified

THE IF FUNCTION

As you can see on this final page of a worksheet, which has not yet been used to create an estimate, all cost lines are still at zero. Even so, the presence of the IF function is visible. It appears in the formula bar, the expanded white box just above the worksheet itself.

The function has been set up to check whether two sums are equal: The first sum, shown on line 360, was arrived at by adding up the four categories of costs shown on line 358—the totals for labor, material, subs, and services for the entire project. The second sum was arrived at by adding up the individual costs, one by one, of all the items in each of the divisions of the estimate; for example, G9 through K48 for all the costs in General Requirements (not visible here since they are on the first page of the estimate and the page shown here is the final page of the estimate).

If the two sums are equal, the IF function signals "verified," as it does here in the highlighted box on line 362. Should the sums not be equal, the IF function would signal "no way pal."

and precision of your estimating process. Here I will provide an overview of the four that I have found to be especially beneficial:

- A stand alone overhead worksheet
- A labor rate sheet
- A direct cost details sheet
- A summary sheet

Once you have added such additional worksheets—all illustrated on the following pages—you will be able to navigate back and forth between them by clicking the sheet tabs discussed earlier. In fact, to facilitate movement back and forth, you can name the tabs for the sheets they lead to. Thus, as you can see on the next page in the illustration of an overhead sheet, you change the initial name of a tab from Sheet 4 to Ov'hd (for Overhead).

An *overhead sheet,* likely the first additional worksheet you will want to create, lists overhead items and their costs and displays overhead subtotals and totals produced by AutoSum. Whenever an overhead cost changes—and overhead does change regularly as costs for everything from insurance and bonds through Internet service and medical plans slide up and (sometimes) down—you can make the change on a line in your overhead worksheet. Excel, again making use of AutoSum formulas you have embedded in your worksheet, will then provide you with the new totals you need for figuring overhead markups.

One caution here: The total shown on the overhead sheet is a total for projected annual costs. There are a variety of ways you can link your overhead sheet to your comprehensive direct costs worksheet and/or the estimate and bid summary sheet we will discuss in a moment. With such linkage, your workbook would automatically mark up direct costs to recapture a portion of the annual cost. However, I will not discuss those ways. If I were to discuss them, it would only be to add that I advise against using them.

As stressed in an earlier chapter, I believe the allocation of overhead charges, like that of profit charges, should be determined on a job-by-job basis with forethought, particularly as regards the likely length of time a project will take and the portion of company capacity it will consume. By automating such a decision you are handing off what is essentially an executive or leadership decision to your computer and its estimating program. I think you will likely be better off keeping the decision, like other executive decisions, to yourself. As good as it is, Excel is not the right tool to use for such decisions. Your brain is.

On the other hand, three additional worksheets—those for labor rates and direct cost details, and the estimate and bid summary—can be productively linked to your comprehensive direct costs worksheet.

Here, however, we are moving to tools that require relatively sophisticated Excel skills to create and use. I would advise holding off on them until you are fully in command of both your comprehensive direct costs worksheet and your overhead sheet. When you are ready, creation (with step-by-step guidance from the videos) of the three additional linked worksheets will turbo charge your Excel estimating.

Labor rate worksheet. This is my favorite worksheet. If you create a labor rate sheet, you can, via the creation of formulas, instruct your direct costs worksheet to (1) select labor rates for various tasks from the labor rate sheet and (2) use those selected rates for figuring labor cost for items in the project. As a result, you don't have to put labor rates into your direct costs worksheet over and over. You just have to keep your labor rate sheet up to date. Your direct costs worksheet will automatically retrieve the correct labor rate for each item of work from the labor rate sheet and incorporate it in your estimate.

Furthermore, once you have added linked labor rate sheets to your workbook, if you must revise your costs for labor—because your workers' comp insurance rate changed, because you gave your crew a raise, because you want to increase the margin of safety in your labor rates, or for other reasons—you only need to make the changes once. You make them in your labor rate sheet. It will automatically figure your updated rates. And your updated

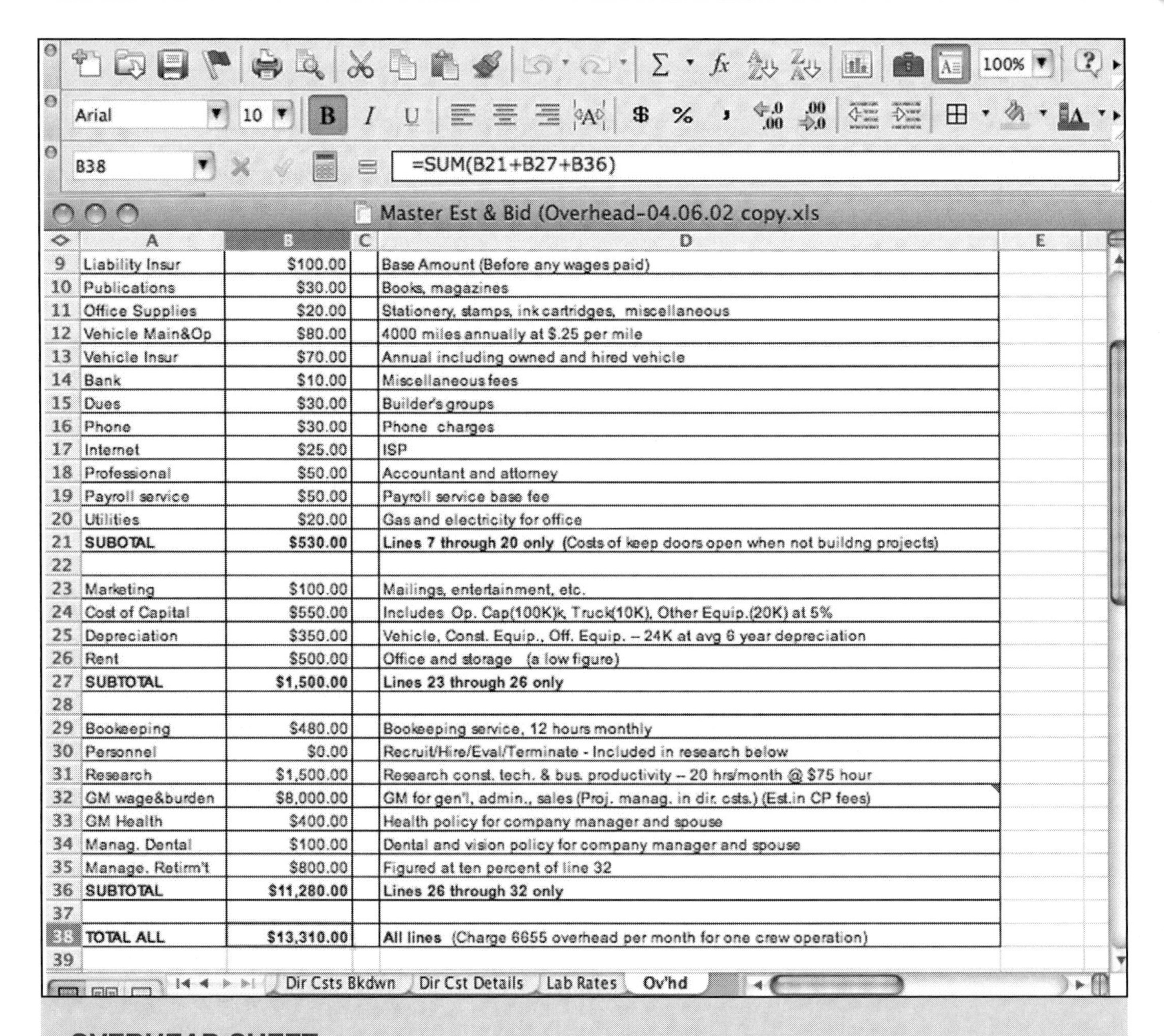

	A	B	C	D
9	Liability Insur	$100.00		Base Amount (Before any wages paid)
10	Publications	$30.00		Books, magazines
11	Office Supplies	$20.00		Stationery, stamps, ink cartridges, miscellaneous
12	Vehicle Main&Op	$80.00		4000 miles annually at $.25 per mile
13	Vehicle Insur	$70.00		Annual including owned and hired vehicle
14	Bank	$10.00		Miscellaneous fees
15	Dues	$30.00		Builder's groups
16	Phone	$30.00		Phone charges
17	Internet	$25.00		ISP
18	Professional	$50.00		Accountant and attorney
19	Payroll service	$50.00		Payroll service base fee
20	Utilities	$20.00		Gas and electricity for office
21	**SUBOTAL**	**$530.00**		**Lines 7 through 20 only** (Costs of keep doors open when not buildng projects)
22				
23	Marketing	$100.00		Mailings, entertainment, etc.
24	Cost of Capital	$550.00		Includes Op. Cap(100K)k, Truck(10K), Other Equip.(20K) at 5%
25	Depreciation	$350.00		Vehicle, Const. Equip., Off. Equip. – 24K at avg 6 year depreciation
26	Rent	$500.00		Office and storage (a low figure)
27	**SUBTOTAL**	**$1,500.00**		**Lines 23 through 26 only**
28				
29	Bookeeping	$480.00		Bookeeping service, 12 hours monthly
30	Personnel	$0.00		Recruit/Hire/Eval/Terminate - Included in research below
31	Research	$1,500.00		Research const. tech. & bus. productivity – 20 hrs/month @ $75 hour
32	GM wage&burden	$8,000.00		GM for gen'l, admin., sales (Proj. manag. in dir. csts.) (Est.in CP fees)
33	GM Health	$400.00		Health policy for company manager and spouse
34	Manag. Dental	$100.00		Dental and vision policy for company manager and spouse
35	Manage. Retirm't	$800.00		Figured at ten percent of line 32
36	**SUBTOTAL**	**$11,280.00**		**Lines 26 through 32 only**
37				
38	**TOTAL ALL**	**$13,310.00**		**All lines** (Charge 6655 overhead per month for one crew operation)
39				

OVERHEAD SHEET

The three subtotals and the final total shown here were all calculated using AutoSum. In the formula bar above the worksheet, you can see the formula produced by AutoSum for the last line in the worksheet, Total All. Note also that the overhead sheet was accessed by clicking on the tab at the bottom of the worksheet that is labeled "Ov'hd."

rates will then be automatically incorporated in the cost of every item in your direct costs worksheet. Compare that to having to go back and refigure, one by one, the labor cost of every item in a project.

Direct cost details worksheet. A direct cost detail sheet provides labor hours and material costs per foot for assemblies. Via formulas, you can instruct your comprehensive direct costs worksheet to gather both the labor hours and material costs for an assembly from the details sheet. Thereby, you can automate the calculation of labor hours and material costs for quantities of work in a job. You simply enter a quantity of

Arial 10

F11 — 2 crw with 10% risk margin

Master Est & Bid (Labor)-04.06.02 copy.xls

	A	B	C	D	E	F	G	H
9								
11		Lead Carp	Carpenter	Apprentice	Laborer	2 crw with 10% risk margin	Lead, Carp, App. W/ 10% risk margin	Lead w/ 2 apprentices w/ 10% risk margin
12	Wage	$35.00	$26.00	$18.00	$14.00			
13	Work Comp Insur-16%/44%	$5.60	$4.16	$7.92	$6.16			
14	Liability Insur/ 6%	$2.10	$1.56	$1.08	$0.84			
15	Social Security/ 7.6%	$2.66	$1.98	$1.37	$1.06			
16	Medicare/1.2%	$0.42	$0.31	$0.22	$0.17			
17	Fed Unemploy Ins/3.5%	$1.23	$0.91	$0.63	$0.49			
18	State Unemploy Ins/1%	$0.35	$0.26	$0.18	$0.14			
19	Personnel Costs							
20	Payroll Costs	$0.11	$0.01	$0.05	$0.03			
21	Training/5%	$1.75	$1.30	$0.90	$0.70			
22	Consumables/8%	$2.80	$2.08	$1.44	$1.12			
23	Medical	$2.00	$2.00	$2.00	$2.00			
24	Dental&Vision	$0.00	$0.00	$0.00	$0.00			
25	Vacation	$0.00	$0.00	$0.00	$0.00			
26	Retirement/10%	$3.50	$2.60	$1.80	$1.40			
27	Total	$57.51	$43.17	$35.59	$28.11	$51.20	$49.97	41.06
28								

Dir Csts Bkdwn | Dir Cst Details | Lab Rates | Ov'hd

Ready — Sum=0

LABOR RATE SHEET

In the labor rate sheet displayed here, I have made cell F27 (highlighted in yellow) active by clicking on it. F27 indicates that $51.20 is the average hourly labor cost per person for a 2-crew. That amount is arrived at by first adding the cost of a lead (B27) to the cost of an apprentice (D27), then dividing the total by two and finallly multiplying the result by 1.1 in to provide a 10% safety margin for the labor cost estimate. With one other step, not taken here, a formula using those three steps would show in the formula bar at the top of the scrreen.

work into the direct costs worksheet, and Excel will refer to your details sheet and figure the labor hours and material costs for that quantity.

For example, one of the direct cost details in the illustration on the next page is for a standard, one-story T-foundation. That detail can be linked via formulas to a comprehensive direct costs worksheet. When you enter quantities in the direct costs worksheet, they are automatically multiplied by the labor hours and material costs in the detail. Thereby, you are given your total labor hours and material costs for the T-foundation. Say you are estimating costs for 100 feet of foundation. You type 100 into the quantity column of your direct costs worksheet. As the illustration on the page 375 displays, the formulas in the labor and material cells of the worksheet will refer to the cost

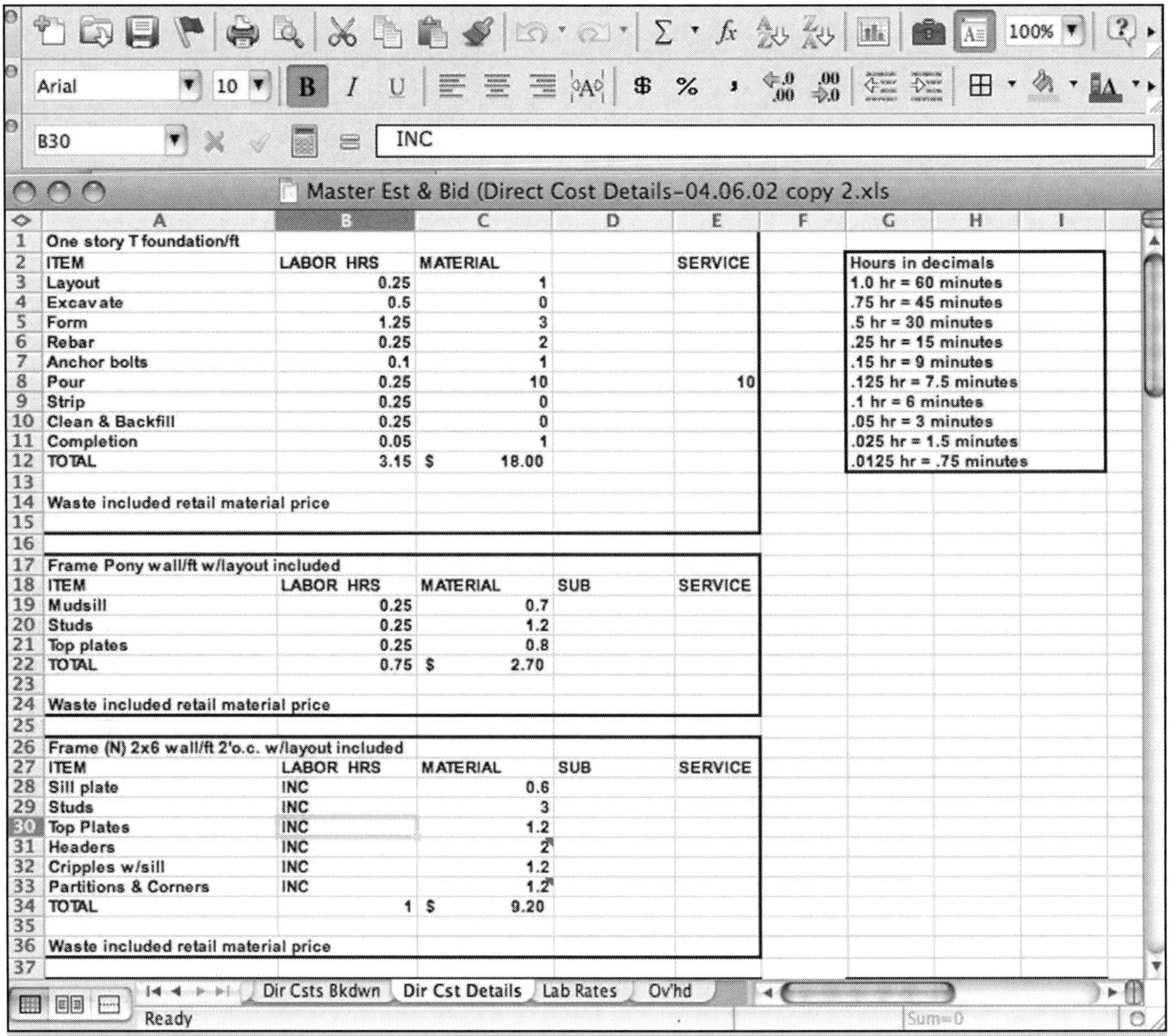

Arial 10 B30 INC 100%

Master Est & Bid (Direct Cost Details-04.06.02 copy 2.xls

	A	B	C	D	E	F	G
1	One story T foundation/ft						
2	ITEM	LABOR HRS	MATERIAL		SERVICE		Hours in decimals
3	Layout	0.25	1				1.0 hr = 60 minutes
4	Excavate	0.5	0				.75 hr = 45 minutes
5	Form	1.25	3				.5 hr = 30 minutes
6	Rebar	0.25	2				.25 hr = 15 minutes
7	Anchor bolts	0.1	1				.15 hr = 9 minutes
8	Pour	0.25	10		10		.125 hr = 7.5 minutes
9	Strip	0.25	0				.1 hr = 6 minutes
10	Clean & Backfill	0.25	0				.05 hr = 3 minutes
11	Completion	0.05	1				.025 hr = 1.5 minutes
12	TOTAL	3.15	$ 18.00				.0125 hr = .75 minutes
13							
14	Waste included retail material price						
15							
16							
17	Frame Pony wall/ft w/layout included						
18	ITEM	LABOR HRS	MATERIAL	SUB	SERVICE		
19	Mudsill	0.25	0.7				
20	Studs	0.25	1.2				
21	Top plates	0.25	0.8				
22	TOTAL	0.75	$ 2.70				
23							
24	Waste included retail material price						
25							
26	Frame (N) 2x6 wall/ft 2'o.c. w/layout included						
27	ITEM	LABOR HRS	MATERIAL	SUB	SERVICE		
28	Sill plate	INC	0.6				
29	Studs	INC	3				
30	Top Plates	INC	1.2				
31	Headers	INC	2				
32	Cripples w/sill	INC	1.2				
33	Partitions & Corners	INC	1.2				
34	TOTAL	1	$ 9.20				
35							
36	Waste included retail material price						
37							

Dir Csts Bkdwn | Dir Cst Details | Lab Rates | Ov'hd

Ready Sum=0

DIRECT COST DETAILS SHEET

A direct cost detail worksheet is not essential to the efficient use of Excel for estimating. But as you gain experience with the program, you may find you can speed up estimating by creating cost details for the standard items of construction that you produce over and over.

detail sheet and calculate the hours (3.1 * 100 = 310 hours).

When you do elect to add cost detail sheets to your workbook, you must, of course, keep the costs in the sheet up to date. That is a chore. I have, therefore, found it practical to maintain cost details only for standard construction details—such as one-story T-foundations and framing of conventional 24" o.c. walls—that occur over and over.

Summary sheet. As a final step in the creation of your Excel estimating and bidding program, you may wish to create a summary sheet such as the one shown on page 376. You can, of course, transfer figures for division totals from your comprehensive direct costs

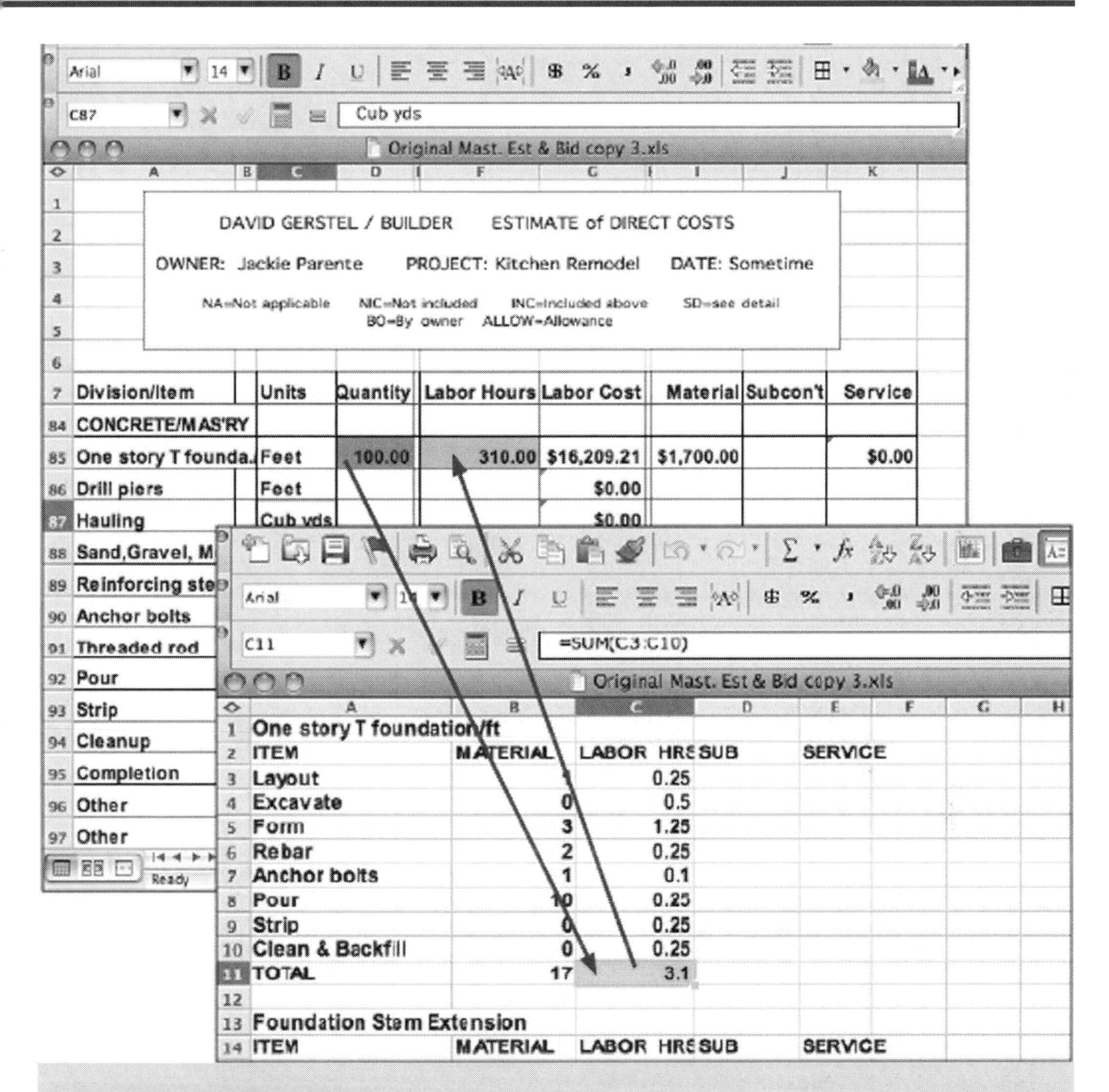

Division/Item	Units	Quantity	Labor Hours	Labor Cost	Material	Subcon't	Service
CONCRETE/MAS'RY							
One story T founda.	Feet	100.00	310.00	$16,209.21	$1,700.00		$0.00
Drill piers	Feet			$0.00			
Hauling	Cub yds			$0.00			
Sand,Gravel, M							
Reinforcing ste							
Anchor bolts							
Threaded rod							
Pour							
Strip							
Cleanup							
Completion							
Other							
Other							

One story T foundation/ft

ITEM	MATERIAL	LABOR HRS	SUB	SERVICE
Layout	1	0.25		
Excavate	0	0.5		
Form	3	1.25		
Rebar	2	0.25		
Anchor bolts	1	0.1		
Pour	10	0.25		
Strip	0	0.25		
Clean & Backfill	0	0.25		
TOTAL	17	3.1		

Foundation Stem Extension

ITEM	MATERIAL	LABOR HRS	SUB	SERVICE

LINKED WORKSHEETS

Here a direct costs worksheet is linked to a direct cost details sheet. In the direct costs worksheet a formula is embedded in the cell F85 in the labor hours column. That formula is instructing the direct costs worksheet to multiply the quantity of 100 feet of one-story T-foundation (highlighted in orange) and multiply it by 3.1 labor hours per foot (highlighted in yellow) from the linked details sheet. As a result, 310 (green) appears in the labor hours column of the direct costs sheet. That is the number of labor hours required to build the 100 feet of foundation. The labor cost of $16,209.21 is produced via a separate link, this one to a labor rate sheet like that illustrated earlier in this chapter.

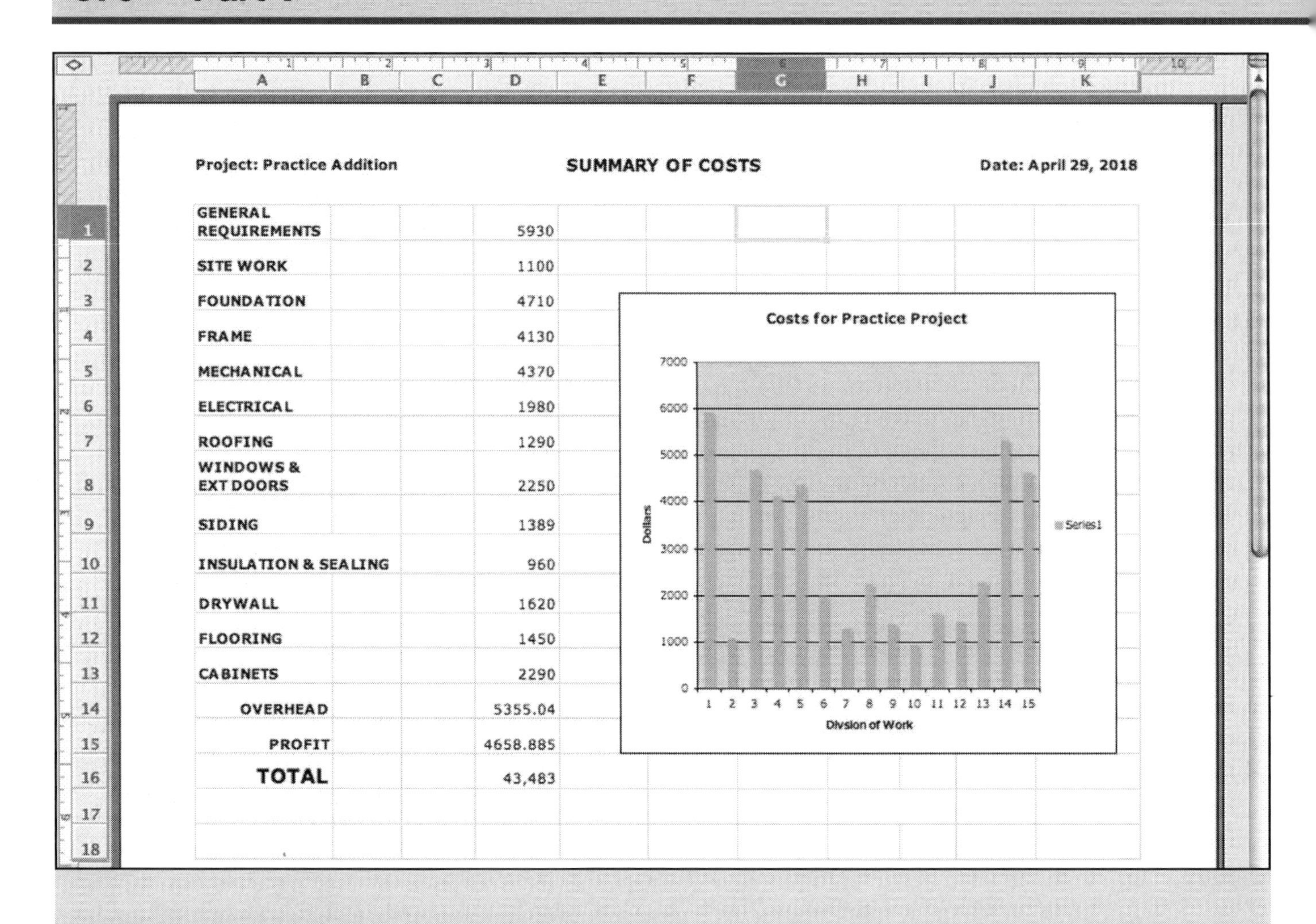

Project: Practice Addition | SUMMARY OF COSTS | Date: April 29, 2018

1	GENERAL REQUIREMENTS	5930
2	SITE WORK	1100
3	FOUNDATION	4710
4	FRAME	4130
5	MECHANICAL	4370
6	ELECTRICAL	1980
7	ROOFING	1290
8	WINDOWS & EXT DOORS	2250
9	SIDING	1389
10	INSULATION & SEALING	960
11	DRYWALL	1620
12	FLOORING	1450
13	CABINETS	2290
14	OVERHEAD	5355.04
15	PROFIT	4658.885
16	TOTAL	43,483

SUMMARY SHEET

After estimating the costs of a project, you can display those costs, division by division, along with your markups for overhead and profit, in a summary of your bid.

If you like, you can add a chart to the summary by clicking on the chart icon in your Excel toolbar. As the videos (and likely the Help included with your edition of Excel) will show you, creation of a chart can be accomplished with a few clicks or keystrokes.

Excel offers a mind-boggling variety of chart possibilities. A simple vertical bar chart such as the one shown here, lists dollar amounts vertically while horizontally listing row numbers that correspond to the rows in the worksheet. (For example, number 15 in the chart corresponds to row 15, the row for profit, in the worksheet). Thereby the chart gives you and your clients a graphic view of just where the costs in the project are concentrated.

worksheet to a summary sheet one by one (just as you would do with a word-processed checklist/spreadsheet). However, as you advance your skills you can again take advantage of Excel's ability to link worksheets and have Excel make the transfer for you. That is, you can instruct the summary sheet to refer to the direct costs worksheet and bring over all the division totals, then add them up using AutoSum. Nice!

Congratulations!

If you have worked your way through *Nail Your Numbers* from start to finish and have traveled the path to skilled estimating and bidding, you deserve kudos. You've put in a lot of work. You have become familiar with the specialized language of estimating and brushed up on your math. You have learned to choose jobs and qualify clients, plans, and specifications systematically; and to methodically build takeoffs. You've built thorough and well-organized task lists that cover every step in the estimating and bidding process from site inspections to the presentation of bids to clients. You have designed, formatted, and built a comprehensive spreadsheet/checklist that greatly reduces the danger of overlooking items during estimating.

Moreover, you have compiled a record of your company's labor productivity rates. You no longer have to rely on guesswork—or, just as bad, averaged numbers grabbed out of a cost catalogue—to estimate labor costs. That, right there, puts you into an elite group of builders.

But you've even gone beyond all that. You have considered several different ways of adding markups for overhead and profit to your estimates of direct costs in order to arrive at realistic selling prices. Going further still, you've learned to protect your selling price along with your client's financial well-being by creating a transparent and consistent contract and change order process. You have even moved your estimating onto Excel and built a powerful workbook that has hugely increased your efficiency via the use of formulas, functions, and templates.

All in all, rather than shortcutting and making the all too common excuses for sloppy estimating and bidding, you have put in the hard miles required to build the sort of strong system that is at the very heart of any sustainable building business.

You are so good at estimating and bidding—along with being a reliable, capable, and respected builder—that you are now able to command pay for your estimating work. You are out of the free bidding trap that squeezes the joy out of so many builders and the life out of their construction companies.

So, congratulations. And good luck. We all need it. When people ask me the key to my success, I usually answer, "Luck." I definitely have enjoyed some dumb luck. But another kind of good fortune does not fall from the sky. It happens in the moment when preparation meets opportunity. You've done the preparation. You are poised to take advantage of opportunity when it comes your way. With some dumb luck, it will. I'll be rooting for you.

GLOSSARY

A & E – Abbreviation for architect and engineer.
ABs – Abbreviation for anchor bolts.
Acute Triangle – A triangle with all angles less than 90°.
Addendum – An addition to a set of plans or a project manual. Plural, addenda.
American Institute of Architects (AIA) – A professional organization of architects. Not all architects are members.
American Society of Testing and Materials (ASTM) – An international organization that publishes technical standards for a wide range of materials and products.
Area – The size of a surface.
Assemblies – A unit of construction that includes several related items of work such as the sills, studs, headers, and sheathing that make up a wall frame assembly.
Associated General Contractors (AGC) – A professional association of builders who do commercial, industrial, and institutional construction.
Assumptions – Specifications assembled by an estimator to compensate for ambiguities or omissions in a set of plans and/or a project manual.
Audit – A systematic review.
Automate – In the context of construction estimating, moving from estimating with paper and pencil to estimating with computer software.
"B" Question – The budget question.
Bid – A price with accompanying assumptions for building a project.
Bond – A guarantee that agreed-upon work will be performed.
Budget – The amount of money allocated for a project.
Builder – Strictly, someone who develops real estate, whether an apartment building, a single house, or a commercial project. More loosely, someone who builds, whether speculatively as a developer or for clients.
Buying a Job – Bidding for a project with little or no markup for profit or even at a price that is less than the sum of your direct and overhead costs.
Callbacks – Requests from an owner to a builder to return to a project after completion and correct defective work.
Change Order – A written extension of a construction contract in which a builder agrees with an owner to do additional work for additional charges or to delete work and reduce charges.
Change Order Artistry – The practice of leaving items that will be needed to complete a project out of an estimate with an eye to earning profit from them during construction.
Civil Engineer – An engineer who provides drawings and specifications for site improvements such as drainage and utilities.
Cladding – Exterior skin of a building such as wood siding, shingles, stucco, or steel.
Clarifications – See *Assumptions.*
Client – A customer.
Cloud – In technology, an off-site data storage facility.
Clouds – In design, cloud-like circles on a drawing that indicate an area in which a change has been made.
Company Rent – An informal term for overhead charges.
Computer Assisted Design (CAD) – Design done with a computer program rather than with paper and pencil.
Constant – A number such as *pi* that is the same for all cases.
Construction Consulting – Services provided during the design and bidding phase of a project. See *Preconstruction Services.*
Construction Specifications Institute (CSI) – The publishers of MasterFormat.
Consumables – The supplies, from pencils through tape measures, saw blades, string, plastic sheeting, and much else, that construction crews use up in the course of their work.
Contingency – A dollar figure in an estimate that provides for unknown or uncertain costs.
Cost Books – Books that claim to provide average costs for construction of items and assemblies called for in building plans and specifications.
Cost Catalogue – See *Cost Books.*
Cost Planning – A preconstruction service designed to maximize the benefits of a project while keeping it within budget.
Cost Reference Book – See *Cost Books.*
Crew Supplies – See *Consumables.*
Crew-day – A day's worth of work by a crew.
Crew-week – A week's worth of work by a crew.
Decimal – A period in a number signifying that digits to the left represent whole numbers—ones, tens, hundreds and larger—and digits to the right represent fractions—tenths, hundredths and smaller. Example: 1234.56.
Deconstruction – Taking apart a building to maximize reuse and recycling of materials rather than demolishing and dumping them.
Details – Generally, specific components of a building. Also, drawings within engineering or architectural plans that provide enlarged views of particular items of construction.
Direct Costs – The costs of constructing a project that are incurred at the job site.
Direct Overhead – See *General Requirements.*
Door and Window Schedule – A list of doors and windows, usually arrayed on a matrix, that specifies manufacturer, size, glazing, and other details.
Elements – See *Items.*

Elevation – Architectural term for "view." An exterior wall elevation gives a view of the exterior surface of a wall.
Employee – A person who works under your supervision, or the supervision of one of your employees, and receives a paycheck. (For the exact legal distinction between an employee and a subcontractor, consult with the appropriate federal and state agencies and/or your business attorney.)
Estimate – An approximation (a very close approximation in the case of a good estimate) of the direct costs of construction for a project.
Estimator – A person who produces estimates. Also, the software used in producing estimates.
Finish Schedule – A list, usually arrayed on a matrix, of the finish materials to be used in a construction project.
Fixed-price Contract – In contrast to a time and materials contract, provides for a project to be built for a predetermined price (as modified by change orders).
Flow Chart – A chart that visually displays the schedule for construction of a project.
FSC – Forest Stewardship Council.
Function – A mathematical formula for relating input to output.
Examples: Hours of work to install an item (input) multiplied by labor rate (function) gives cost of labor for the installation (output).
Five hours to install a pedestal sink (input) multiplied by $70/hour (function) equals $350 (output).
Gantt Chart – See *Flow Chart.*
General Building Contractor – A contractor who takes on responsibility for an entire project, doing the work either with his crew (his "own forces") or via subcontractors.
General Conditions – An extension of a contract, specifically the detailed extension of certain lengthy and complex contracts created by the AIA. (Note: The term is sometimes, though not in this book, also used in place of "General Requirements," a division of the work necessary to construct a project.)
General Requirements – Services and work necessary for constructing a project that do not become a permanent part of the project. Examples: Scaffolding, delivery of materials, storing and handling of materials, project management.
Grinders – Customers who will squeeze a builder for everything they can get while paying as little as possible for it.
Gross Profit Margin – Overhead and profit combined and then expressed as a percentage of the total selling price for a project.
GRs – An abbreviation for General Requirements.
HDs – Abbreviation for hold downs, hardware used to connect frames to foundations
Hidden Conditions – Work necessary for the completion of a project, such as correction of defective framing, that is hidden from view at the time of an estimate.
Historical Labor Productivity Record – A record of a crew's productivity for a particular item of work. Also known more simply as a labor productivity record or even labor record.
Holdback – Money an owner of a project deducts from payments to a contractor and retains till completion of the project. Also called "Retention."
Home Office Overhead – See *Overhead.*
Hunt and Peck – Two-fingered typing.
HVAC – Heating, Ventilation, and Air Conditioning.
Indirect Overhead – See *Overhead.*
Integrated Project Delivery (IPD) – An alternative to competitive bidding, characterized by collaboration from the inception of design through the completion of construction between a project's owner, design team, and builder.
Interior Designers – Designers who are typically focused on finishes, including furnishings, and have little knowledge of construction.
Items – Components of a project. Also called "elements" and "parts."
K – An abbreviation for a thousand. For example, $10K means $10,000.
KD – An abbreviation for "kiln-dried."
Labor Burden – Costs for an employee above and beyond wages. Examples: Social Security. Unemployment taxes. Consumables. Training.
Labor Rate – Charge for labor at a job site. Includes wages, labor burden, and sometimes other charges.
Lead – A first contact about a potential project.
Lead Sheet – A form for recording information about a potential project.
Liquidated Damages – Money due from a contractor to an owner for damage done during a project. Example: Delay in reaching completion of the project, resulting in the owner suffering loss of rents.
Lowball – Bidding projects at unduly low prices with the hope of surviving in a down market or of making a profit by levying change orders during construction.
Markup – A percentage or sum added to direct costs to recapture overhead or profit.
MasterFormat – The format, i.e., organization, of divisions, phases, assemblies, and items of work published by the Construction Specifications Institute (CPI).
Matrix – A sheet divided into grids that is used for the display of information.

Mobilization – Setting up for a task or tasks.
MPE – An abbreviation for mechanical, plumbing, electrical.
NA – Not applicable.
NAHB – National Association of Home Builders.
NARI – National Association of the Remodeling Industry.
Negotiated Bid – A vague term describing a process whereby an owner makes an agreement with a builder to work out a contract for construction of a project.
NIC – Not included.
Obtuse Triangle – A triangle that has an angle greater than 90°.
Overhead – In contrast to direct costs, which are the costs incurred for construction at a job site, overhead expenses are those incurred for the off-job-site costs of running a company.
Parts – See Items.
PDF – An abbreviation for Portable Document Format, a file format used for design documents that allows them to be easily distributed by computer.
Percentage – A way of expressing a fraction. Example: ¼ = 25%.
Pi – A mathematical constant used in figuring the area of circles and the volume of cylinders.
Place – In math, the location of a digit relative to a decimal point that determines that digit's value. Example: 8.35. The three is in the tenths place while the five is in the hundredths place.
Plan Views – In architectural drawings, views from above that show the arrangement of spaces (rooms) in a project.
Plans – Architectural and engineering drawings for a project.
Precedence – Of greater importance or having priority. Example: Information in a project manual for a project can take precedence over, i.e., be given priority over, information in the plans for the project.
Preconstruction services – Services including estimating, construction consulting, and value engineering that precede construction of a project.
Preliminary Drawings – Drawings that are roughly halfway to completion.
Prevailing Wage – Usually the wage paid to unionized workers.
Price at a Discount – Bidding low with relatively little or no markup for profit and perhaps not even for overhead. See *Buying a Job.*
Profit – In actual practice, charges to cover the unknown expenses that will occur during or after the completion of a project. More hopefully, a reward for taking and skillfully managing the risks of being in the construction business.
Project Manual – A manual that provides the technical specifications, bidding requirements, and contract documents for a project.
Project Overhead – See *General Requirements.*
Proof – A procedure intended to ensure that the result of a mathematical computation (such as adding up a column of numbers) has produced an accurate result.
Pythagorean Theorem – A mathematical formula that allows you to determine the length of one side of a right triangle when the other two are known. It is useful for figuring rafter lengths.
Qualify – To evaluate a client and project for a bid.
Quantify – To figure an amount.
Quantity Survey – See *Takeoff.*
Recap Sheet – A sheet summarizing all the divisions of work in a project together with their costs, and which may also provide space for charges for overhead, profit, and bonds along with the total selling price of the project.
Recapture – Recover expenses, as in charging enough to recapture overhead costs.
Reflected Ceiling Views – A view of a ceiling as it would be seen if reflected on the flooring beneath it.
Request for Information – A request, usually to a project designer, to clarify details, resolve contradictions, or fill in blank areas in plans and specifications.
Retention – See *Holdback*.
RFI – An abbreviation of Request for Information.
Right Triangle – A triangle that includes a 90° angle.
Round Off – To shorten a number to the nearest number with fewer decimal places. Example: 5.46 is rounded off to 5.5 (not 5.4).
Schedules – A list or matrix that displays information such as door and window manufacturers and sizes.
Schematic Drawings – A first stage of architectural drawings that show the function and general form of a project.
Scope of Work – The work called for in a contract.
Section – A cutaway view or slice through a building that reveals how it is put together.
Selling Price – A term some contractors use in place of "bid" to describe the price for which they are offering to build a project.
Selling, General, and Administrative Expense (SGA) – A more formal term for "Overhead."
Setup – See *Mobilization.*
Slipperies – An informal term used by the author to designate items in an estimate that tend to escape attention.
Sonotube – A tube made of high-strength cardboard that is used in the formwork for concrete columns.
Specifications – Notes on a plan or in a project manual that describe the quality of the work called for in a project.
Specs – Shorthand for project manual.
Sponsor – A broad term covering clients (customers) for both private

and public projects.
Staging Area – A space, either right at a job site or nearby, for storing material and equipment.
Standard of Care – A legal term signifying the expectations for performance during construction of a project.
Stick by Stick – Estimating one item or small unit of work at a time.
Stipulation – A requirement.
Structural Engineer – The engineer responsible for designing the structural components, i.e., foundation and frame, of a project.
Subcontractor – An independent individual or company that a builder hires to do a portion of project, such as plumbing, drywall, or tile.
Take Off – To quantify the amount of work in a project.
Takeoff – A summary of the quantities of work in a project.
Template – An estimate worksheet that can be used repeatedly for similar projects.
Ten-Fingered Typing – Using all ten fingers to type rather than "hunting and pecking" with only two fingers.
Thermal Boundary – The envelope of a building, enhanced with insulation and sealant, which prevents heat transfer.
Time and Material Contracts – A variety of contracts under which owners agree to pay for the labor and material used on a project at a predetermined rate rather than contracting for the project to be built at a fixed price.
Time Sensitive – As used in this book, a term used to refer to costs, such as rental of a sani-can, that increase as a project increases in length.
Trapezoid – A four-sided figure with only one pair of opposite sides parallel.
Unit Cost – A cost of an item or assembly that includes labor and material.
Unit Price – A cost of an item or assembly that includes direct costs, overhead and profit.
Value Engineering – Redesigning items, assemblies, systems, or space in a project to maximize benefit while controlling costs.
Volume – In estimating, the measure of space inside a box-shaped or cylindrical form.
WAG – Estimator slang meaning "wild-assed guess."
Wages – Pay to employees (as opposed to payment of subcontractors' bills).
What-ifs – Alternate scenarios for construction, typically considered for the purpose of managing the cost of a project.
Working Capital – Money available to pay a company's ongoing operating expenses.
Working Drawings – Drawings intended for use on a job site during construction.
Your Own Forces – Your crew.

RESOURCES

Estimating and Bidding Books

Estimating for the General Contractor and ***Bidding for the General Contractor*** by Paul Cook. These provide an exceptionally clear explanation and analysis of estimating and bidding procedures. The books were published in 1982 and 1985, so they predate the technology we now use. But they remain valuable because what is critical in estimating and bidding is to get your procedures right. If they are not right, computerizing them won't help. It might make matters worse. Even as he shares the insights gleaned from years of producing estimates, Cook gets the procedures close to perfect, and his procedures can easily be adapted to newer technology. An upside to the books' having been published so far back is that they are available on the web for a few dollars each. At that price, buying Mr. Cook's books online from a used bookseller is like bumping into a painting by Norman Rockwell at a flea market and picking it up for the small bills in your wallet.

Construction Estimating & Bidding: Theory, Principles, Process from the Associated General Contractors. While the book is a bit sophisticated for start-up builders, if you are an established builder trying to strengthen your estimating and bidding systems, spending the hundred and fifty bucks or so that it costs may prove one of the best investments you will ever make.

Construction Estimating Complete Handbook by Adam Ding. Contains lengthy checklists you can draw upon to build your own spreadsheet/checklist. It will be especially useful for that purpose if you move from residential to commercial construction. The handbook also contains an 80-page digest of Excel that is useful as a source of answers to specific questions about Excel operations. Additionally, Ding offers useful tools for various phases of bidding such as comparing and evaluating subcontractor bids.

Estimating Building Costs by Wayne DelPico. Manages, unfortunately, to be overly detailed at times, as when it offers lengthy descriptions of garage door parts, while being sketchy in important respects—such as its failure to provide real-life examples of the sound practices it preaches. However, it does include instructive drawings and tables and is well organized and well written. By reading selectively, you can avoid information you do not need and pick up useful tips, know-how, and general principles.

Excel for Dummies. Poorly edited and not a good choice for learning Excel, but a useful reference when you need answers to specific questions about Excel operations. You want the version for the edition of Excel you are using (2016, 2013, 2010, etc.)

Plan Reading & Material Takeoff by Wayne DelPico. If you need to sharpen your plan reading and takeoff skills, DelPico's book should get you up to speed. It shares the weaknesses and strengths of DelPico's other book, described above. Used copies of both DelPico books can be purchased inexpensively online.

The Number Yard by Adam Cook. If you need to refresh or sharpen your basic math skills, Adam Cook's concise and clear narrative will provide you with good-humored help.

Integrated Project Delivery from the California Council of the AIA. The IPD manual describes in detail an alternative to competitive bidding that features close collaboration from the inception of a project between the owner, design team, and builder. Unfortunately, the manual is so badly written that it is at best tiresome reading and at worst hard to follow. But it makes a powerful case against competitive bidding and does offer a chart that clearly illustrates the IPD alternative. One builder hands it out to architects as a way of encouraging them to move away from competitive bidding and toward his cost planning approach

Construction cost data books from RSMeans. I recommend avoiding them as a source of construction costs for actual estimates and bids. They are, however, useful, as a source of divisions, assemblies, and items for building up your spreadsheet/checklist. New editions are pricey. However, editions only a few years old, which are as good a checklist source, can be purchased for a few bucks.

Construction Business Books

Broken Buildings, Busted Budgets by Barry B. LePatner. Offers an insightful analysis of the construction industry and its dysfunctions. (Disclosure: I may be partial to the book because the author quotes from or paraphrases my own writings dozens of times).

Construction Contactor's Survival Guide by Thomas Schleifer. Distills ten primary reasons builders fail. The most accomplished builder I have ever known read it yearly prior to his retirement.

Crafting the Considerate House by David Gerstel. Chronicles the design and construction of a resource-efficient, affordable, comfortable, and attractive family. home and offers numerous ideas for value engineering residential projects.

Elements of Building by Mark Kerson. EOB is organized as a series of concise commentaries rather than as an instructional narrative. The author intended it for startup builders. It's popularity suggests it is of substantial value to them; though when I first saw it, I had doubts as to whether it would be since it does not make use of graphics — not charts, checklists, forms, or photos. I do believe it is valuable for established builders. Kerson's distillation of lessons learned from his decades in the building business encourage the reader to revisit fundamental issues and can help even the most seasoned builder stay on track. I have kept a copy on my desk for a some years. Often I open it at random and read a few pages. Always I am stimulated to re-examine my ideas about the building business.

Markup and Profit by Michael Stone. For Stone, markup and profit is not just an important issue; it is holy ground, to be protected with religious fervor, given priority over all other considerations, including the well-being of one's crew. You will not find in Stone's book any photos of beloved projects, no joy in the craft of building, no suggestion that some jobs are worth doing even if there's little profit in them (in fact, he views the idea with contempt). Stone is all business, all the time. About the business of building he has much to say that is thought-provoking, useful, and founded on his decades of experience working as a consultant to builders all over the country. Even if you choose to operate from different values, you can learn much from Stone's book.

Managing the Small Construction Business. A collection of articles from the *Journal of Light Construction.* I especially recommend the articles by Martin King, Deva Rajan, Larry Hayden, and Sal Alfano, long-time editor of JLC.

Running a Successful Construction Company by David Gerstel. Builders have been recommending this book to one another for 25 years and several hundred thousand of them have picked up either a new or used copy. One commented, "I value the book because it is concise and confident, yet Gerstel shares his missteps and learning experiences as well as his successes." When I read the book myself, I mostly see room for improvement. But given reader responses, I have to assume it remains helpful to many builders.

Web Sites

Paul Eldrenkamp, a builder in the Boston area, is a clear thinker and an engaging writer. To access his thinking on estimating and bidding, you may have to do a little digging, because his writings on business practices are mixed in with posts from other people in his company (Byggmeister) and with his own articles on environmentally considerate construction. But believe me, the digging will be well worth your while. Start digging at www.byggmeister.com. Alternately, you can begin at jlconline.com/author/paul-eldrenkamp

Fernando Pagés Ruiz has ranged widely across the light-frame world, producing everything from small residential remodels to housing developments. His building business blog, which you will find on the *Fine Homebuilding* web site, ranges just as widely, covering subjects from the effective use of smartphone apps for project management to developers' tales of boom, bust, despair, and recovery. The blog is thought-provoking while offering useful and practical advice.

Math is Fun Mathisfun.com Though intended for schoolteachers, not builders, it provides clear explanations of the math needed for estimating.

MasterFormat Divisions. archtoolbox.com./representation/specifications Scroll down to see both the old and the new MasterFormat Divisions

Videos

For excellent instructional videos about Excel, please see the sidebar titled "A Low-Cost Approach to Learning Excel" in Chapter Nineteen.

Index

A

B

C

D

E

T

V

W

Acknowledgements

A great many people -- builders, specialty contractors, engineers, designers, and attorneys -- have generously taken time from their busy days to discuss the craft and challenges of estimating and bidding with me, and have often done so in recorded interviews. By and large they granted those interviews after I promised to refrain from identifying them by name when I quoted or paraphrased their insights, their revelations of their companies' practices, and their candid descriptions of their failures and frustrations. Therefore, I cannot thank them here by name. However, they know who they are. I am grateful to them. If this book is valuable to its readers, it is in large part because of the contributions of the construction professionals who have shared their deep knowledge with me.

During the writing and production of *Nail Your Numbers,* I have been fortunate to enjoy the guidance, criticism, and support of superb editors and a remarkable graphics designer. Jackie Parente has once again served as my astute and deeply knowledgeable developmental editor. Michael and Pam Rosenthal of P&M Editorial Services provided unusually comprehensive and thoughtful copy editing. And Deb Tremper of Six Penny Graphics supported me with awesome skill and efficiency in working through aesthetic and technical challenges in the design of both the cover and interior of the book.

I am particularly grateful to the San Francisco based architect John McClean who gave me extensive and ultimately decisive advice on the cover design.

Peter Chapman of *Fine Homebuilding* books; Mark Kerson, the author of *Elements of Building* (see Resources); George Kiskaddon of the Builders Booksource; and Fernando Pages, the creator of the *Fine Homebuilding* business blog, and have all buoyed my spirits with much appreciated encouragement during the years it has taken me to write and produce *Nail Your Numbers.*

Finally, and above all, I am grateful to my wife Sandra, to whom this book is dedicated, for her unflagging support and encouragement.

Made in United States
Orlando, FL
28 October 2022